EXPERIMENTS IN

GROB BASIC ELECTRONICS

FRANK PUGH
Santa Rosa Junior College
Santa Rosa, California

WES PONICK
Hewlett-Packard Company
Network Measurement Division
Santa Rosa, California

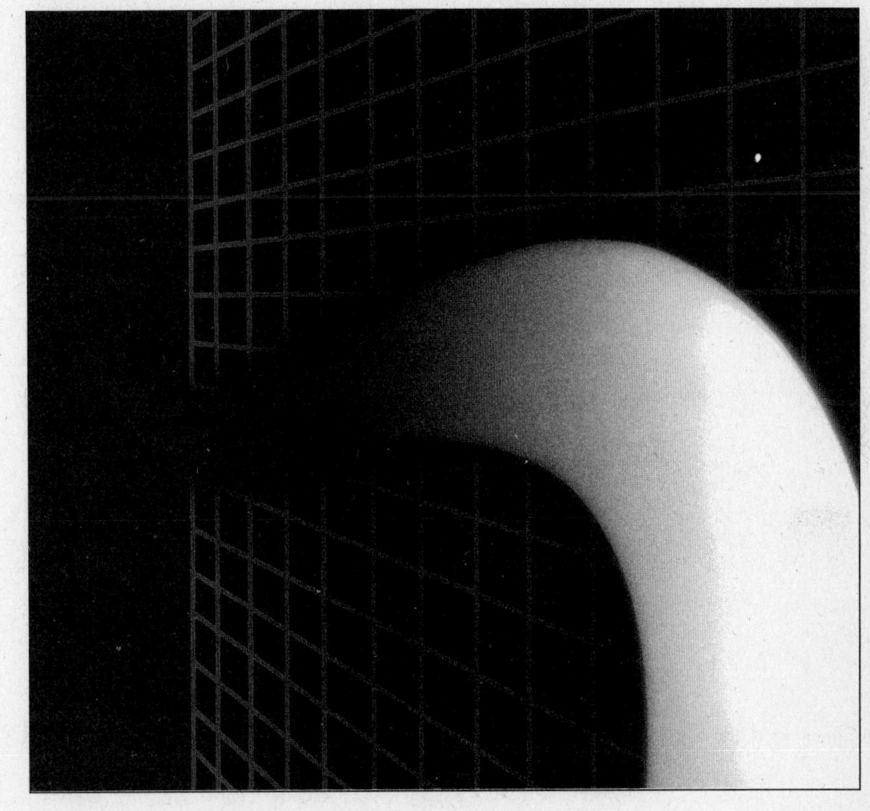

THIRD EDITION

GLENCOE

Macmillan/McGraw-Hill

Lake Forest, Illinois Columbus, Ohio Mission Hills, California Peoria, Illinois

This textbook has been prepared with the assistance of Publishing Advisory Service.

Experiments in Grob Basic Electronics, Third Edition

Send all inquiries to:
GLENCOE DIVISION
Macmillan/McGraw-Hill
936 Eastwind Drive
Westerville, OH 43081

ISBN 0-02-800764-6

Printed in the United States of America.

1 2 3 4 5 6 7 8 9 POH 99 98 97 96 95 94 93 92 91

CONTENTS

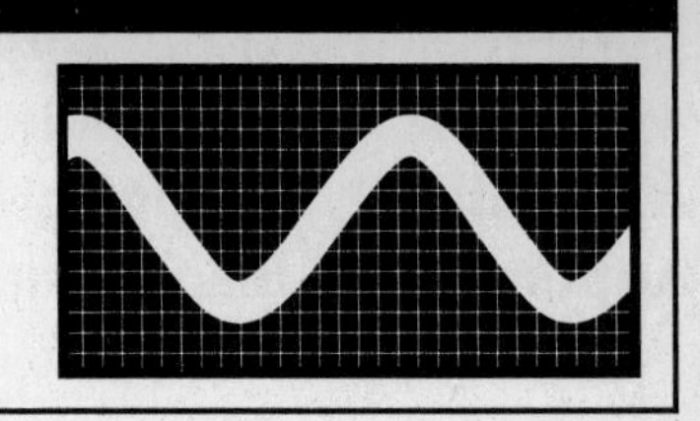

PREFACE

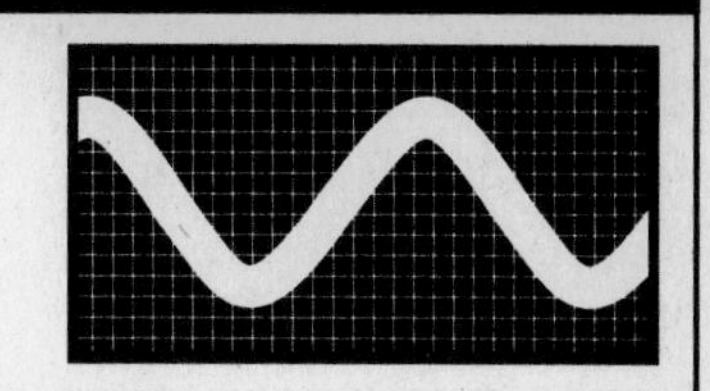

Experiments in Grob Basic Electronics is a lab manual for the beginning electronics student who does not have any previous experience in electricity or electronics. It has been developed especially for use with *Grob Basic Electronics,* Seventh Edition, and the experiments are coordinated with that text in sequence and in technical usage. There are 52 experiments, ranging from an introduction to electronic equipment and components to a primer on operational amplifiers. The emphasis throughout is on basic concepts. Although the experiments build on one another, with simple concepts being developed before more complex subjects are introduced, minor modifications in the sequence can be made as necessary to suit the requirements of an individual electronics program. All experiments have been student-tested at Santa Rosa Junior College.

Each experiment, which takes approximately three hours to perform, is organized as follows. First, the basic principles are explained in detail; then, the student is encouraged to apply electronics theory to troubleshooting; and finally, the student writes a comprehensive report. Techniques for good technical report preparation are discussed in the first group of experiments. Additional information about preparing reports is provided in Appendix D.

The authors would like to thank Tom Power, instructor and founder of the Department of Electronics at Santa Rosa Junior College, for his unflagging encouragement. We also wish to thank David Newton of the ITT Technical Institute, Nashville, and Melvin Duvall of Sacramento City College for their assistance with the preparation of this laboratory manual. In addition, we thank Jeanne Ponick, Ann Pugh, our children, and all the students at SRJC for their patience and support.

Frank Pugh
Wes Ponick

EXPERIMENT 1

INTRODUCTION TO EQUIPMENT AND COMPONENTS

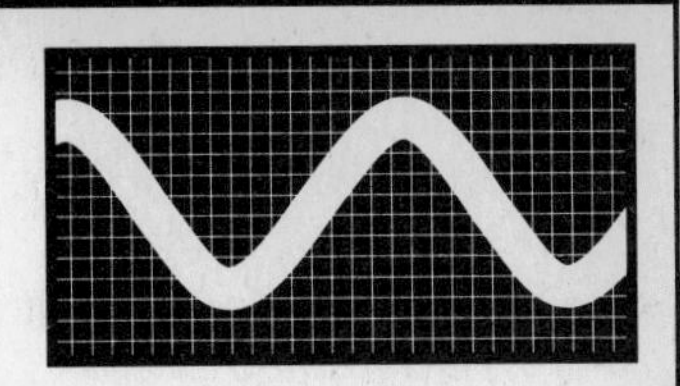

OBJECTIVES

At the completion of this experiment, you will:

- Be familiar with the basic equipment in an electronics lab.
- Be familiar with the basic components used in electronics.
- Know the resistor color code.

SUGGESTED READING

Chapters 1 and 2, *Basic Electronics,* B. Grob, Seventh Edition

INTRODUCTION

This first experiment is simply an introduction to the electronics lab and the equipment and components you will find there. To use an electronics lab, it is necessary to become familiar with some typical equipment and components. On the following pages, there are descriptions of the types of equipment you will be using.

Familiarize yourself with the equipment here and in your lab. If you know what an ohmmeter, voltmeter, and ammeter are before you actually use them, you will be well-prepared. After studying the material, answer the questions in the quiz. Then, learn how to read the resistor color code, and fill in the correct answers to Tables 1-1 and 1-2. Turn in the quiz and the tables for grading.

Experiments 1 and 2 are often combined for the first week or two of study. Remember, this first experiment can be done outside the lab. However, Experiment 2 must be done in the lab.

EQUIPMENT

Every electronics lab, and every school, has different types of equipment. However, there are some basic meters that are common to all electronics labs. In one form or another, your school will have these meters. They may appear to be different only because they are made by different manufacturers.

Simple Meters

Ohmmeter: Used to measure resistance (in ohms). Remember that resistance is the opposition to current flow in a circuit. Resistors usually have color stripes on them to identify their values. One way to be sure your resistor is the correct value is to measure it with an ohmmeter. Then you will know its value in ohms.

Ammeter: Used to measure current (in amperes). Remember that current is an electric charge in motion. Ammeters usually measure the current in milliamperes (0.001 A), because most electronics lab courses do not require large amounts of current.

Voltmeter: Used to measure voltage (in volts). Remember that voltage refers to units of potential difference. Most voltmeters use more than one range, or scale, because voltage, unlike current, is often measured in larger values.

Multimeters

VOM (VOLT-OHM-MILLIAMMETER): A multipurpose meter used to measure dc and ac voltage, dc current, and resistance.

VTVM (VACUUM TUBE VOLTMETER): Another type of multipurpose meter used to measure voltage and resistance only.

DMM (DIGITAL MULTIMETER): A digital multipurpose meter, also known as a DVM (digital voltmeter).

Other Equipment

DC Power Supply: A source of potential difference, like a battery. It supplies voltage and current, and it can be adjusted to provide the required voltage for any experiment.

Soldering Iron: A pointed electric appliance that heats electric connections so that solder (tin and lead) will melt on those connections. Used to join components and circuits.

Breadboard/Springboard/Protoboard: Used to assemble basic circuits, by either soldering, inserting into springs, or joining together in sockets. These boards are the tools of designers, students, and hobbyists.

Component Familiarization (Symbols)

-\/\/\/- The symbol for resistance or a resistor.

—|||— The symbol for a battery (cells) or a power supply.

\+ Positive symbol (associated with the color red).

– Negative symbol (associated with the color black).

—|(— Capacitor (or capacitance) symbol. Used in later study.

⌒⌒⌒ Inductor (or inductance) symbol. Used in later study.

Leads are simply insulated wires used to join the meters to the circuits, the power supply to the circuit, etc. They are conductors and have no polarity of their own. Leads have different types of connectors on the ends, such as alligator clips, banana plugs, and BNC connectors.

Ω = ohms (unit of resistance), Greek letter omega

∞ = infinity (used to indicate infinite resistance)

Common Symbols for Multipliers

Lowercase k = kilo = 1000 or 1×10^3
Uppercase M = mega = 1,000,000 or 1×10^6
Lowercase m = milli = 0.001 or 1×10^{-3}
Greek letter mu (μ) = micro = 0.000 001 or 1×10^{-6}

PROCEDURE

Resistor Color Code

Answer questions 1 to 15.
Study Fig. 1-1, then complete Tables 1-1 and 1-2 by filling in the blanks with the appropriate value or color.

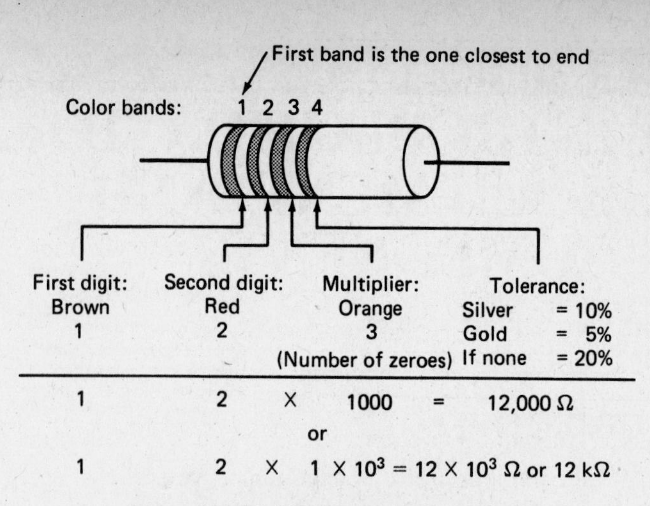

Remember: Multiplier means adding zeros.
Here, add three zeros.

Note: If the multiplier (third band) is gold, multiply by 0.1. Here, a gold band multiplier would mean 1.2 Ω. If the multiplier is black, multiply by 1.

COLOR CODE

Color	Value	Color	Value
Black	0	Green	5
Brown	1	Blue	6
Red	2	Violet	7
Orange	3	Grey	8
Yellow	4	White	9

Fig. 1-1. How to read color bands (stripes) on carbon resistors. Resistors come in various shapes and sizes. This is only one common form. The larger the physical size, the greater the wattage rating. Also, some resistors have five color stripes (see Appendix).

NAME ______________________________ DATE ______________

RESULTS FOR EXPERIMENT 1

QUESTIONS

Fill in the blanks (1–15) with the letter of the correct answer.

____ **1.** An instrument used to measure potential difference.

____ **2.** An instrument used to measure current.

____ **3.** An instrument used to measure resistance.

____ **4.** A passive component that opposes the flow of current.

____ **5.** An instrument used to heat solder and join components.

____ **6.** A source of dc voltage other than a battery.

____ **7.** A symbol for dc voltage source.

____ **8.** A symbol for resistance.

____ **9.** A color used to represent negative polarity.

____ **10.** An item used to temporarily build circuits on.

____ **11.** A Greek letter used to represent a unit of resistance.

____ **12.** An English letter used to represent 1000.

____ **13.** An English letter used to represent 0.001.

____ **14.** An instrument capable of measuring both voltage and current.

____ **15.** A symbol for infinity.

a. ammeter
b. —/\/\/—
c. power supply
d. voltmeter
e. —|||—
f. black
g. ohmmeter
h. breadboard
i. soldering iron
j. resistor
k. VOM/VTVM
l. m
m. k
n. Ω
o. ∞

REPORT

TABLE 1–1. Resistor Color Codes

First Digit Band 1	Second Digit Band 2	Multiplier Band 3	Tolerance Band 4	Resistor Value
Red	Brown	Brown	Gold	________
Brown	Brown	Black	Gold	________
Green	Blue	Red	Silver	________
Blue	Green	Yellow	Silver	________
Red	Red	Orange	Silver	________
Orange	White	Brown	Gold	________
Blue	Green	Black	Silver	________
Brown	Black	Red	Gold	________
Yellow	Violet	Green	Gold	________
Brown	Black	Orange	Silver	________
Orange	Orange	Orange	Silver	________
Brown	Black	Gold	Gold	________
White	Blue	Red	Silver	________
Brown	Black	Yellow	Silver	________
Brown	Green	Green	Gold	________

TABLE 1–2. Resistor Color Codes

Band 1 Color	Band 2 Color	Band 3 Color	Band 4 Color	Resistor Value
________	________	________	________	680 kΩ, 5%
________	________	________	________	10 kΩ, 10%
________	________	________	________	100 kΩ, 5%
________	________	________	________	3.3 MΩ, 5%
________	________	________	________	1.2 kΩ, 10%
________	________	________	________	820 Ω, 10%
________	________	________	________	47 kΩ, 5%
________	________	________	________	330 Ω, 10%
________	________	________	________	470 kΩ, 5%
________	________	________	________	560 Ω, 10%
________	________	________	________	1.5 MΩ, 10%
________	________	________	________	220 Ω, 5%
________	________	________	________	56 Ω, 10%
________	________	________	________	12 kΩ, 5%
________	________	________	________	560 kΩ, 5%

EXPERIMENT 2

HOW TO USE BASIC LAB EQUIPMENT

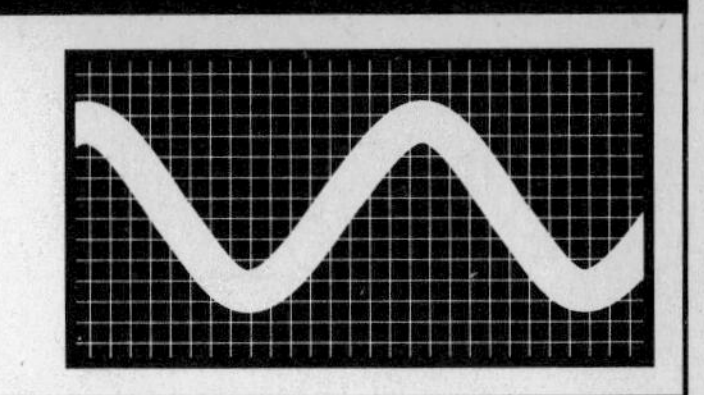

OBJECTIVES

At the completion of this experiment, you will be able to:

- Read linear and nonlinear meter scales.
- Measure resistance by using an ohmmeter (VTVM, DVM, or VOM).
- Measure voltage by using a voltmeter (VOM, DVM, or VTVM).

SUGGESTED READING

Chapters 1 and 2, *Basic Electronics,* B. Grob, Seventh Edition

INTRODUCTION

This experiment will help prepare you to use the meters in your lab. You will concentrate only on voltage and resistance measurements. It is extremely important to build a strong foundation for laboratory work. Therefore, read this experiment before following the procedure in the lab.

Voltmeters and ohmmeters vary with each situation and manufacturer. However, the meters described in this experiment are generic and should apply in most cases. Most of the explanations are given in the procedure. (Current measurements will be discussed in Experiment 3).

EQUIPMENT

VTVM, VOM, or DVM as specified by the lab for ohmmeter and voltmeter functions.
DC power supply

COMPONENTS

30 Resistors, all 0.5 W unless indicated otherwise:

(1) 10 Ω
(1) 56 Ω
(1) 100 Ω
(1) 220 Ω
(1) 390 Ω
(1) 470 Ω
(1) 680 Ω
(1) 4.7 kΩ
(1) 5.6 kΩ (0.5 W or less)
(3) 10 kΩ (0.5 W or less)
(1) 22 kΩ (0.5 W or less)
(1) 33 kΩ (0.5 W or less)
(2) 47 kΩ (0.5 W or less)
(1) 68 kΩ (0.5 W or less)
(1) 820 Ω
(1) 1 kΩ
(1) 1.2 kΩ
(1) 1.5 kΩ
(1) 2.2 kΩ
(1) 3.3 kΩ
(1) 86 kΩ (0.5 W or less)
(2) 100 kΩ (0.5 W or less)
(1) 220 kΩ (0.5 W or less)
(1) 470 kΩ (0.5 W or less)
(1) 1.2 MΩ (0.5 W or less)
(1) 3.3 MΩ (0.5 W or less)

PROCEDURE

1. Reading simple (linear) voltmeter scales: Refer to Fig. 2-1. There are eight typical voltmeter scales shown (*a* to *h*). Notice that each meter face has two scales. Also, notice that the scales are linear, that is, equally divided. Try to read these scales. Remember that it is the range switch that determines which scale is used. For example, meter A is on the 10-V range. Therefore the reading is 6.5 V. If meter A were on the 100-V range, the reading would be 65 V. Be sure you understand that the range indicates the greatest value you can measure on the scale. Fill in the voltage readings for meters *b* through *h*. Check your answers at the end of this experiment.

2. VTVM (Vacuum Tube Voltmeter): Refer to Fig. 2-2. Shown here is a generic representation of a typical VTVM. Before the development of digital multimeters (DMM), the VTVM was one of the most widely used multimeters in the electronics industry. Although the VTVM may be more difficult to read than a digital meter, it is a much better meter to use as a learning tool because the VTVM requires more operator adjustments in order to properly use it, and it also requires the interpretation of scales.

If you have a VTVM in your lab, refer to it along with Fig. 2-2 and continue. The VTVM must be plugged into an ac wall outlet. It can measure resistance and dc or ac voltage only. It does not measure current (amperes).

Find the following items on the VTVM and check them off as you locate them:

Function Switch: This switch turns the VTVM on. Actually, it is a rotary switch and should be in the OFF position when not in use. However, VTVMs require some warm-up time and are often left on during lab hours. The function switch allows you to

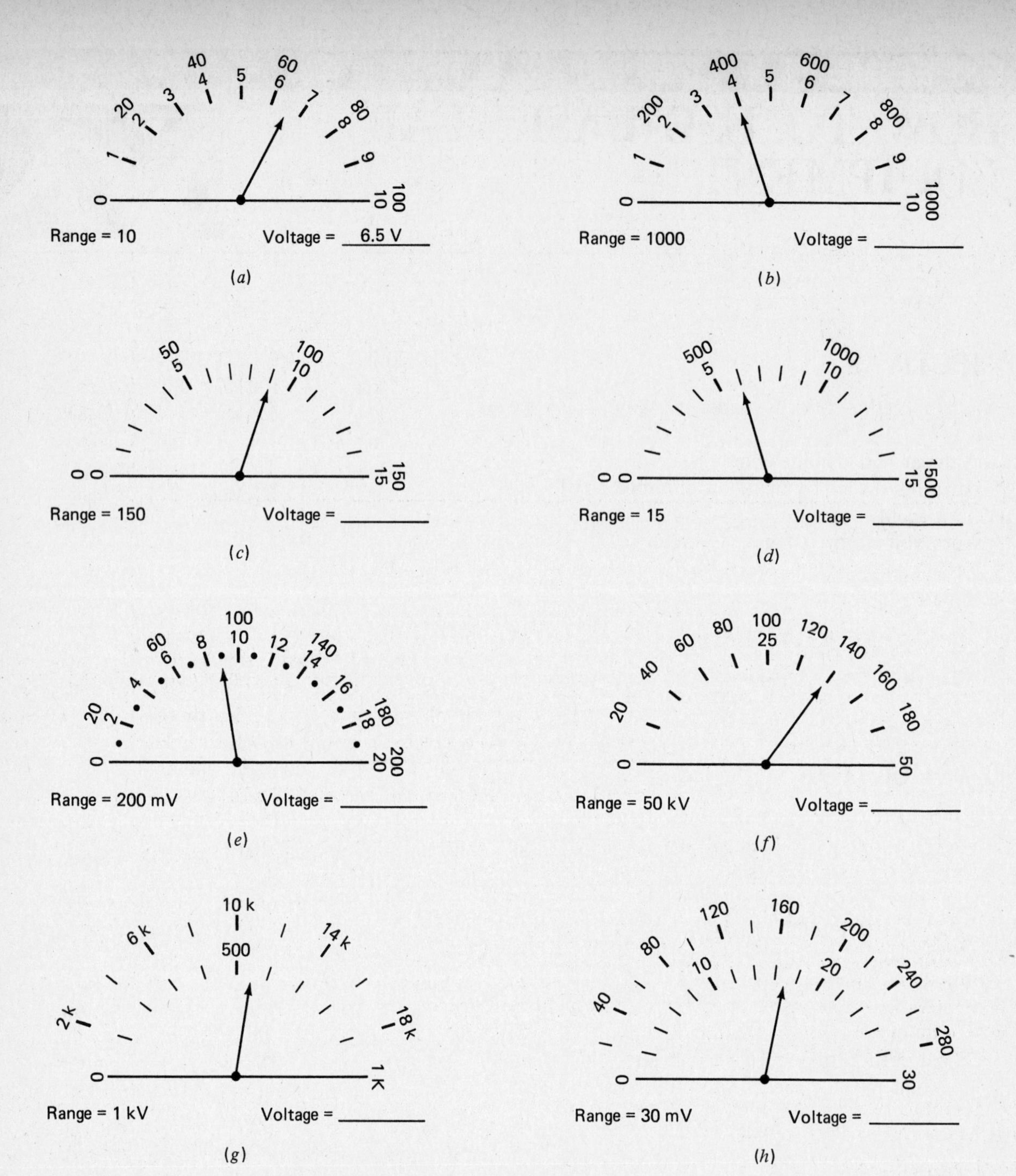

Fig. 2-1. Practice voltmeter reading (simple linear scales). *Note: Range* means the full-scale deflection or the greatest value on a particular scale.

choose the function you want from the VTVM. Thus, setting the function switch turns the VTVM into either an ohmmeter or a voltmeter (AC or +DC or −DC). You will probably use it as either an ohmmeter or a +dc voltmeter during your beginning studies.

Range Switch: This switch determines the scale that will be used. Notice that Fig. 2-2 shows eight different range settings. Each range setting corresponds to the function switch setting. For example, "15 V, R × 100" can be used for either a voltage or a resistance (ohms) measurement. In this case, if the function switch is on OHMS, the range would automatically be R × 100, not 15 V.

Scales (Ohms and DC/RMS): The OHMS scale, marked Ω for resistance, is on top. It is a nonlinear scale (not equally divided). It goes from zero to infinity (∞). In this case, Fig. 2-2 shows a reading of 2200 Ω. If the function switch were on +DC, the DC/RMS scale would be used, and the reading would be approximately 10.3 V. Remember that the full-scale deflection is 15 V and not 1.5 V, because the range switch determines the full-scale deflection. You, the operator, must make the mental adjustment when you read the meter scale. This takes practice.

Zero Adjustment: This knob is a potentiometer that

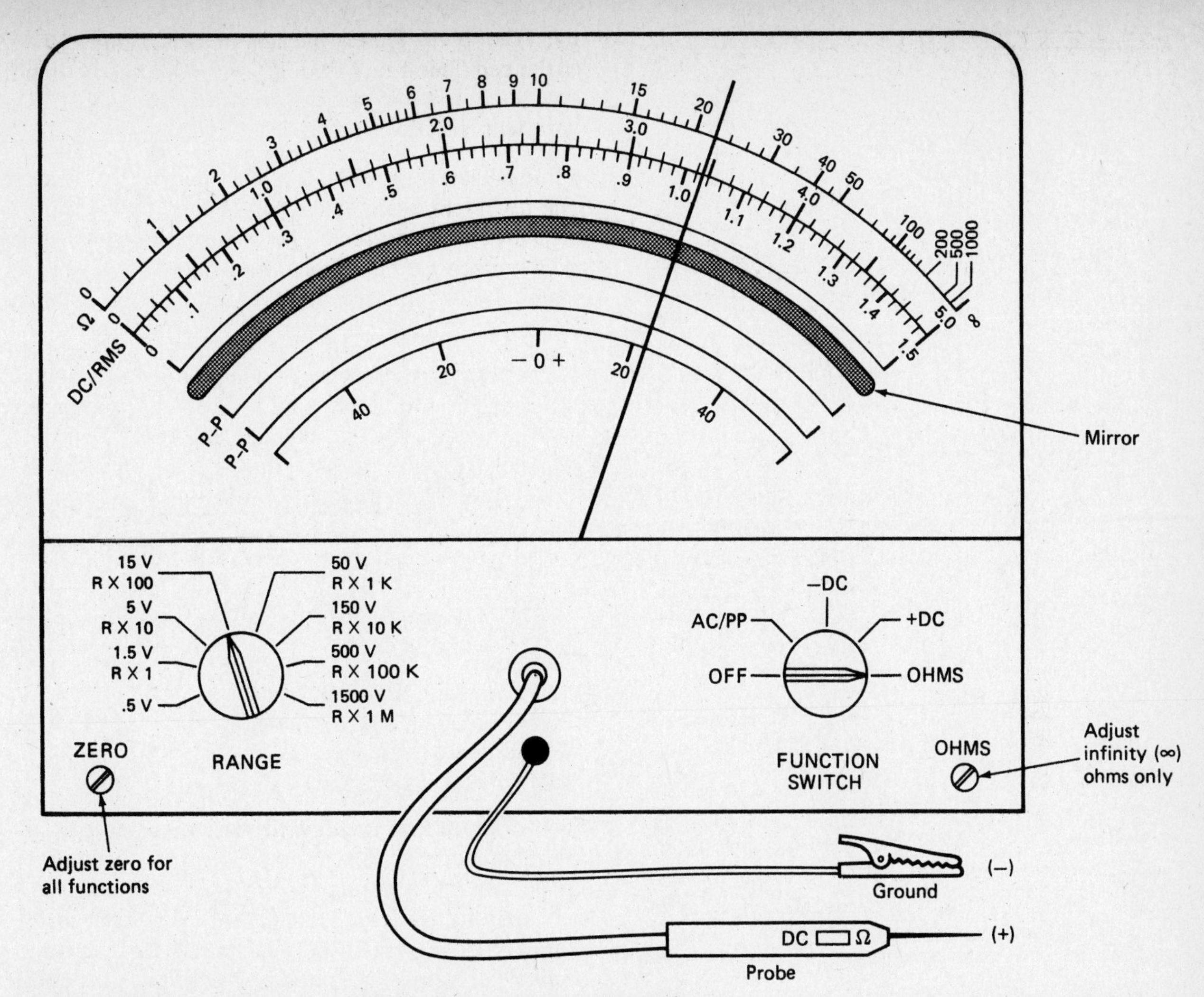

Fig. 2-2. VTVM (typical generic type). The top scale is nonlinear (0 to infinity), used for resistance (Ω). The remaining scales are linear (0 to 5, 0 to 1.5, etc.), used for voltage. Note that most VTVMs have a mirror strip between scales to prevent parallax errors. These visual errors are similar to those made by passengers in a car trying to read the speedometer from an angle. To avoid parallax errors, line up the needle with the mirror reflection and with the division line on the scale.

allows you to position the needle directly on the zero line before making a measurement.

Ohms Adjustment: This knob, also a potentiometer, allows you to line up the needle on the other end of the scale, on the infinity line. It is only used for resistance measurements (ohmmeter function).

Ground Lead: This lead is usually connected first. It is the negative (−), or ground, side. First, connect this ground strap or lead to one end of the component you are measuring. Then, make connection with the probe (+).

Probe: Often called the *positive lead,* this pointed probe has a switch on it for measuring either ohms or dc voltage. After the ground lead is connected, it is easy to use this probe to make contact with the component you are measuring.

3. VOM (Volt-Ohm-Milliammeter): Refer to Fig. 2-3. Shown here is a generic representation of a typical VOM. VOMs are portable because, unlike VTVMs, they operate on batteries. VOMs are usually less expensive than digital multimeters or VTVMs. However, the VOM is still a good meter to use as a learning tool.

If you have a VOM in your lab, refer to it along with Fig. 2-3 and continue. The VOM can measure voltage, resistance (ohms), and current (amperes). Depending upon the model, VOMs can measure milliamps or even as much as 10 to 12 A.

Find the following items on the VOM and check them off as you go along.

Function/Range Switch: This rotary switch should be in the OFF position when not in use; otherwise, the VOM's battery may become depleted. Not only does this switch allow you to choose the desired functions (ohmmeter, voltmeter, ammeter), but it also indicates the range. Notice that, in Fig. 2-3, the

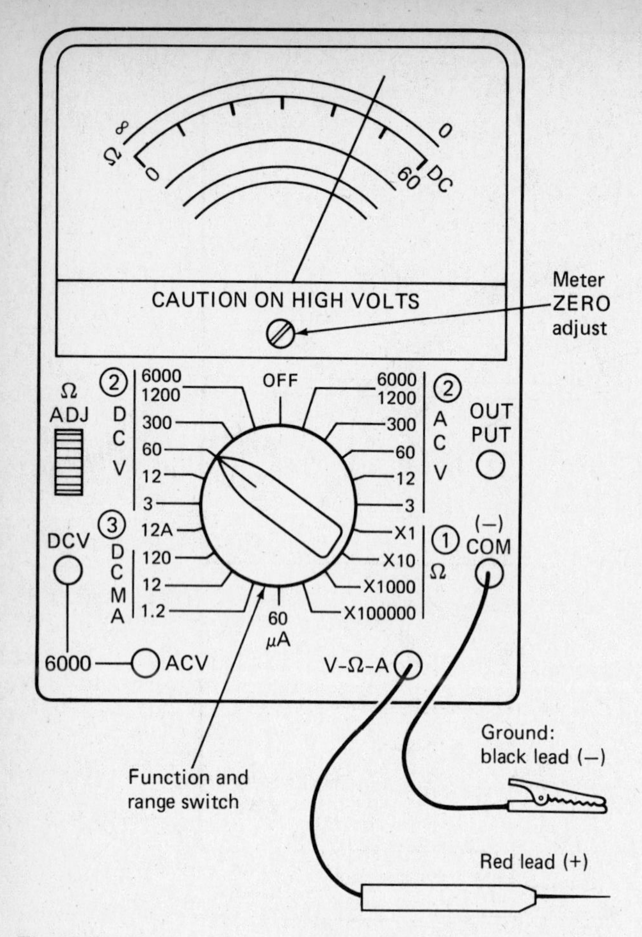

Fig. 2-3. VOM. Linear and nonlinear scale (ohms). A generic VOM measures ① resistance, ohms (Ω), ② ac and dc voltage, and ③ current, amperes (A). The VOM shown is set for ②, 0 to 60 V dc.

switch is set to measure dc voltage with a range of 0 to 60 V. VOMs often seem more difficult to use than VTVMs because they have so many switches and inputs located in a small area.

Ohms Adjust: This is the zero adjustment for measuring resistance (ohms). Use it to line up the needle directly on the zero line when using the VOM as an ohmmeter when the leads are shorted.

COM: This is the common or negative input. That is, it is the place where the black ground lead is plugged into the meter. Remember that VOMs, like most electronic instruments, have two inputs: a negative and a positive.

VΩA: This is the positive input. Plug the other lead (red, if available) into this jack. Note that the words *jack, input, terminal,* and *plug-in* are often used interchangeably. You will become used to this terminology as you continue. This plug-in is used whenever measuring volts, ohms, or amperes (dc milliamperes).

Other Inputs: Do not be concerned with the other inputs at this time. You will not be using them. However, notice that the zero adjust is not marked on the VOM itself. This is because it is not adjusted (balanced) like the VTVM. If your VOM needle is not on the zero line in the dc volts function, adjust it here with a screwdriver.

Scales: Like the VTVM, the VOM has several scales. The nonlinear scale at the top is used for resistance measurements in ohms. The linear scale below is usually used for dc voltage. And the milliamperes scale (current measurements) may be another scale altogether, depending upon the manufacturer.

4. Summary of Meters: As you continue through this manual, you will mostly be making three basic measurements: current, voltage, and resistance. Therefore, do not be discouraged if the variation in meter types or terminology seems to be overwhelming.

You may be using one of the following:

Voltmeter
Ammeter
Ohmmeter
VOM (volt-ohm-milliammeter)
VTVM (vacuum tube voltmeter)
DVM (digital voltmeter)
DMM (digital multimeter)

Each of the above meters will vary depending upon the manufacturer. But you only need to concentrate on the meters in your particular lab. Therefore, you may wish to ask for a demonstration, or, if available, you may wish to read the manufacturer's operating manual for your meter. In any case, remember these things:

The nonlinear scale is for measuring resistance in ohms.
The linear scale is for voltage and current measurements.
The words *ground* or *negative* usually refer to the same thing: one side of the meter connection.
The zero or infinity adjustments are used to calibrate the meter for use as an ohmmeter.
Only voltage and resistance measurements are required for this experiment. Current will be measured in Experiment 3.

5. Reading Ohmmeter and Voltmeter Scales (VTVM Type): Refer to Fig. 2-4. There are 12 typical VTVM (similar to VOM) measurements. The nonlinear top scale is used for resistance, and the linear scale beneath that is used for voltage. Notice that the DC OR RMS (rms = root mean square, and is essentially the same as dc voltage) scale is actually many scales in one. Also, the range switch not only determines which scale to use, 0 to 5.0 or 0 to 1.5, but it also determines if you need to convert the 0 to 5.0 scale to 0 to 50.0, etc. For measurements 5 to 12 fill in the correct values for resistance or voltage. Check your answers on page 14.

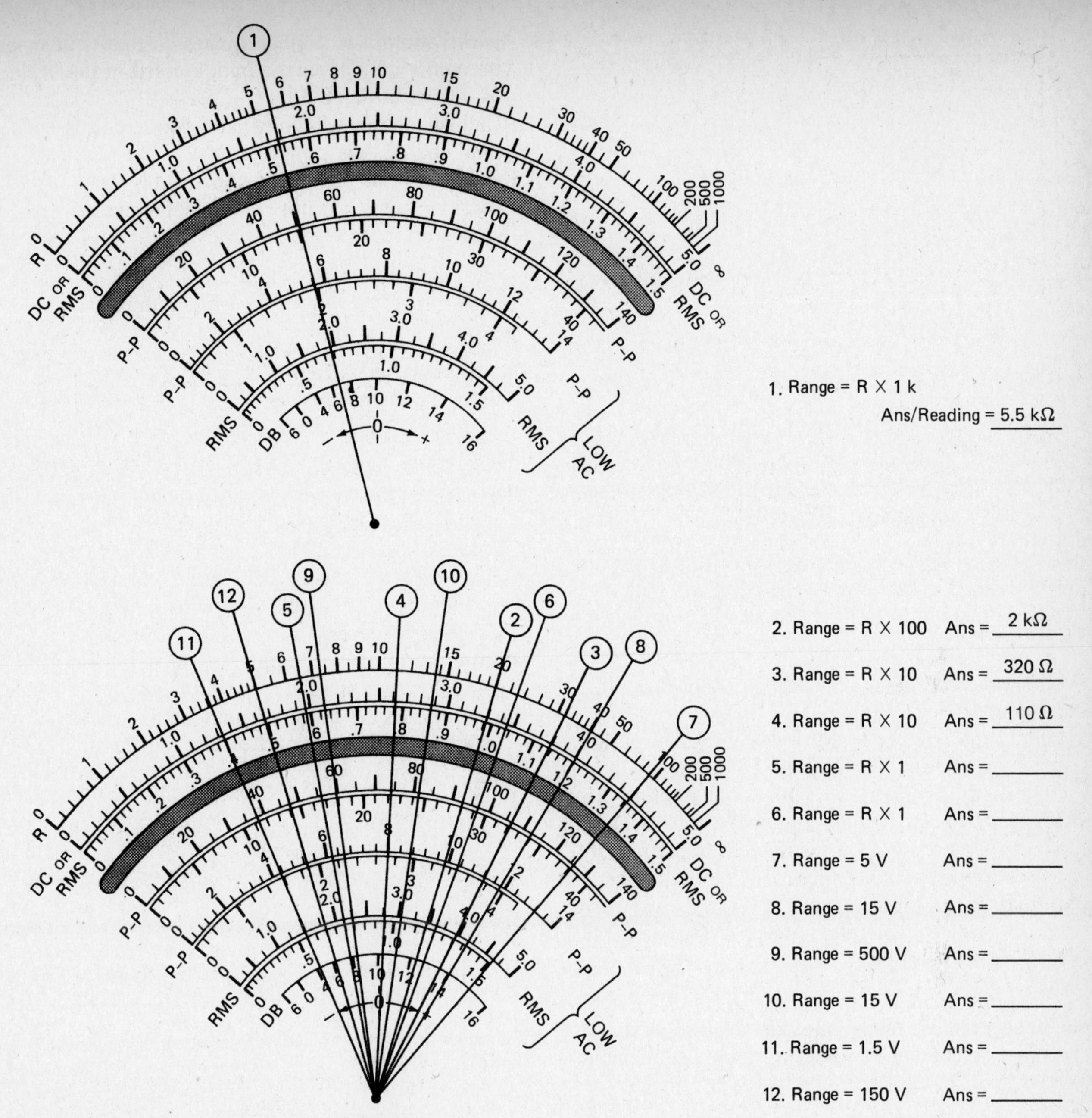

1. Range = R × 1 k
   Ans/Reading = 5.5 kΩ

2. Range = R × 100 Ans = 2 kΩ
3. Range = R × 10 Ans = 320 Ω
4. Range = R × 10 Ans = 110 Ω
5. Range = R × 1 Ans = ______
6. Range = R × 1 Ans = ______
7. Range = 5 V Ans = ______
8. Range = 15 V Ans = ______
9. Range = 500 V Ans = ______
10. Range = 15 V Ans = ______
11. Range = 1.5 V Ans = ______
12. Range = 150 V Ans = ______

Fig. 2-4. Ohmmeter and voltmeter scales (VTVM type).

6. Measuring Resistance Using a VTVM: (A VOM can be substituted.) Follow the procedure below and check off each step as you continue.

Turn on the VTVM by setting the function switch to the OHMS position. Notice that the needle will go toward the infinity (∞) line. This is because infinite resistance is the starting point; between the ground lead and the positive lead (probe), there is only air, or infinite resistance.

Switch the probe to the OHMS position.

After allowing the VTVM to warm up for 1 or 2 min, connect both leads together, resulting in a short circuit, or zero resistance. The needle should now go in the other direction, toward zero.

Keeping both leads connected (short-circuited), use the ZERO adjust to align the needle (pointer) with the zero line. Be sure to use the mirror for proper alignment and reduction of parallax error. If the pointer, the zero line, and the mirror reflection are one, you have zeroed the ohmmeter correctly.

Disconnect the two leads. The needle will move toward the infinity line. Now, use the OHMS adjust and align the pointer with the infinity line. Alignment of the pointer, the infinity line, and the mirror reflection of the needle (pointer) will eliminate parallax errors.

Repeat the zero adjustment once more (short-circuit the leads together), and the meter should now be calibrated.

Set the range for R × 10.

Connect the two leads across the 100-Ω resistor as

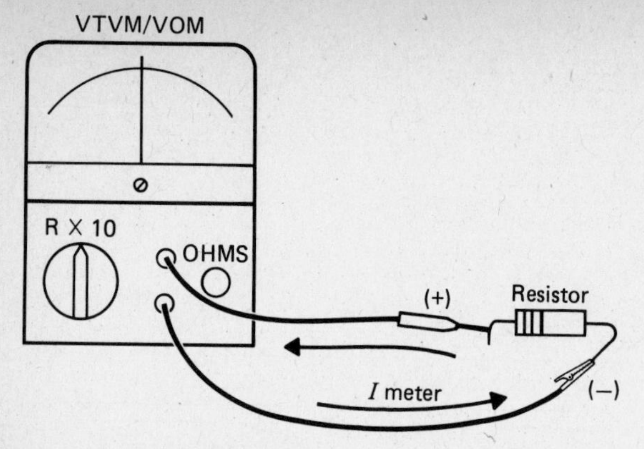

Fig. 2-5. Circuit building aid.

shown in Fig. 2-5. It does not matter which side of the resistor you use; you may use either side. Also, never measure resistance when voltage is applied. You may damage the meter if you do.

Note: When measuring resistance, either end of the resistor can be connected. The resistor's polarity (+ or −) is due to the current passing through the resistor.

You should see the needle resting on or near the number 10.

Multiply the number that the needle is indicating by the range setting. For example, if the needle is resting exactly between 9 and 10, at 9.5, multiply 9.5 × 10. The 10 is for R × 10. The result is 95 Ω. Do not be concerned if your measurement is not exactly 100 Ω. Remember that there may be differences due to manufacturing tolerances. Even a 100-Ω resistor, with 10 percent tolerance (silver, fourth band) may be 90 Ω and still be good.

Now, measure all the resistors listed in the component section. Write the measured value next to the nominal (color band) value in Table 2-1. Remember:

Always attempt to change ranges so that your measurements fall within the middle third of the scale, as shown in Fig. 2-6.

Readjust the OHMS (infinity) and ZERO controls each time you change ranges.

Remember that you cannot be entirely exact when reading scales. Do not be overly concerned if your values are not perfect.

In Fig. 2-6 note that both the 4–5 and the 40–50 divisions are not within the middle third. The choice is yours.

7. Measuring Voltage Using a VTVM: (A VOM or a simple voltmeter can be substituted.) Follow the procedure below and check off each step as you continue.

Turn the VTVM on by setting the function switch to positive dc volts (+DC).

Switch the probe (VTVM only) to dc volts.

Adjust the ZERO control to be sure you start with the needle aligned with the zero division line. The VTVM is now ready to be used as a voltmeter.

Locate the dc power supply. It will be similar to the generic power supply shown in Fig. 2-7.

Locate the ON/OFF switch. It may be a separate switch, or it may be part of the voltage control.

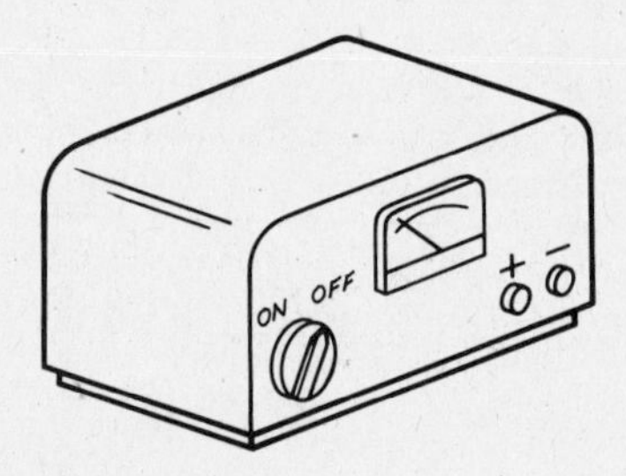

Fig. 2-7. DC power supply.

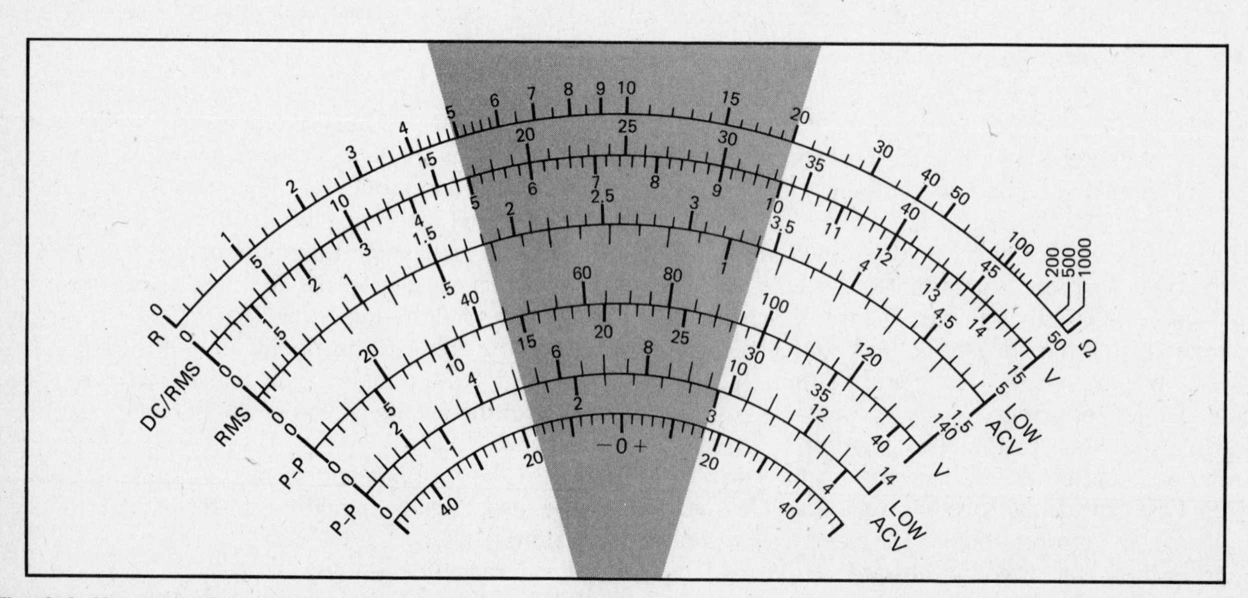

Fig. 2-6. Use the middle third of scale for readings, if possible.

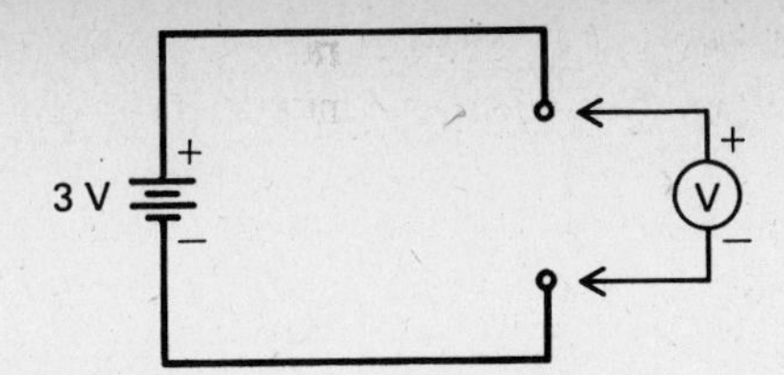

Fig. 2-8. Simple schematic of a dc voltage measurement. (A schematic is a type of electronic road map or blueprint.) The 3 V on the left means that the power supply is set at 3 V. The V in a circle on the right indicates a voltmeter (VTVM or VOM).

Be sure to locate the positive (+) and negative (−) terminals of the power supply. These are similar to the ends (+ and −) of a battery.

Set the power supply range switch (if any) to the setting that will allow you to adjust for up to 5 V (for example, 10 V will allow you to adjust for 10 V or less).

Now that you are ready to measure voltage, connect the VTVM (voltmeter) directly to the power supply as shown in Figs. 2-8 and 2-9.

Remember, connect the negative ground lead to the negative output of the power supply. Connect the probe, or positive side, of the VTVM voltmeter to the positive side of the power supply.

Now, slowly increase the voltage control. The needle (VTVM) should deflect upscale. If not, check your connections. Stop at +3 V.

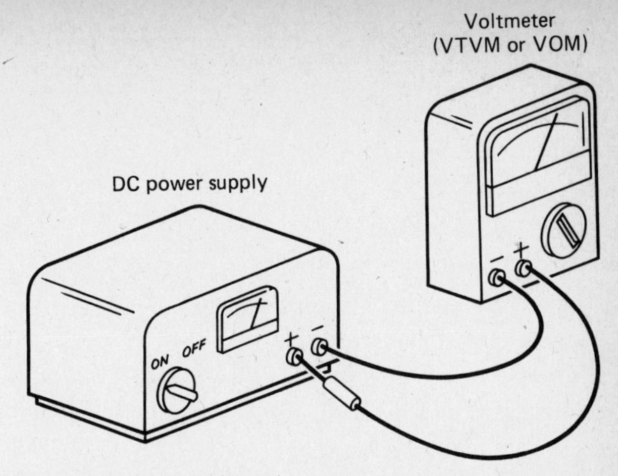

Fig. 2-9. Circuit building aid. Measuring the output of a dc power supply by using a voltmeter (VTVM or VOM).

Does your meter indicate +3 V? If so, you have probably been successful.

Once you feel comfortable, you may wish to try your skill with a known voltage. For example, measure a flashlight battery (1.5 V). You should read 1.5 V on your meter. Remember, always observe the proper polarity (+ and −).

NAME ______________________________ DATE ____________

RESULTS FOR EXPERIMENT 2

QUESTIONS

Answer TRUE (T) or FALSE (F) to the following:

____ **1.** It is always necessary to allow VTVMs to warm up prior to use.

____ **2.** The VTVM can measure current.

____ **3.** An ohmmeter will show zero ohms when the leads are not connected together (open circuit).

____ **4.** Linear scales are used for resistance measurements.

____ **5.** It is necessary to adjust infinity (∞) and zero ohms whenever changing ranges on an ohmmeter.

____ **6.** When measuring voltage, the negative lead is connected to the positive side of the voltage source or battery.

____ **7.** Shorting the leads together on an ohmmeter results in zero ohms.

____ **8.** Parallax is an error resulting from reading meter scales from an angular view.

____ **9.** An ohmmeter cannot be damaged by measuring voltage.

____ **10.** The range switch is only used for voltage measurements.

REPORT

Turn in the following:

1. Your data table (Table 2-1) of resistance measurements.
2. The answers to questions 1 to 10.
3. A one-page summary of voltage- and resistance-measuring techniques. Write a paragraph on each. Discuss only those things that are important for making safe and accurate measurements.

TABLE 2–1. Resistance Measurements Using the Ohmmeter

Nominal Value, Ω	Measured Value, Ω	Nominal Value, Ω	Measured Value, Ω

Answers to Fig. 2-1

b. 400 V
c. 90 V
d. 6 V
e. 90 mV
f. 35 kV
g. 550 V
h. 17 mV

Answers to Fig. 2-4

5. 6.5 Ω
6. 23 Ω
7. 4.5 V
8. 12 V
9. 210 V
10. 8.6 V
11. 0.4 V
12. 50 V

EXPERIMENT 3

OHM'S LAW

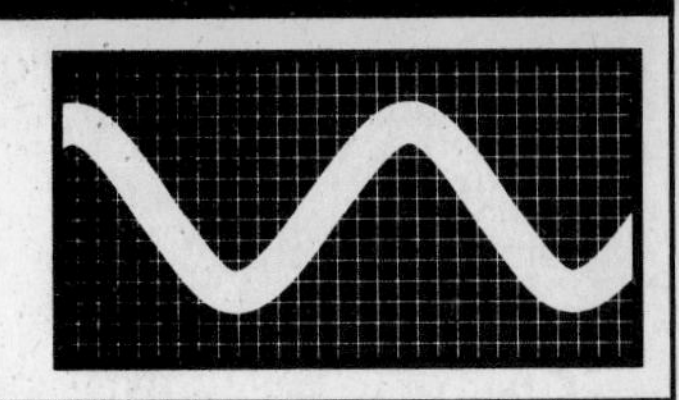

OBJECTIVES

At the completion of this experiment, you will be able to:

- Validate the Ohm's law expression, where

$$V = I \times R$$

$$I = \frac{V}{R}$$

$$R = \frac{V}{I}$$

SUGGESTED READING

Chapter 3, *Basic Electronics,* B. Grob, Seventh Edition

INTRODUCTION

Ohm's law is the most widely used principle in the study of basic electronics. In 1828, George Simon Ohm experimentally determined that the amount of current I in a circuit depended upon the amount of resistance R and the amount of voltage V. If any two of the factors V, I, or R are known, the third factor can be determined by calculation: $I = V/R$, $V = IR$, and $R = V/I$. Also, the amount of electric power, measured in watts, can be determined indirectly by using Ohm's law:

$$P = \frac{V^2}{R} \quad \text{or} \quad P = I^2R$$

Ohm's law is usually represented by the equation $I = V/R$. Because it is a mathematical representation of a physical occurrence, it is important to remember that the relationship between the three factors (I, V, and R) can also be expressed as follows:

Current is directly proportional to voltage if the resistance does not vary.
Current is inversely proportional to resistance if the voltage does not vary.

This experiment provides data that will validate the relationships between current, voltage, and resistance.

Measuring Current with an Ammeter (VOM)

You have already learned how to use an ohmmeter to measure resistance and a voltmeter to measure voltage in Experiment 2. For both those meters, it was necessary to place the two (+ and −) leads *across* the component you measured. However, an ammeter is actually connected *in* the circuit—in the path of current (electron) flow. Therefore, the leads are used to connect the ammeter into a circuit only after you break the circuit open. Refer to Fig. 3-1.

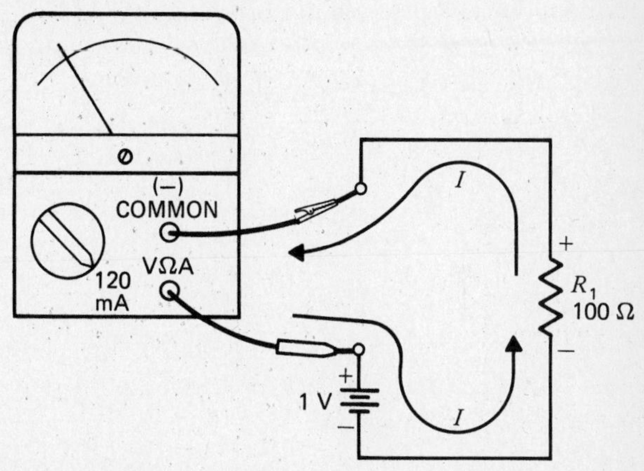

Fig. 3-1. Ammeter (VOM) connected in a circuit to measure current.

Notice that the VOM range/function switch is set to 120 mA (full-scale deflection). Also notice that the electron flow (current) leaves the negative side of the battery, flows through the resistor, and enters the negative side of the ammeter (VOM). Then it exits the ammeter and returns to the positive side of the battery.

Let us simplify Fig. 3-1 by replacing the ammeter with the symbol A. Now, the same circuit of Fig. 3-1 can be seen in the schematic of Fig. 3-2.

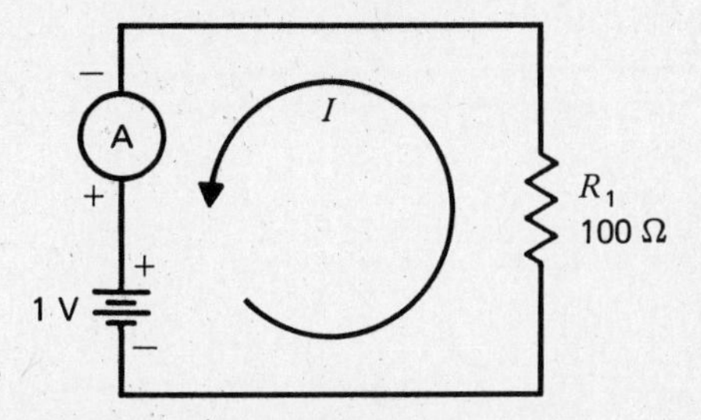

Fig. 3-2. Schematic representation of Fig. 3-1.

Here, it should be understood that to connect the ammeter, the circuit was broken between the positive side of the battery and the resistor. Also, to determine where the polarities are, notice that wherever electron flow enters a component, that side of the component is then considered negative. Where electron flow leaves the component, closer to the positive side of the battery source, it is considered positive. Thus, the resistor and ammeter, in Fig. 3-2, are shown where electron flow determines polarity. Of course, electron flow must enter the common, or negative, side of the ammeter and exit from the positive side (VΩA).

Of course, power is turned off before making any connections.

Finally, an easy way to protect an ammeter, or any meter, is to allow one lead (positive, for example) to act as a switch, as shown in Fig. 3-3. Here, the positive lead (VΩA) of the ammeter can be touched lightly to the connection so that you can be sure the proper polarity (meter goes upscale) is there. Lightly tap or touch the positive lead to the place you are going to connect it. Watch the meter needle; it should go upscale without pegging (attempting to go out of range).

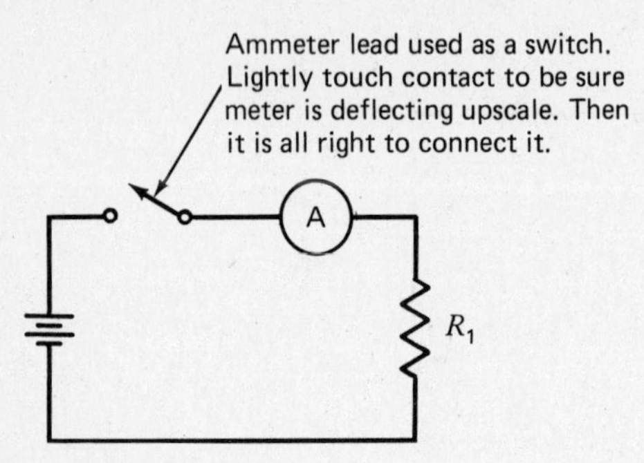

Fig. 3-3. Ammeter in circuit.

EQUIPMENT

Ohmmeter
Voltmeter
Ammeter
Power supply
Connecting leads
Circuit board

COMPONENTS

Resistors (all 0.5 W):

(1) 100 Ω (1) 820 Ω
(1) 330 Ω (1) 1 kΩ
(1) 560 Ω

PROCEDURE

1. Connect the circuit of Fig. 3-4*a* with the power supply turned off. The polarity of the power supply, the voltmeter, and the ammeter must be correct in order to avoid damage to the equipment.

Note: Use the circuit building aid of Fig. 3-4*b* if you have trouble.

2. Refer to Table 3-1. Calculate the current for the first voltage setting (1.5 V, 100 Ω nominal value). Record the value in Table 3-1 under the heading Calculated Current.

3. Check your circuit connection by tracing the path of electron flow. Be sure the ammeter reading is correct for the value of calculated current. Also, be sure the voltmeter range is correct.

4. Before turning the power supply on, turn the voltage adjust knob to zero. Now, turn the power supply on. Use the voltmeter to monitor the power supply and adjust for 1.5 V applied voltage.

Note: The ammeter can also be monitored. Unless you are using a digital meter, the pointer or needle should deflect upscale.

5. Read the value of measured current and record the results in Table 3-1.

6. Repeat procedures 2 to 5 for each value of applied voltage listed in Table 3-1.

Note: It is not necessary to turn the power supply off unless you are disconnecting the circuit in order to change meter ranges.

7. After recording all the measured and calculated values of current for Table 3-1, turn off the power supply. Be sure that the voltage adjust is set to zero. Disconnect the circuit. The calculated values of power can be done later. Indicate which formula you used.

8. Refer to Table 3-2. Use an ohmmeter to measure and record the values of each resistor listed in Table 3-2.

Note: Remember to adjust the ohmmeter for zero and infinity, depending upon the type you are using.

9. Connect the circuit of Fig. 3-5. Keep the power supply turned off and begin with the first value of resistance in Table 3-2, $R = 1000\ \Omega$ (1 kΩ).

Note: This is the same basic circuit as Fig. 3-1. However, the ammeter is located in a different part of the circuit's current path. Use the same precautions as you did with Fig. 3-4.

10. Refer to Table 3-2. Calculate the current for the value of resistance listed in Table 3-2 (5.0 V, 1 kΩ nominal value). Record the calculated value in Table 3-2 under the heading Calculated Current.

11. Check your circuit. Be sure that the ammeter range is correct for the value of calculated current. Also, be sure that the voltmeter range is correct.

12. Turn on the power supply and adjust for 5.0 V applied voltage. Monitor the meters as you did for the circuit of Fig. 3-4.

13. Read the value of measured current and record the results in Table 3-2.

14. Turn the power supply off. Replace the resistor with the next value of resistance listed in Table 3-2.

15. Repeat procedures 10 to 14 for each value of resistance listed in Table 3-2.

16. Turn the power supply off. Disconnect the circuit.

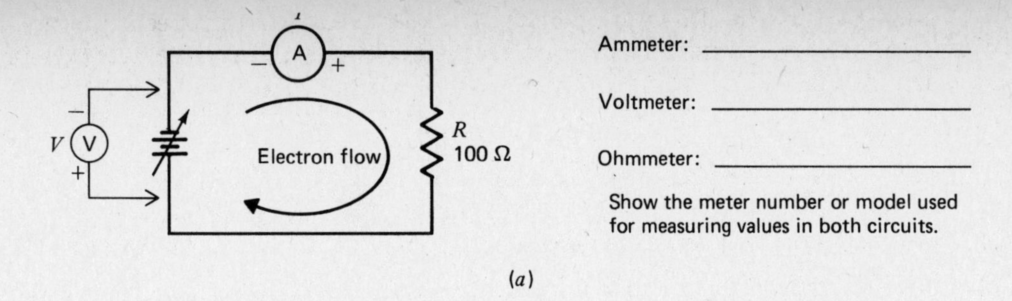

Ammeter: ______________________

Voltmeter: ______________________

Ohmmeter: ______________________

Show the meter number or model used for measuring values in both circuits.

(*a*)

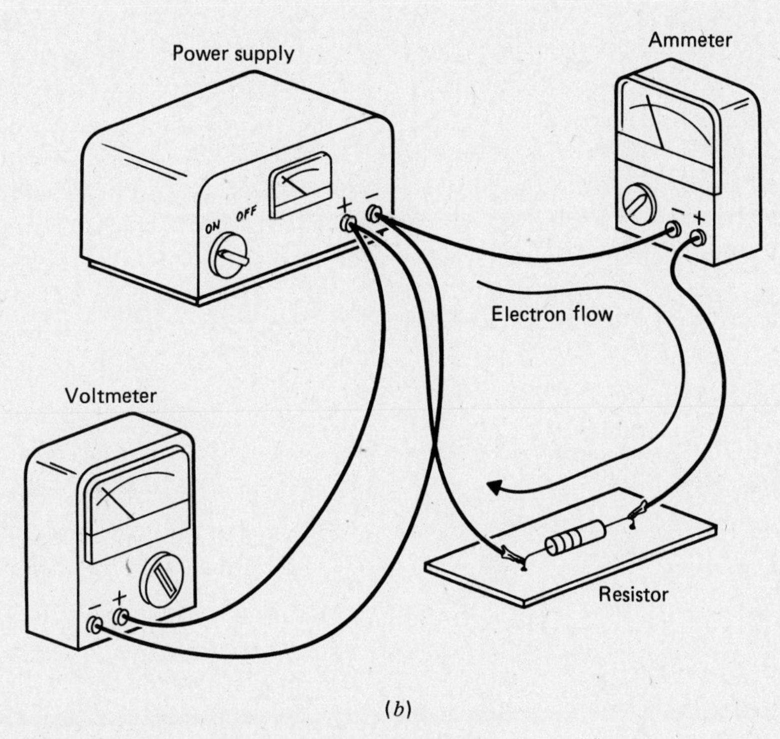

(*b*)

Fig. 3-4. (*a*) Ohm's law schematic. (*b*) Circuit building aid. (Note that arrows indicate electron flow or current path.)

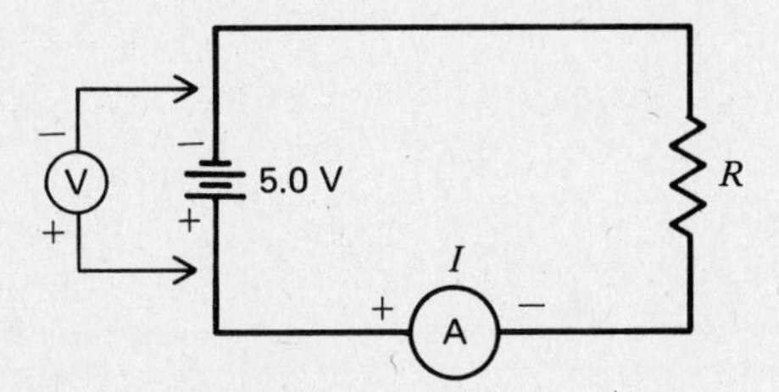

Fig. 3-5. Ohm's law circuit.

NAME ______________________________ DATE ____________

RESULTS FOR EXPERIMENT 3

QUESTIONS

Choose the correct answer.

_____ **1.** If the circuit of Fig. 3-1 had 10 kΩ of resistance, the amount of applied voltage necessary to produce 1 mA would be:
A. 1000 V **B.** 100 V **C.** 10 V **D.** 1 V

_____ **2.** In the circuit of Fig. 3-1, if the applied voltage was increased, the amount of power would be:
A. Decreased **B.** Increased **C.** Stayed the same

_____ **3.** Compared to a voltmeter, the ammeter in Fig. 3-1 is:
A. Drawn differently and connected differently **B.** Drawn the same and connected differently **C.** Drawn differently and connected the same **D.** Drawn the same and connected the same

_____ **4.** Referring to the circuit of Fig. 3-2, if the voltage was doubled for each step but the resistance was halved, the current (Table 3-2) would:
A. Increase by twice as much **B.** Decrease by one-half **C.** Decrease by one-fourth **D.** Increase by four times as much

_____ **5.** According to Ohm's law and the data gathered in the experiment:
A. The less resistance, the less current with constant voltage **B.** The more current, the less voltage with constant resistance **C.** The less resistance, the more current with constant voltage **D.** The less current, the more voltage with constant resistance.

_____ **6.** If the terminals (negative and positive) of the power supply in Fig. 3-2 were reversed, it would be necessary to:
A. Reverse the terminals of the ammeter and the resistor **B.** Reverse the terminals of the voltmeter and the ammeter **C.** Reverse the terminals of the voltmeter and the resistor **D.** Reverse the terminals of the resistor only

_____ **7.** Because the ammeter is connected in the same way as the resistor, would you expect the ammeter's resistance to be:
A. Very large **B.** Very small **C.** Medium value

_____ **8.** To obtain a current value of 30 mA, the amount of voltage and resistance necessary would be:
A. $V = 100$ V, $R = 333$ Ω **B.** $V = 15$ V, $R = 500$ Ω **C.** $V = 3$ V, $R = 100$ Ω **D.** A, B, and C

REPORT

Organize your written report as follows:

1. Cover sheet, also called a blank style sheet, first.
2. Results: Data Tables 3-1 and 3-2.
3. Schematics with the numbers or models of the equipment used.
4. Answers to questions 1 to 8.

Note: Discuss the results after you analyze the data. For this report, discuss voltage, current, and resistance as they are related according to Ohm's law and your data. Write short concise sentences. Remember, the *Discussion* is the most important part of a report; it relates your understanding of the experiment.

The *Purpose* is usually a statement about what you are trying to validate. The *Procedure* is the lab manual title, page number(s) where the procedure is found, and any changes made to the procedure. The *Results* are listed as any data tables or graphs by number. The *Conclusion* is a brief summary of the experiment with respect to the Purpose. Be neat and accurate. Answer the questions assigned, and attach the answers on a separate sheet of lined 8½ × 11 in. paper. Type or write the report in ink (black or blue). A sample report is included in the Appendix.

TABLE 3–1. Ohm's Law

Applied Voltage, V	Nominal Resistance, Ω	Measured Current, mA	Calculated Current,* mA	Calculated Power†
1.5	100	______	______	______
2.5	100	______	______	______
3.5	100	______	______	______
4.5	100	______	______	______
6.0	100	______	______	______
7.0	100	______	______	______
8.0	100	______	______	______
9.0	100	______	______	______

*Formula: Applied voltage/nominal resistance

†Formula: ______________

TABLE 3–2. Ohm's Law

Applied Voltage, V	Nominal Resistance, Ω	Measured Resistance, Ω	Measured Current, mA	Calculated Current,* mA
5.0	1000	______	______	______
5.0	820	______	______	______
5.0	560	______	______	______
5.0	330	______	______	______
5.0	100	______	______	______

*Formula: Applied voltage/measured resistance

EXPERIMENT 4
SERIES CIRCUITS

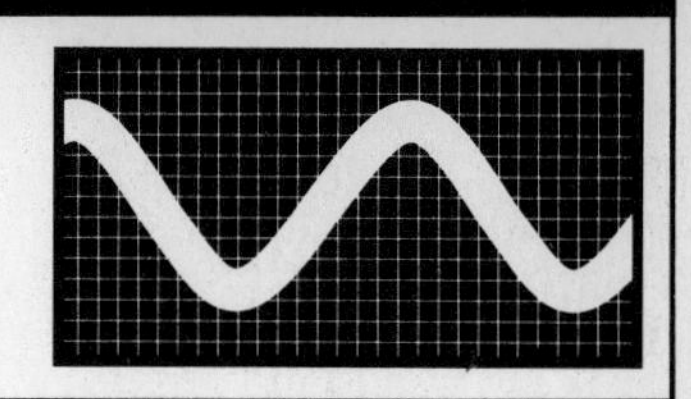

OBJECTIVES

At the completion of this experiment, you should be able to:

- Recognize the basic characteristics of series circuits.
- Compare the mathematical relationships existing in a series circuit.
- Compare the mathematical calculations to the measured values of a series circuit.

SUGGESTED READING

Chapter 4, *Basic Electronics,* B. Grob, Seventh Edition

INTRODUCTION

An electric circuit is a complete path through which electrons can flow from the negative terminal of the voltage source, through the connecting wires or conductors, through the load or loads, and back to the positive terminal of the voltage source.

If the circuit is arranged so that the electrons have only one possible path, the circuit is called a *series circuit.* Therefore, a series circuit is defined as a circuit that contains only one path for current flow. Figure 4-1 shows a series circuit with several resistors.

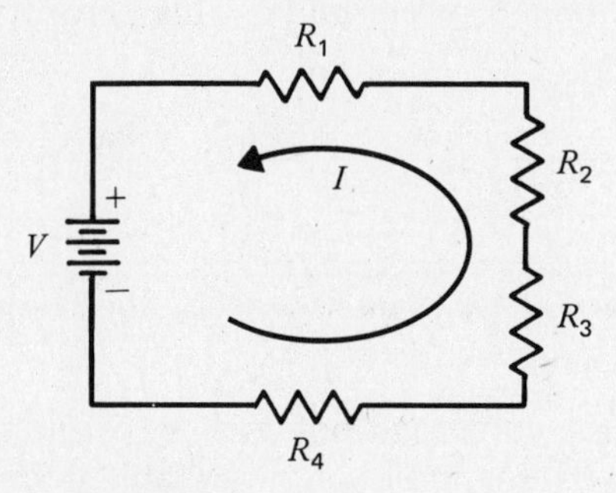

Fig. 4-1. Series circuit.

One of the most important aspects of a series circuit is its relationship to current. Current in a series circuit is determined by Ohm's law, which defines a proportional relation between voltage and the total circuit resistance. This relationship between total voltage and total circuit resistance results in the current being the same value throughout the entire circuit. In other words, the measured current will be the same value at any point in a series circuit.

Figure 4-2 shows a series circuit consisting of a battery and two resistors. The battery is labeled V_t and provides the total voltage across the circuit. The resistors are labeled R_1 and R_2. The resistance values are $R_1 = 100\ \Omega$ and $R_2 = 2.7\ \text{k}\Omega$. The power supply has been adjusted to a level of 20 V.

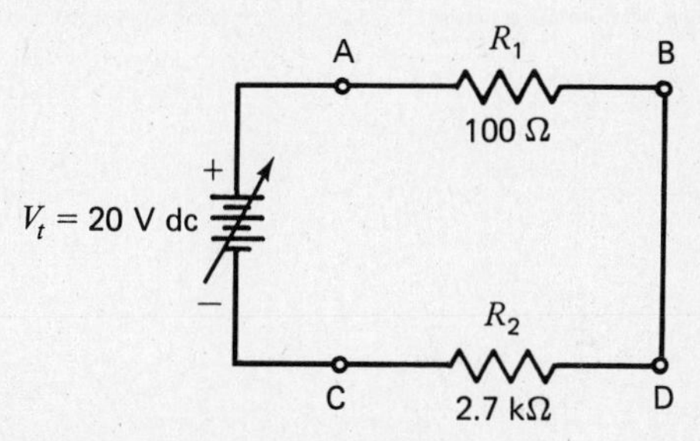

Fig. 4-2. Series circuit for analysis.

Figure 4-2 shows several points labeled A, B, C, and D. These are the points at which current could be measured. To measure the current at these points, it is necessary to break into the circuit and insert an ammeter in series. Remember, in a series circuit the amount of current measured will be the same at each of these points.

The total current, labeled I_t, flowing throughout the series circuit depends upon two factors: the total resistance R_t and the applied total voltage V_t. The applied total voltage was given in Fig. 4-2 such that

$$V_t = 20\ \text{V}$$

To determine the current, the total resistance R_t must be calculated by

$$R_t = R_1 + R_2$$

The total resistance for the circuit shown in Fig. 4-2 is

$$R_t = 100\ \Omega + 2700\ \Omega = 2800\ \Omega$$

For circuits that have three (3) or more series resistors, the above formula may be modified as shown to include the additional resistors:

$$R_t = R_1 + R_2 + R_3 + \cdots$$

When solving for the current I, Ohm's law states that

$$I = \frac{V}{R}$$

When solving for current I_t,

$$I_t = \frac{V_t}{R_t} = \frac{20\ \text{V}}{2800\ \Omega} = 0.00714\ \text{A} = 7.14\ \text{mA}$$

Applied voltage will be dropped proportionally across individual resistances, depending upon their value of resistance. As shown in Fig. 4-3, the *IR* voltage drops can be solved as follows. The sum of the voltage drops will be the total applied voltage. This is stated by

$$V_t = IR_1 + IR_2$$

or $V_t = V_{r_1} + V_{r_2}$

where

$$V_{r_1} = I_t R_1 = (7.14\ \text{mA})(100\ \Omega) = 0.71\ \text{V}$$

and

$$V_{r_2} = I_t R_2 = (7.14\ \text{mA})(2700\ \Omega) = 19.27\ \text{V}$$

Therefore,

$$V_t = 0.71\ \text{V} + 19.27\ \text{V} = 19.98\ \text{V} = 20\ \text{V (approximately)}$$

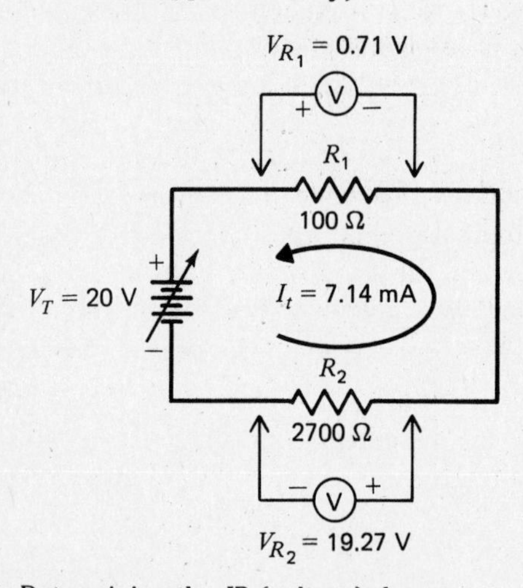

Fig. 4-3. Determining the *IR* (voltage) drops.

EQUIPMENT

Breadboard
DC power supply
Voltmeter
Ammeter
Ohmmeter

COMPONENTS

Resistors (all 0.5 W):

(1) 150 Ω (1) 10 Ω
(1) 15 Ω

PROCEDURE

1. With an ohmmeter, measure each resistor value for the resistors required for this experiment. Connect the resistors in series and measure the total resistance R_t. Record the results in Table 4-1.

2. Using the information in the introduction and Fig. 4-4, calculate the total resistance R_t, the series current I_t, and the *IR* voltage drops across R_1, R_2, and R_3. Record the results in Table 4-2.

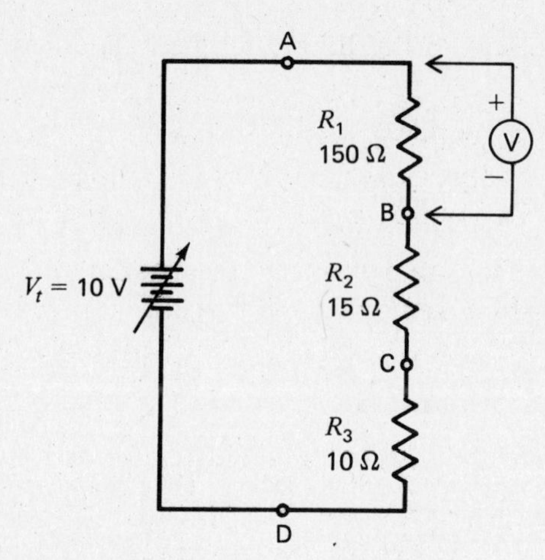

Fig. 4-4. Series circuit.

3. Connect the circuit in Fig. 4-4 and turn on the dc power supply. Using a voltmeter, adjust the power supply level to 10 V dc.

4. Measure the *IR* voltage drops across each resistance as indicated in Fig. 4-4. For example, the voltmeter is shown connected across R_1 to measure V_{R_1}. Record the results in Table 4-2.

5. After recording the measured voltage drops, compare the measured and calculated values. Use the following formula for determining this percentage:

$$\% = \frac{\text{difference between meas. and calc. values}}{\text{calc. values}} \times 100$$

If any error is greater than 20 percent, then repeat calculations or measurements.

6. Measure the current at points A, B, C, and D in Fig. 4-4. Record this information in Table 4-3.

Note 1: To measure current, you must actually break the circuit at these points and insert your ammeter (or current-measuring resistor, if you are not using an ammeter) in accordance with the information presented in the introduction.

Note 2: Check with your instructor in case you need to leave this circuit connected. When all results have been recorded, call your instructor over to your lab station to verify your results.

NAME ______ DATE ______

RESULTS FOR EXPERIMENT 4

QUESTIONS

Choose the correct answer.

_____ **1.** In a series circuit, total current I_t is equal to:
A. $R_t \times I_t$ **B.** $V_t \times R_t$ **C.** V_t/R_t **D.** R_t/V_t

_____ **2.** In a series circuit, the current is:
A. Different through every resistor in series **B.** Always the same through every resistor in series **C.** Calculated by using Ohms law as $I = V \times R$ **D.** Found only by using the voltmeter

_____ **3.** The total voltage in a series circuit is:
A. Equal to total resistance **B.** Found by adding the current through each resistor **C.** Equal to the sum of the series *IR* voltage drops **D.** Found by using an ohmmeter

_____ **4.** In a series circuit with 10 V applied:
A. The greater the total resistance, the less the total current **B.** The greater the total current, the greater the total resistance **C.** The *IR* voltage drops will each equal 10 V **D.** The sum of the *IR* voltage drops will equal 10 V

_____ **5.** When an *IR* voltage drop exists in a series circuit:
A. The polarity of the resistor is equal to positive **B.** The polarity of the resistor is equal to negative **C.** The polarity of the resistor is less than the total current on both sides **D.** The polarity of the resistor is positive on one end and negative on the other because of current flowing through it

REPORT

When completing the report, discuss your results by answering the following three questions as the most significant aspects:

1. Does $V_t = I_{R_1} + I_{R_2} + I_{R_3}$?
2. Is the measurement current the same at all parts of the series circuit?
3. Does $R_t = R_1 + R_2 + R_3$?

Write your own conclusion by summarizing the concept of series circuits.

Report Notes

The test circuit (Fig. 4-5) appears along with Tables 4-1 to 4-3. Use them as part of your report. Tear them out here and staple them to your report. You will be responsible for completing the remaining sections of the report.

Note: All necessary paper has been provided for this experiment. The following experiments will gradually include less and less given material until the report becomes your full responsibility.

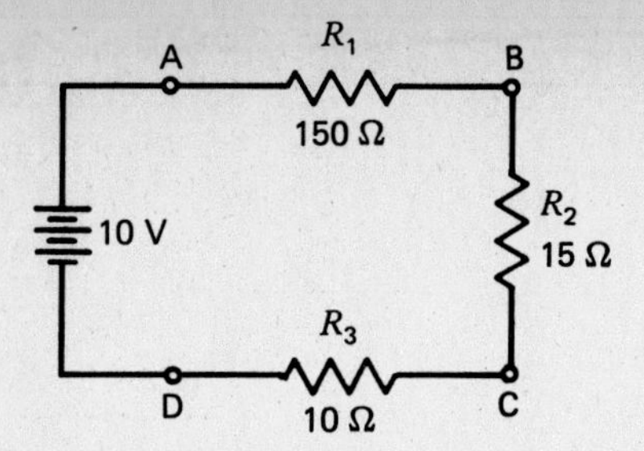

Fig. 4-5. Series circuit.

TABLE 4–1

Nominal Resistance, Ω	Measured, Ω
$R_1 = 150$	
$R_2 = 15$	
$R_3 = 10$	
$R_t = 175$	

TABLE 4–2

	Calculated	Measured	% Error
R_t			
I_t			
V_{R_1}			
V_{R_2}			
V_{R_3}			

TABLE 4–3

Point	Current, mA
A	
B	
C	
D	

EXPERIMENT 5

SERIES-AIDING AND SERIES-OPPOSING VOLTAGES

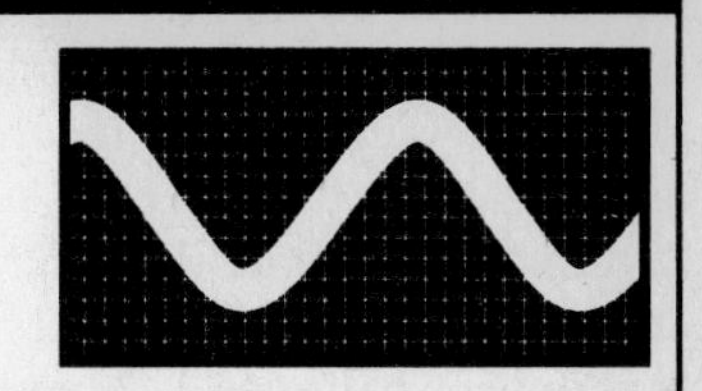

OBJECTIVES

At the completion of this experiment, you will be able to:

- Define the terms series-aiding and series-opposing.
- Construct a series-aiding and a series-opposing circuit.
- Measure current and voltage and anticipate correct polarity connections.

SUGGESTED READING

Chapter 4, *Basic Electronics,* B. Grob, Seventh Edition

INTRODUCTION

In many practical applications, a circuit may contain more than one voltage source. Voltage sources that cause current to flow in the same direction are considered to be *series-aiding,* and their voltages add. Voltage sources that tend to force current in opposite directions are said to be *series-opposing,* and the effective voltage source is the difference between the opposing voltages. When two opposing sources are inserted into a circuit, current flow would be in a direction determined by the larger source. Examples of series-aiding and series-opposing sources are shown in Fig. 5-1*a* and *b.*

Series-aiding voltages are connected such that the currents of the sources add, as shown in Fig. 5-1*a*. The 10 V of V_1 produces 1 A of current flow through the 10-Ω resistance of R_1. Also, the voltage source V_2, of 5 V, creates 0.5 A of current flowing through the 10-Ω resistance of R_1: The total current would then be additive and be 1.5 A.

When voltage sources are connected in a series-aiding fashion, where the negative terminal of one source is connected to the positive terminal of the next, voltages V_1 and V_2 are added to find the total voltage.

$$V_{\text{total}} = 5\text{ V} + 10\text{ V} = 15\text{ V}$$

The total current can then be determined by

$$I_{\text{total}} = 15\text{ V}/10\Omega = 1.5\text{ A}$$

Series-opposing voltages can be subtracted, as shown in Fig. 5-1*b*. The currents that are generated are opposing each other. These voltages can still be algebraically added, keeping in mind their algebraic sign. Here, V_2 is smaller than V_1, and the difference between them is +5 V, resulting in an I_{total} of 0.5 A.

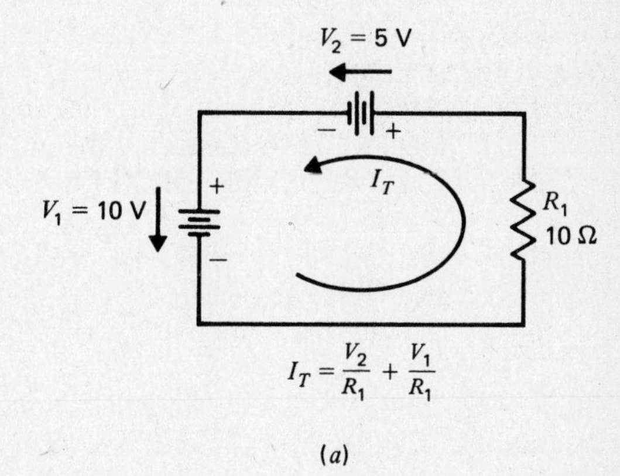

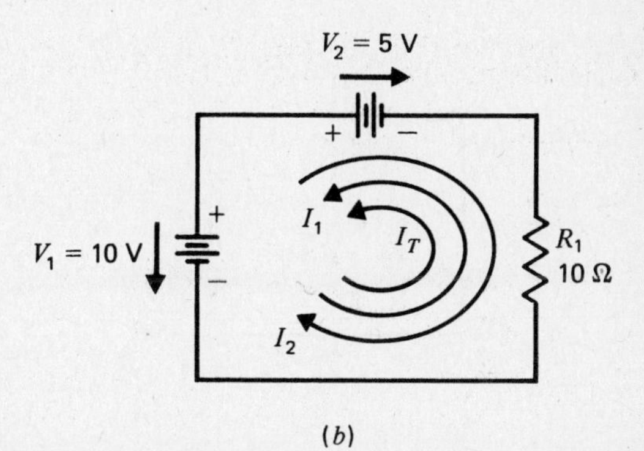

Fig. 5-1. (*a*) Series-aiding voltages. (*b*) Series-opposing voltages.

EQUIPMENT

VOM
Test leads

COMPONENTS

(4) D cells (1.5-V batteries)

PROCEDURE

1. Connect the negative lead of the meter to the

negative terminal of the battery, and connect the positive lead of the meter to the positive terminal of the battery. Identify and number each battery 1 through 4. Measure and record in Table 5-1 the voltage of each of the dry cells supplied to you.

2. Connect batteries 1 and 2 as series-aiding. Measure and record in Table 5-2 their total voltage.

3. Connect batteries 1, 2, and 3 as series-aiding. Measure and record in Table 5-2 their total voltage.

4. Repeat this process, measuring and recording the voltage of four batteries in series-aiding, in Table 5-2.

5. Connect two batteries in parallel. (Be sure to connect the negative terminal of one battery to the negative terminal of the other, and connect the positive terminal of one battery to the positive terminal of the other. In this way, we have connected them together.) Measure and record this voltage in Table 5-3.

6. Connect three batteries in parallel. Measure and record this voltage in Table 5-3. Then connect four batteries in parallel. Measure and record this voltage in Table 5-3 also.

7. Connect the batteries in a series-parallel arrangement as shown in Fig. 5-2. Measure and record in Table 5-4 the voltage from point A to point B.

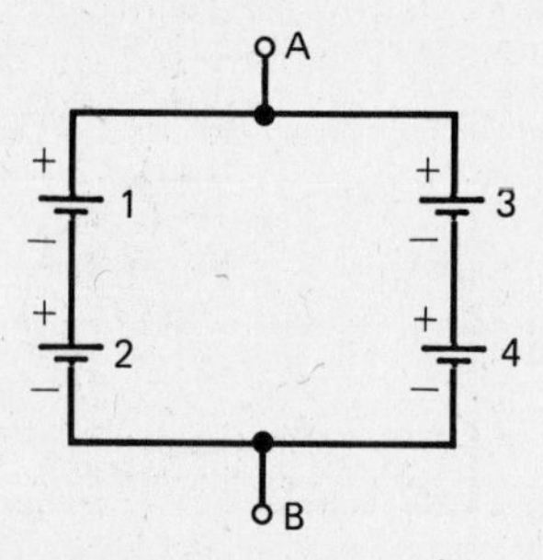

Fig. 5-2. Series-parallel batteries.

8. Connect the circuit as shown in Fig. 5-3. Measure and record in Table 5-4 the voltage from point C to point D.

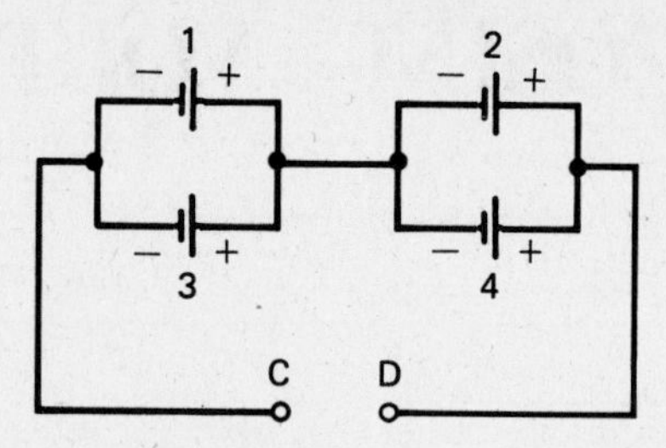

Fig. 5-3. Series-aiding circuit.

9. Connect the series-aiding–series-opposing arrangements shown in Fig. 5-4*a* to *c*. Measure and record in Table 5-5 the voltage across each circuit. Before connecting the voltmeter, determine which are the probable positive and negative terminals.

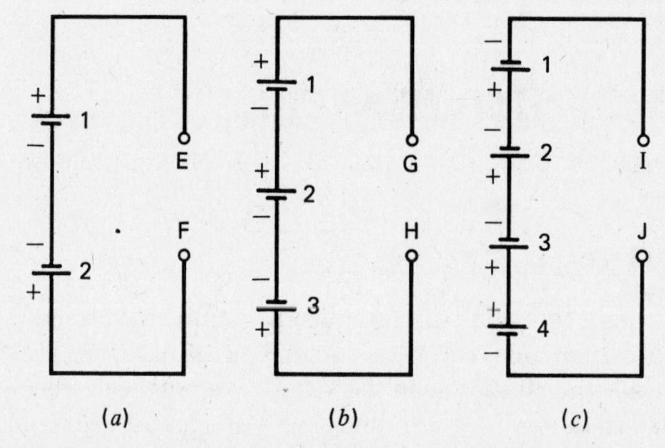

Fig. 5-4. (*a*) Series-opposing circuit. (*b*) Series-aiding and -opposing circuit. (*c*) Series-aiding and -opposing circuit.

NAME ____________________ DATE ____________

RESULTS FOR EXPERIMENT 5

QUESTIONS

1. Name four precautions which must be observed in measuring voltages.

2. What would happen to a dry cell or battery if the positive and negative terminals were short-circuited?

3. What arrangement of six dry cells gives the maximum voltage?

4. Draw a practical arrangement of ten 1.5-V dry cells to give a battery of 7.5 V.

5. Explain the difference in connection between two dry cells connected in series-aiding and in series-opposing.

REPORT

Write a complete report. Describe how combinations of positive and negative voltages add and oppose each other.

TABLE 5–1

Battery	Measured Voltages
1	________
2	________
3	________
4	________

TABLE 5–2

Series-Aiding Battery	Measured Voltages
1 + 2	________
1 + 2 + 3	________
1 + 2 + 3 + 4	________

TABLE 5–3

Parallel-Arrangement Battery	Measured Voltages
1 + 2	________
1 + 2 + 3	________
1 + 2 + 3 + 4	________

TABLE 5–4

Circuit	Voltages	Terminals
Fig. 5-2	________	A to B
Fig. 5-3	________	C to D

TABLE 5–5

Circuit	Voltages	Terminals
Fig. 5-4*a*	________	E to F
Fig. 5-4*b*	________	G to H
Fig. 5-4*c*	________	I to J

EXPERIMENT 6

PARALLEL CIRCUITS

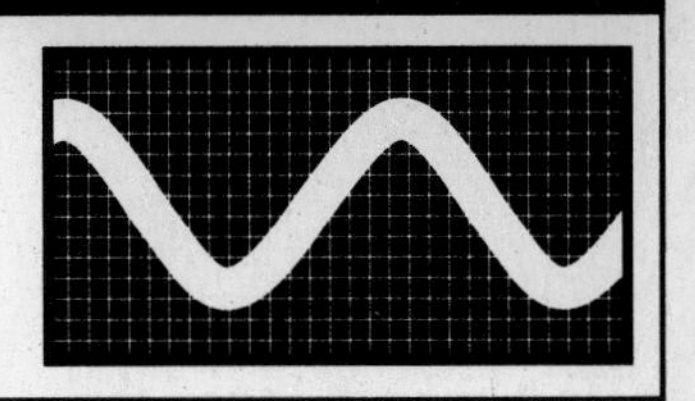

OBJECTIVES

At the completion of this experiment, you will be able to:

- Identify a parallel circuit.
- Accurately measure current in a parallel circuit.
- Use Ohm's law to verify measurements taken in a parallel circuit.

SUGGESTED READING

Chapter 5, *Basic Electronics,* B. Grob, Seventh Edition

INTRODUCTION

A parallel circuit is defined as one having more than one current path connected to a common voltage source. Parallel circuits, therefore, must contain two or more load resistances which are not connected in series. Study the parallel circuit shown in Fig. 6-1.

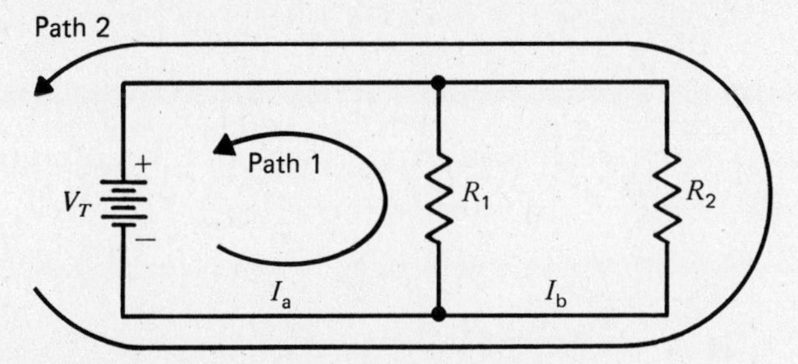

Fig. 6-1. Parallel circuit with two branches.

Beginning at the voltage source V_T and tracing counterclockwise around the circuit, two complete and separate paths can be identified in which current can flow. One path is traced from the source through resistance R_1 and back to the source; the other is traced from the source through resistance R_2 and back to the source.

The source voltage in a series circuit divides proportionately across each resistor in the circuit. In a parallel circuit, the same voltage is present across all the resistors of a parallel bank. In other words,

$$V_T = V_{R_1} = V_{R_2} = \cdots$$

The current in a circuit is inversely proportional to the circuit resistance. This fact, obtained from Ohm's law, establishes the relationship upon which the following discussion is developed. A single current flows in a series circuit.

In summary, when two or more electronic components are connected across a single voltage source, they are said to be *in parallel.* The voltage across each component is the same. The current through each component, or branch, is determined by the resistance of that branch and voltage across the bank of branches. Adding a parallel resistance of any value increases the total current. The total resistance of a circuit (R_T) can be found by dividing the total voltage applied (V_T) by the total current (I_T).

When measuring currents and resistances in the following procedure, some important precautions should be observed. When measuring currents, be sure to install the ammeter in a series configuration, with the ammeter connected in series with the individual branch resistances through which the current is flowing. Also, before measuring branch resistances, be sure to turn off all voltage sources.

EQUIPMENT

DC power supply, 0–10 V
Ammeter
Voltmeter
Protoboard or springboard
Test leads

COMPONENTS

Resistors (all 0.5 W):

(2) 1 kΩ (1) 2.2 kΩ

PROCEDURE

1. Measure the resistance values of R_1, R_2, and R_3 and record in the required locations of Tables 6-1 to 6-3, where the nominal values are:

$$R_1 = 1.0\ \text{k}\Omega$$
$$R_2 = 1.0\ \text{k}\Omega$$
$$R_3 = 2.2\ \text{k}\Omega$$

2. Connect the circuit as shown with R_1 only in Fig. 6-2. Adjust the power supply voltage to 10 V.
3. Using an ammeter, measure and record in Table 6-1 the current through points a and b.
4. Using a voltmeter, measure and record in Table 6-1 the voltage across R_1.
5. To this circuit, add R_2 across (meaning that R_2 is

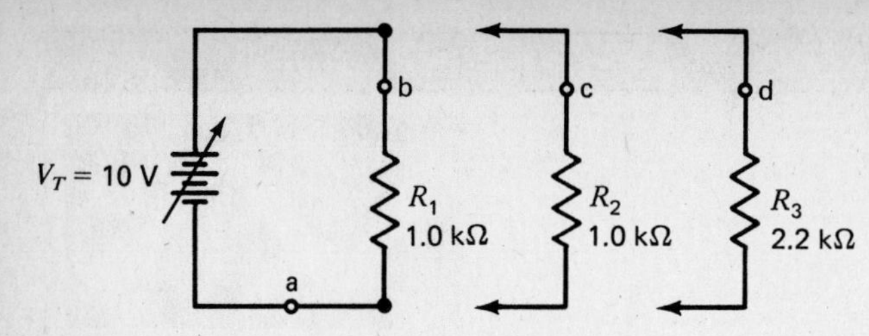

Fig. 6-2. Three-branch parallel circuit.

connected in parallel) R_1. With an ammeter, measure and record in Table 6-2 the current through points a, b, and c.

6. Using a voltmeter, measure and record in Table 6-2 the voltages across R_1 and R_2.

7. Finally, add R_3 across R_2. With an ammeter, measure and record in Table 6-3 the current through points a, b, c, and d.

8. Using a voltmeter, measure and record in Table 6-3 the voltages across R_1, R_2, and R_3.

9. Calculate the total resistances and the current for each of the three previous circuits and record this information in Tables 6-1 to 6-3.

NAME DATE

RESULTS FOR EXPERIMENT 6

QUESTIONS

1. What is a parallel circuit? What circuit characteristics indicate that a parallel circuit condition exists?

2. In the circuit of Fig. 6-2, determine the power being dissipated by each resistor using the values of current determined in procedure step 7. Use the formula I^2R = power (watts).

3. Are the voltages the same across each resistor in a parallel circuit?

4. Are the currents the same through each resistor in a parallel circuit?

5. Suppose in procedure step 7 that R_3 developed a short-circuited condition. How would the current flowing through each resistor change? Would the voltage drops across each resistor change? How?

REPORT

Write a complete report. Describe the three most significant aspects of the experiment.

TABLE 6–1

	Measured	Calculated
R_1	______	
R_T		______
V_{R_1}	______	
I_a	______	______
I_b	______	______

TABLE 6–2

	Measured	Calculated
R_1	______	
R_2	______	
R_T		______
V_{R_1}	______	
V_{R_2}	______	
I_a	______	______
I_b	______	______
I_c	______	______

TABLE 6–3

	Measured	Calculated
R_1	______	
R_2	______	
R_3	______	
R_T		______
V_{R_1}	______	
V_{R_2}	______	
V_{R_3}	______	
I_a	______	______
I_b	______	______
I_c	______	______
I_d	______	______

EXPERIMENT 7

SERIES-PARALLEL CIRCUITS

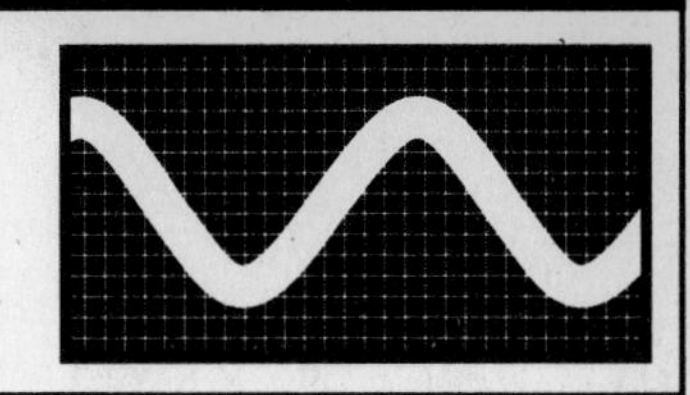

OBJECTIVES

At the completion of this experiment, you will be able to:

- Identify a series-parallel circuit.
- Accurately measure voltages and current present in a series-parallel circuit.

SUGGESTED READING

Chapter 6, *Basic Electronics,* B. Grob, Seventh Edition

INTRODUCTION

Consider the circuit shown in Fig. 7-1. Resistors connected together as shown are often called a *resistor network.* In this network, resistors R_2 and R_3 are considered to be in a parallel arrangement. Also, this parallel arrangement is in series with R_1. To calculate the total resistance R_T, the equivalent resistance of R_2 and R_3 must be determined, where

$$R_{eq} = \frac{R_2 \times R_3}{R_2 + R_3}$$
$$= 150\ \Omega \text{ (approx.)}$$

Therefore, the equivalent resistance of this parallel arrangement is approximately equal to 150 Ω.

The total resistance is now easily found by adding the two resistances R_1 and R_{eq}, where

$$R_T = R_1 + R_{eq}$$

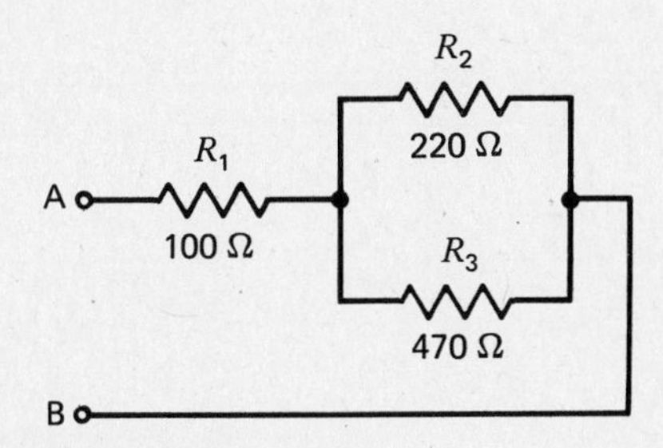

Fig. 7-1. Series-parallel circuit.

EQUIPMENT

Ohmmeter
DC power supply
Ammeter
Voltmeter
Protoboard or springboard
Leads

COMPONENTS

Resistors (all 0.5 W unless indicated otherwise):

(1) 68 Ω	(2) 150 Ω
(3) 100 Ω	(1) 220 Ω
(1) 120 Ω	(1) 470 Ω
(1) 150 Ω 1 W	

PROCEDURE

1. Measure and record the actual resistance values shown in Table 7-1.

2. Connect the circuit shown in Fig. 7-2, where $V = 10$ V, $R_1 = 100\ \Omega$, $R_2 = 220\ \Omega$, and $R_3 = 470\ \Omega$.

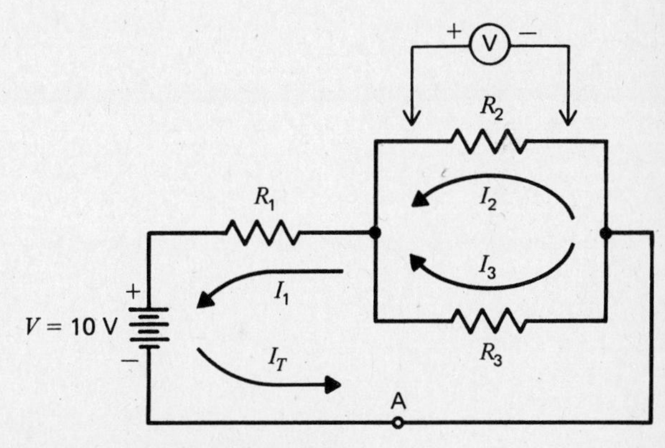

Fig. 7-2. Series-parallel circuit.

3. Calculate, measure, and record V_1, V_2, V_3, I_1, I_2, I_3, and I_T (at point A) in Table 7-2.

4. Measure and record in Table 7-2 the total applied voltage.

5. With the voltage source removed, measure, calculate and record R_T in Table 7-2.

6. Calculate the percentage of error for Table 7-2.

7. Construct the circuit shown in Fig. 7-3, where

$V_T = 10$ V	$R_4 = 120\ \Omega$
$R_1 = 150\ \Omega/1$ W	$R_5 = 150\ \Omega$
$R_2 = 100\ \Omega$	$R_6 = 100\ \Omega$
$R_3 = 100\ \Omega$	$R_7 = 68\ \Omega$

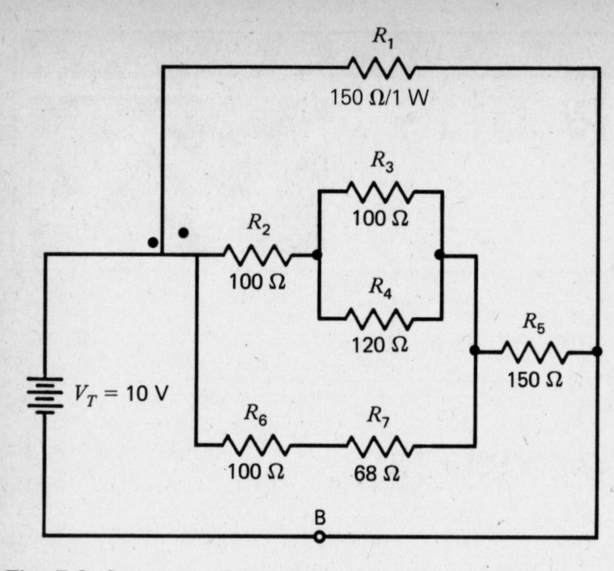

Fig. 7-3. Series-parallel circuit.

8. Calculate, measure, and record V_1, V_2, V_3, V_4, V_5, V_6, V_7, I_1, I_2, I_3, I_4, I_5, I_6, I_7, and I_T (at point B) in Table 7-3.

9. Measure and record in Table 7-3 the total applied voltage V_T.

10. With the voltage source removed, measure, calculate and record R_T in Table 7-3.

11. Calculate the percentage of error for Table 7-3.

NAME DATE

RESULTS FOR EXPERIMENT 7

QUESTIONS

1. In your own words, explain what a series-parallel circuit is.

2. In the circuit shown in Fig. 7-2, determine the power being dissipated by each resistor. Use the formula VI = power (watts).

3. In Fig. 7-2, are the voltages the same across each resistor?

4. In Fig. 7-2, are the currents the same through each resistor?

5. Suppose in Fig. 7-3 that R_3 developed a short-circuited condition. How would the current flowing through each resistor change? Would the voltage drop across each resistor change? How?

REPORT

Write a complete report. Discuss how Ohm's law applies, and compare the calculated and measured data. Write a conclusion that summarizes the circuit operation.

TABLE 7–1 (values for Fig. 7-3)

Nominal Resistance	Measured Resistance
$R_1 = 150\ \Omega$	________
$R_2 = 100\ \Omega$	________
$R_3 = 100\ \Omega$	________
$R_4 = 120\ \Omega$	________
$R_5 = 150\ \Omega$	________
$R_6 = 100\ \Omega$	________
$R_7 = 68\ \Omega$	________

TABLE 7–2

	Calculated	Measured	% Error
V_1			
V_2			
V_3			
V_T			
I_1			
I_2			
I_3			
I_T			
R_T			

TABLE 7–3

	Calculated	Measured	% Error
V_1			
V_2			
V_3			
V_4			
V_5			
V_6			
V_7			
V_T			
I_1			
I_2			
I_3			
I_4			
I_5			
I_6			
I_7			
I_T			
R_T			

EXPERIMENT 8

ADDITIONAL SERIES-PARALLEL CIRCUITS

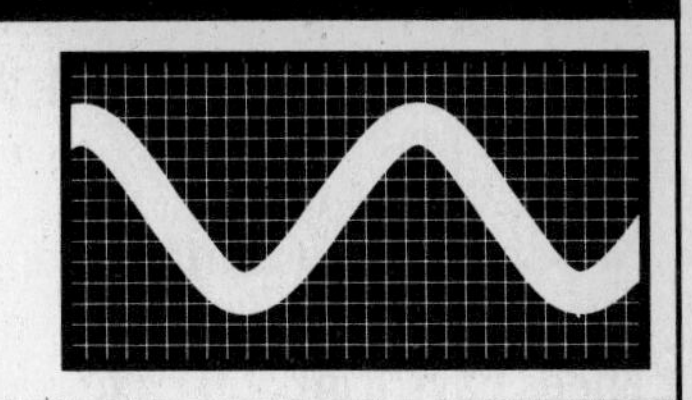

OBJECTIVES

At the completion of this experiment, you will be able to:

- Redraw and simplify a series-parallel circuit.
- Write an equation for the total resistance.
- Determine which resistances have the greatest effect on R_T.

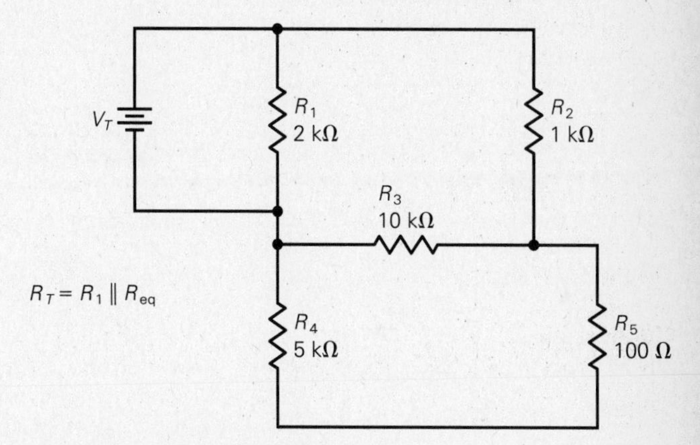

Fig. 8-1. Resistive network.

SUGGESTED READING

Chapter 6, *Basic Electronics,* B. Grob, Seventh Edition

INTRODUCTION

This experiment is a continuation of the previous experiment on series-parallel circuits. The same concepts discussed in that experiment apply here. The following points summarize the concepts of a series-parallel circuit:

- The R_T of two parallel branches equals the product divided by the sum

$$R_T = \frac{R_1 \times R_2}{R_1 + R_2}$$

- Adding resistance in series increases the total resistance and decreases the total current.
- Adding resistance in parallel increases the total current and decreases the total resistance.
- The equivalent resistance R_{eq} of a series-parallel network must be determined before the total resistance R_T can be found.

Consider the circuit shown in Fig. 8-1. This resistive network has a total resistance of

$$R_T = R_1 \parallel R_{eq}$$

The two vertical lines in the equation are used as a symbol to indicate resistances in parallel. Thus, $R_T = R_1 \parallel R_{eq}$ can be literally taken to mean

$$R_T = R_1 \text{ in parallel with } R_{eq}$$

To determine R_T, the circuit must be reduced or redrawn so that all the resistors, except R_1, can be combined into one equivalent resistance R_{eq}. This is because R_1 is in parallel with the equivalent total resistance of R_2 through R_5.

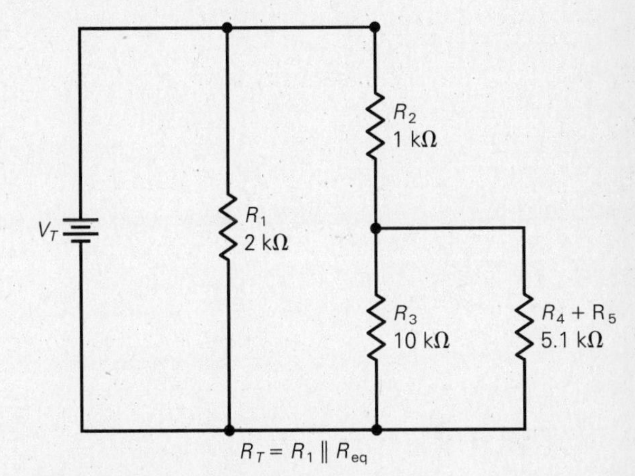

Fig. 8-2. Redrawn resistive network.

Although the circuit can be redrawn in more than one way, there is a method that can be used easily: Look for the resistors farthest from the power source, and combine them, working backward toward the source. The result of this method is

$$R_{eq} = [(R_4 + R_5) \parallel R_3] + R_2$$

The network can be redrawn as shown in Fig. 8-2. Notice that the redrawn figure makes it easier to write the equation. In this case, you can begin with the sum of $R_5 + R_4$ in parallel with R_3. This would be

$$R_4 + R_5 = 5.1\ \text{k}\Omega.$$

Then, by using the product-over-the-sum method

$$\frac{R_3 \times 5.1\ \text{k}\Omega}{R_3 + 5.1\ \text{k}\Omega} = \frac{10\ \text{k}\Omega \times 5.1\ \text{k}\Omega}{10\ \text{k}\Omega + 5.1\ \text{k}\Omega} = \frac{51.0\ \text{k}\Omega}{15.1\ \text{k}\Omega} = 3.38\ \text{k}\Omega$$

Next, add 3.38 kΩ to 1 kΩ R_2 so that $R_{eq} = 4.38$ kΩ. Finally,

$$R_T = R_1 \parallel R_{eq} = \frac{2\text{ k}\Omega \times 4.38\text{ k}\Omega}{2\text{ k}\Omega + 4.38\text{ k}\Omega} = \frac{8.76\text{ k}\Omega}{6.38\text{ k}\Omega} = 1.37\text{ k}\Omega$$

Not only can you write an equation for R_T by redrawing the circuit, but also you can see how certain resistances have a greater or lesser effect on the network. For example, R_5 has almost no effect on R_T because R_5 is less than 10 percent of the resistance in series and is added to R_4. However, if R_4 were removed from the circuit, then R_5 would have a greater effect on the equivalent resistance.

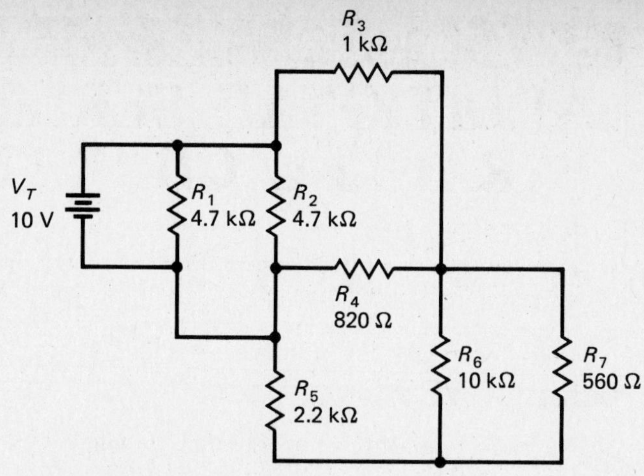

Fig. 8-3. Resistive network.

EQUIPMENT

Ohmmeter
DC power supply
Ammeter
Voltmeter
Protoboard or springboard
Leads

COMPONENTS

Resistors (all 0.5 W):
(2) 4.7 kΩ
(1) 560 Ω
(1) 820 Ω
(2) 10 kΩ
(1) 2.2 kΩ
(1) 1 kΩ

PROCEDURE

1. Measure and record the resistor values shown in Table 8-1.
2. Connect the circuit shown in Fig. 8-3.
3. Measure and record the voltages around the circuit as shown in Table 8-1.
4. Disconnect the power supply V_T, and measure the total circuit resistance. Record the results in Table 8-1.
5. Disconnect R_1 and V_T, and measure and record the resistance R_{eq}.
6. Calculate the current through each resistor, using Ohm's law and the measured voltage drops. Record the results in Table 8-1 and include I_T, the total circuit current.
7. Remove R_6 from the circuit and measure the total circuit resistance. Record the results in Table 8-1.
8. Replace R_2 with a 10-kΩ resistor, and measure the total circuit resistance. Record the results in Table 8-1. (R_6 is still removed.)
9. On a separate sheet of 8 × 11 in. paper, redraw the circuit so that only four resistances represent the simplified circuit. Label all resistances so that the combined resistances are easy to identify. For example, one resistance might be $(R_3 + R_4) \parallel R_2$, etc.
10. Write an equation for the total resistance, and show how to calculate R_T by using R_{eq} and the product-over-the-sum method. Use the same sheet of paper as in step 9 above.

NAME DATE

RESULTS FOR EXPERIMENT 8

QUESTIONS

1. Which resistor in the circuit of Fig. 8-2 has the least effect on R_T and why?

2. What would happen to the circuit of Fig. 8-2 if R_1 were decreased to 10 Ω?

3. What would happen to the circuit of Fig. 8-2 if R_1 were increased to 10 MΩ?

4. Which resistor in the circuit of Fig. 8-2 has the least effect on I_T?

5. Which resistor in the circuit of Fig. 8-2 has the greatest effect on I_T?

REPORT

Turn in all data and the equations. No formal report is required for this experiment.

TABLE 8–1

Resistance Values Nominal or Calculated	Ω Measured	V Measured	I Calculated
$R_1 = 4.7\ \text{k}\Omega$	________	________	________
$R_2 = 4.7\ \text{k}\Omega$	________	________	________
$R_3 = 1\ \text{k}\Omega$	________	________	________
$R_4 = 820\ \Omega$	________	________	________
$R_5 = 2.2\ \text{k}\Omega$	________	________	________
$R_6 = 10\ \text{k}\Omega$	________	________	________
$R_7 = 560\ \Omega$	________	________	________
$R_T =$ ________	________		$I_T =$ ________
$R_{eq} =$ ________	________		
R_T with R_6 removed	________		
R_T with $R_2 = 10\ \text{k}\Omega$	________		

EXPERIMENT 9

OPENS AND SHORTS IN SERIES-PARALLEL CIRCUITS

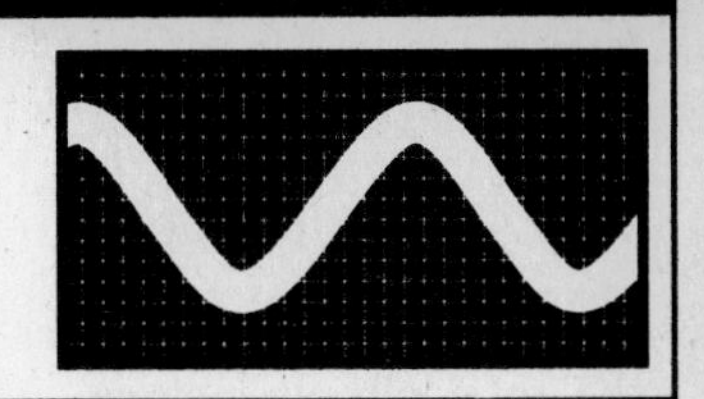

OBJECTIVES

At the completion of this experiment, you will be able to:

- Determine the changes in circuit current and voltage drops resulting from a short circuit.
- Determine the changes in circuit current and voltage drops resulting from an open circuit.

SUGGESTED READING

Chapter 6, *Basic Electronics,* B. Grob, Seventh Edition

INTRODUCTION

A short circuit has practically zero resistance. Its effect, therefore, is to allow excessive current to flow in a circuit, although this is not usually intentional. An open circuit has the opposite effect because an open circuit has infinitely high resistance with practically zero current.

Therefore, if one path in a circuit changes (becomes open or short), the circuit's voltage, resistance, and current in the other paths change as well. For example, the series-parallel circuit shown in Fig. 9-1 becomes a series circuit when there is a short across circuit points A and B.

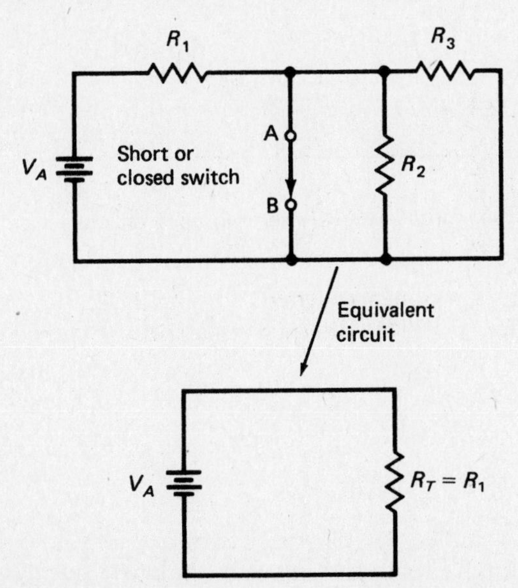

Fig. 9-1. Effect of a short in a series-parallel circuit.

As an example of an open circuit, the series-parallel circuit in Fig. 9-2 becomes a series circuit with just R_1 and R_2 when there is an open between points A and B.

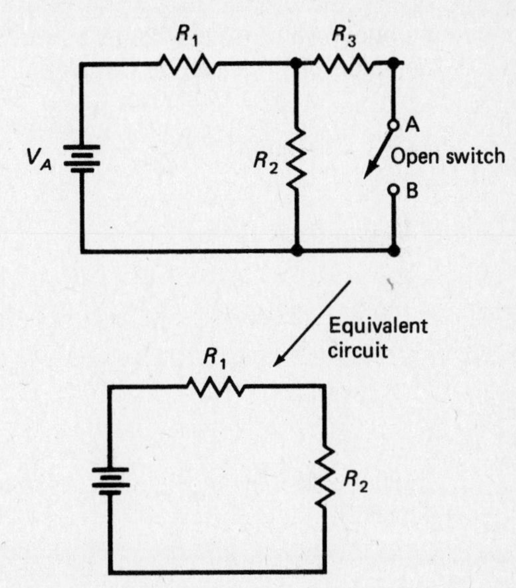

Fig. 9-2. Effects of an open in a series-parallel circuit.

The Short Circuit

You can determine the effect of a short in a series-parallel circuit. For example, in the circuit of Fig. 9-1, a switch is shown between points A and B. This switch represents a possible flaw in construction or operation of an actual circuit.

Refer to Fig. 9-3 where the circuit does not have a short. Specific component values have been assigned to the same circuit as in Fig. 9-1. As shown in Fig. 9-3,

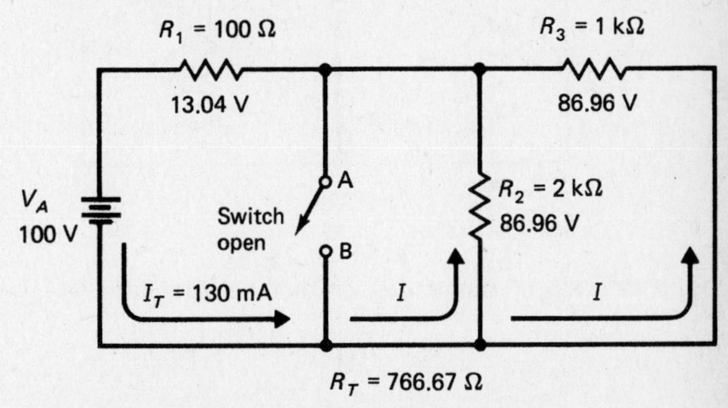

Fig. 9-3. Circuit values where no short exists.

$R_1 = 100\ \Omega$, $R_2 = 1000\ \Omega$, and $R_3 = 2000\ \Omega$. The total resistance R_T can be determined as follows:

$$R_T = R_1 + \frac{R_2 \times R_3}{R_2 + R_3}$$
$$= 766.67\ \Omega$$

Knowing the total resistance is essential in determining the total circuit current. If the applied voltage V_A is 100 V, then

$$I_T = \frac{V_A}{R_T}$$
$$= \frac{100\ \text{V}}{766.67\ \Omega}$$
$$= 0.13\ \text{A} \quad (\text{or } 130\ \text{mA})$$

Another component value that will be interesting to identify is the value of V_1. It can be calculated as

$$V_1 = I_T \times R_1$$
$$= 0.13\ \text{A} \times 100\ \Omega$$
$$= 13.04\ \text{V}$$

The value of V_2 can be found as well:

$$V_2 = V_A - V_1$$
$$= 100\ \text{V} - 13.04\ \text{V}$$
$$= 86.96\ \text{V}$$

The value of V_3 is then

$$V_3 = V_A - V_1$$
$$= 100\ \text{V} - 13.04\ \text{V}$$
$$= 86.96\ \text{V}$$

The voltage values of V_2 and V_3 should be equivalent since R_2 and R_3 form a parallel circuit.

Figure 9-3 shows the calculated effect of a short with a series-parallel circuit. If the circuit is shorted from point A to B, then the effects on resistors R_2 and R_3 are eliminated. The elimination of R_2 and R_3 creates predictable changes in circuit current and the voltage drop of R_1. The circuit of Fig. 9-4 shows the electrical effects. Since R_2 and R_3 are eliminated, the only effective resistance left in the circuit is R_1.

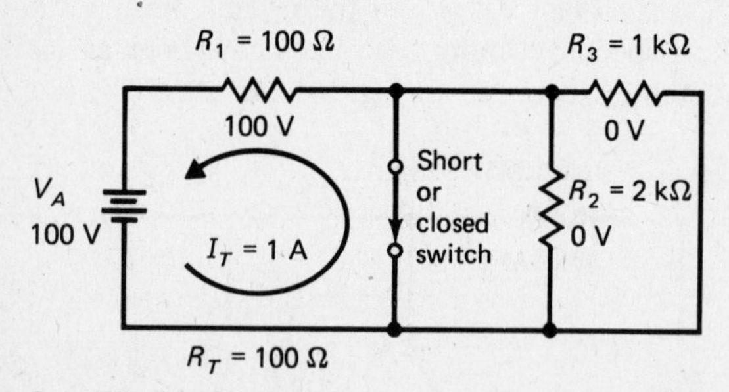

Fig. 9-4. Effects of a short.

The total circuit current I_T can then be calculated as

$$I_T = \frac{V_A}{R_1}$$
$$= \frac{100\ \text{V}}{100\ \Omega}$$
$$= 1\ \text{A}$$

The voltage drop across R_1 is then

$$V_1 = I_T \times R_1$$
$$= 1\ \text{A} \times 100\ \Omega$$
$$= 100\ \text{V}$$

The voltage drop across R_2 is thus

$$V_2 = V_A - V_1$$
$$= 100\ \text{V} - 100\ \text{V}$$
$$= 0\ \text{V}$$

The voltage drop across R_3 is

$$V_3 = V_A - V_1$$
$$= 100\ \text{V} - 100\ \text{V}$$
$$= 0\ \text{V}$$

The voltage drops across R_2 and R_3 should equal 0 V since their resistance values are reduced to $0\ \Omega$ because of the short.

In summary, these circuit changes occur in two ways. First, the circuit currents increase significantly. Second, the voltage drop from the increase in current flow also increases.

The Open Circuit

An open circuit provides practically infinite resistance to the applied voltage V_A. Its overall effect on the circuit would be zero or minimal current flow. You can determine the effect of an open path in a series-parallel circuit. For example, in the circuit of Fig. 9-5, component

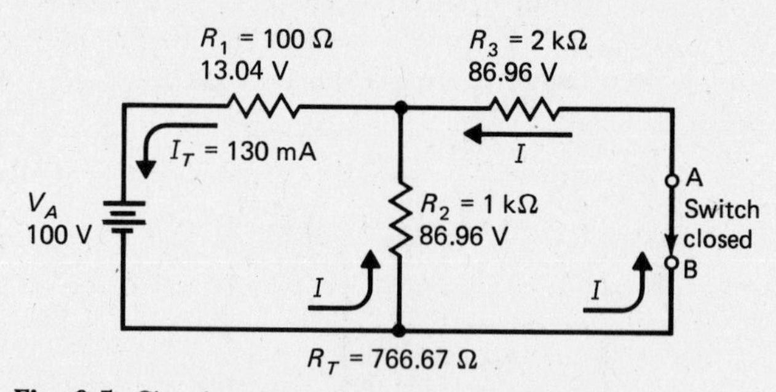

Fig. 9-5. Circuit values where no open exists.

values have been assigned as $R_1 = 100\ \Omega$, $R_2 = 1000\ \Omega$, and $R_3 = 2000\ \Omega$. The total resistance R_T can be determined as follows:

$$R_T = R_1 + \frac{R_2 \times R_3}{R_2 + R_3}$$
$$= 766.67\ \Omega$$

Knowing the total resistance is essential in determining the total circuit current. If the applied voltage V_A is

100 V, then

$$I_T = \frac{V_A}{R_T}$$
$$= \frac{100\ \text{V}}{766.67\ \Omega}$$
$$= 0.13\ \text{A} \quad (\text{or } 130\ \text{mA})$$

Another component value that will be important to identify is the voltage value of V_1. It can be calculated as

$$V_1 = I_T \times R_1$$
$$= 0.13\ \text{A} \times 100\ \Omega$$
$$= 13.04\ \text{V}$$

The voltage of V_2 is

$$V_2 = V_A - V_1$$
$$= 100\ \text{V} - 13.04\ \text{V}$$
$$= 86.96\ \text{V}$$

The voltage of V_3 is

$$V_3 = V_A - V_1$$
$$= 100\ \text{V} - 13.04\ \text{V}$$
$$= 86.96\ \text{V}$$

The values are shown in Fig. 9-5.

If the circuit is open between points A and B, then R_3 is no longer part of the circuit. The removal of R_3 creates a predictable change in the circuit current and voltage drops of R_1 and R_2.

The circuit of Fig. 9-6 shows the overall electrical effects when the switch is opened between points A and B. The circuit current I_T can be calculated after the total resistance R_T is found:

$$R_T = R_1 + R_2$$
$$= 100\ \Omega + 1000\ \Omega$$
$$= 1100\ \Omega$$

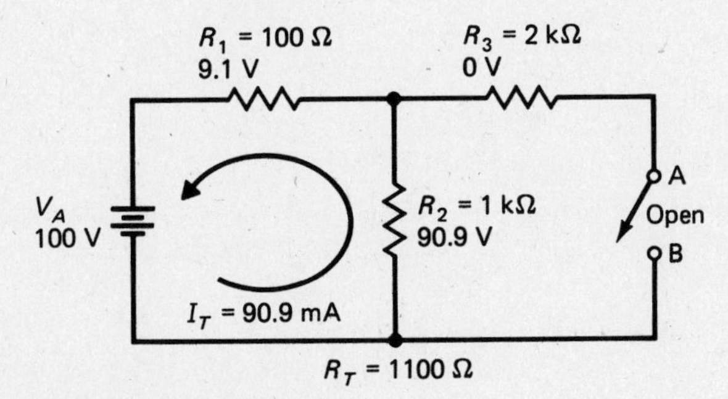

Fig. 9-6. Effects of an open.

Then I_T can be calculated as

$$I_T = \frac{V_A}{R_1}$$
$$= \frac{100\ \text{V}}{1100\ \Omega}$$
$$= 0.0909\ \text{A} \quad (\text{or } 90.9\ \text{mA})$$

The voltage drop across R_1 is then

$$V_1 = I_T \times R_1$$
$$= 0.0909\ \text{A} \times 100\ \Omega$$
$$= 9.09 \text{ or } 9.1\ \text{V}$$

The voltage drop across R_2 is then

$$V_2 = I_T \times R_2$$
$$= 0.0909\ \text{A} \times 1000\ \Omega$$
$$= 90.9\ \text{V}$$

The voltage value of R_3 is 0 V, due to the open circuit.

EQUIPMENT

Voltmeter
Power supply
Protoboard

COMPONENTS

(3) 150-Ω 0.5-W resistors
(1) SPST switch

PROCEDURE

1. Measure and record in Table 9-1 the values of the three resistors used in this experiment.

2. Construct the circuit of Fig. 9-7, and adjust the voltage of the power supply to 10 V.

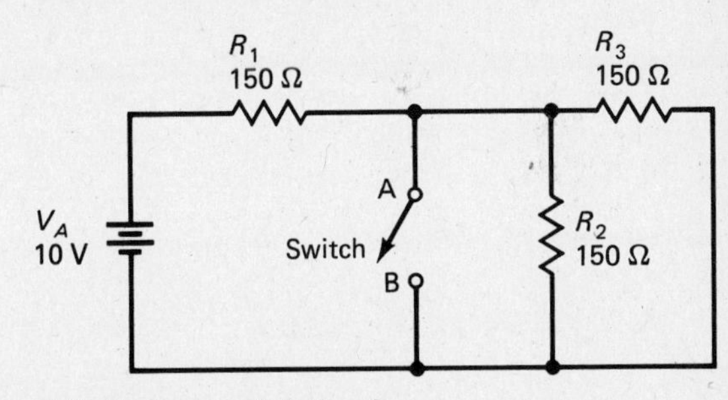

Fig. 9-7. Series-parallel circuit.

3. Calculate and record in Table 9-1, for the circuit of Fig. 9-7, the values of the total resistance R_T; the currents I_{R_1}, I_{R_2}, and I_{R_3}; and the voltages V_{R_1}, V_{R_2}, and V_{R_3}.

4. Measure and record in Table 9-1 the measured values of I_{R_1}, I_{R_2}, I_{R_3}, V_{R_1}, V_{R_2}, V_{R_3}.

5. Connect points A and B.

6. Calculate and record in Table 9-1 the values of R_T, I_{R_1}, I_{R_2}, I_{R_3}, V_{R_1}, V_{R_2}, and V_{R_3}.

7. Measure and record in Table 9-1 the values of I_{R_1}, I_{R_2}, I_{R_3}, V_{R_1}, V_{R_2}, and V_{R_3}.

8. Turn off the power supply, and disconnect the circuit.

9. Measure and record in Table 9-2 the values of the three resistors used in this experiment.

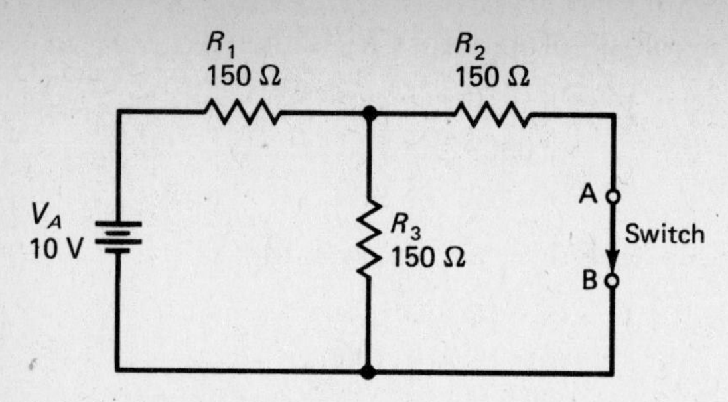

Fig. 9-8. Series-parallel circuit.

10. Construct the circuit of Fig. 9-8, and adjust the voltage of the power supply to 10 V.

11. Calculate and record in Table 9-2, for the circuit of Fig. 9-8, the values of the total resistance R_T; currents I_{R_1}, I_{R_2}, I_{R_3}; and voltages V_{R_1}, V_{R_2}, and V_{R_3}.
12. Measure and record in Table 9-2 the measured values of I_{R_1}, I_{R_2}, I_{R_3}, V_{R_1}, V_{R_2}, and V_{R_3}.
13. Disconnect or open points A and B.
14. Calculate and record in Table 9-2 the values of R_T, I_{R_1}, I_{R_2}, I_{R_3}, V_{R_1}, V_{R_2}, and V_{R_3}.
15. Measure and record in Table 9-2 the values of I_{R_1}, I_{R_2}, I_{R_3}, V_{R_1}, V_{R_2}, and V_{R_3}.
16. Turn off the power supply and disconnect the circuit.

NAME

DATE

RESULTS FOR EXPERIMENT 9

QUESTIONS

1. What are the characteristics of a shorted circuit?

2. What are the characteristics of an open circuit?

3. Compare Tables 9-1 and 9-2. What are the differences and similarities?

REPORT

Write a complete report. Discuss the measured and calculated results. Discuss the three most significant aspects of the experiment and write a conclusion.

TABLE 9–1 (V_A = 10 volts)

Component	Measured	Calculated
Unshorted Circuit, Fig. 9–7		
R_1	______	
R_2	______	
R_3	______	
R_T		______
I_{R_1}	______	______
I_{R_2}	______	______
I_{R_3}	______	______
V_{R_1}	______	______
V_{R_2}	______	______
V_{R_3}	______	______
Shorted Circuit, Fig. 9–7		
R_T		______
I_{R_1}	______	______
I_{R_2}	______	______
I_{R_3}	______	______
V_{R_1}	______	______
V_{R_2}	______	______
V_{R_3}	______	______

TABLE 9–2 (V_A = 10 volts)

Component	Measured	Calculated
Shorted Circuit, Fig. 9–8		
R_1	______	
R_2	______	
R_3	______	
R_T		______
I_{R_1}	______	______
I_{R_2}	______	______
I_{R_3}	______	______
V_{R_1}	______	______
V_{R_2}	______	______
V_{R_3}	______	______
Opened Circuit, Fig. 9–8		
R_T		______
I_{R_1}	______	______
I_{R_2}	______	______
I_{R_3}	______	______
V_{R_1}	______	______
V_{R_2}	______	______
V_{R_3}	______	______

EXPERIMENT 10
KIRCHHOFF'S LAWS

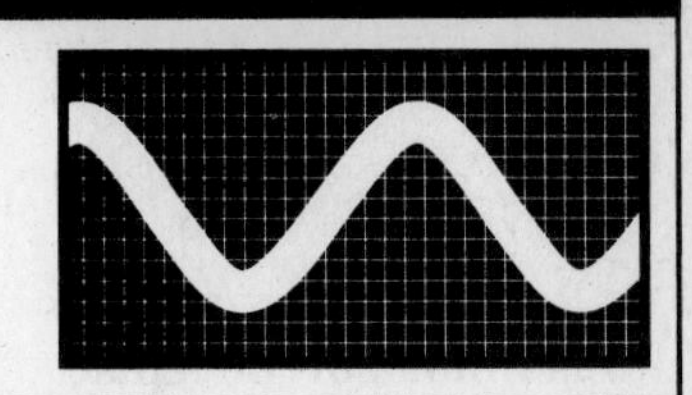

OBJECTIVES

At the completion of this experiment, you will be able to:

- Validate Kirchhoff's current and voltage laws.
- Gain proficiency with lab equipment and technique.

SUGGESTED READING

Chapter 9, *Basic Electronics*, B. Grob, Seventh Edition

INTRODUCTION

In 1847, Gustav R. Kirchhoff formulated two laws that have become fundamental to the studies of basic electronics. They are as follows:

1. The algebraic sum of the currents into and out of any point must be equal to zero.
2. The algebraic sum of the applied source voltages and the *IR* voltage drops in any closed path must be equal to zero.

At first, Kirchhoff's laws seem obvious. That is, it seems obvious to state that whatever goes into a circuit must also equal what comes out of it. However, Kirchhoff's laws are used to analyze circuits that are not simple series or parallel or series-parallel circuits. For example, circuits that contain more than one voltage source and circuits called *unbalanced bridges* often cannot be understood without using Kirchhoff's laws to analyze those circuits. In its simplest form, Kirchhoff's law could be used to analyze the circuit in Fig. 10-1.

Simplified version:

$I_1 = I_2 + I_6 \qquad I_2 = I_3 + I_4 \qquad I_5 = I_2$
$I_6 = I_1 - I_2 \qquad I_3 = I_2 - I_4 \qquad I_7 = I_6 + I_5$
$I_2 = I_1 - I_6 \qquad I_4 = I_5 - I_3 \qquad I_7 = I_1$

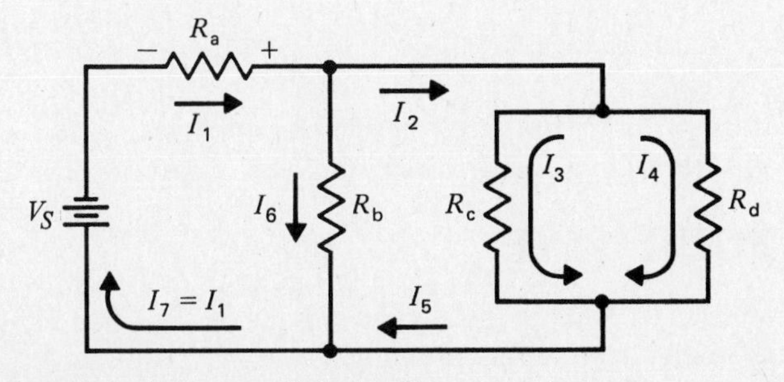

Fig. 10-1. Series-parallel circuit.

Algebraic version:

$+I_1 - I_6 - I_2 = 0 \qquad +I_5 + I_6 - I_7 = 0$
$+I_2 - I_3 - I_4 = 0 \qquad I_{total} = I_1 = I_7$

Note: For the circuit of Fig. 10-1, if $V_S = 20$ V, the sum of the *IR* voltage drops across the series-parallel combination of R_a, R_b, R_c, and R_d will also equal 20 V.

Be sure that you understand the algebraic version of the circuit. It has become standard practice to assign positive and negative values to the currents as follows: The current into any point is positive (+), and the current out of any point is negative (−). Also, this would be true if there were more than one path for the current to enter or leave. Do not confuse this with electron flow.

Although the circuit of Fig. 10-1 is a series-parallel circuit, it was used to demonstrate how Kirchhoff's laws operate. Now consider the circuit of Fig. 10-2. This circuit could not be solved for its currents and *IR* voltage drops without using Kirchhoff's laws because of the two source voltages.

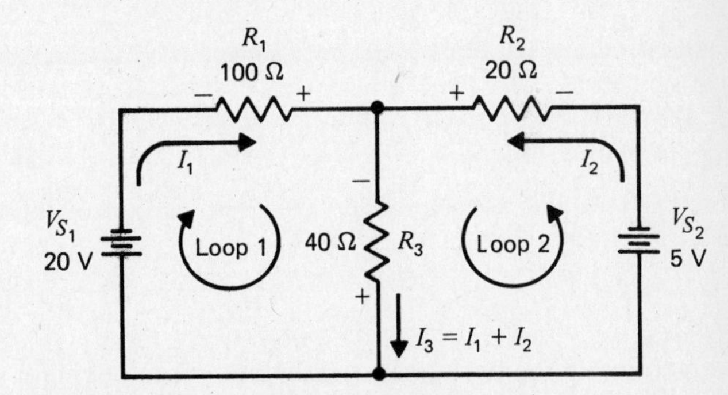

Fig. 10-2. Circuit with two voltage sources.

By using the loop method, Kirchhoff's laws can be used to determine the currents and *IR* voltages in the circuit by calculation. For example,

$$V_{S_1} = 20\text{ V} - V_{R_1} - V_{R_3} = 0 \qquad \text{(loop 1)}$$
$$V_{S_2} = 5\text{ V} - V_{R_2} - V_{R_3} = 0 \qquad \text{(loop 2)}$$

By following the loop around its path, Kirchhoff's law provides a method. Take the source as positive, and take the individual path resistance as negative. The sum is equal to zero. This means that circuits will be divided into separate loops as if the other voltage source and its corresponding path did not exist. The *IR* voltage drops would be calculated as

$$V_{R_1} = I_1R_1 = I_1 \times 100\ \Omega$$
$$V_{R_2} = I_2R_2 = I_2 \times 20\ \Omega$$
$$V_{R_3} = I_3R_3 = (I_1 + I_2)R_3 = (I_1 + I_2) \times 40\ \Omega$$

Refer to the equations for loops 1 and 2. The *IR* voltages would now be replaced in the formulas as follows:

For loop 1,

$$20\text{ V} - 100(I_1) - 40(I_1+I_2) = 0$$
$$-140I_1 - 40I_2 = -20\text{ V}$$

and for loop 2,

$$5\text{ V} - 20(I_2) - 40(I_1+I_2) = 0$$
$$-60I_2 - 40I_1 = -5\text{ V}$$

Note that the final equations are transposed versions that can be simplified further by division:

$$\frac{-140I_1 - 40I_2}{-20\text{ V}} = \frac{-20\text{ V}}{-20\text{ V}} \text{ or } 7I_1 + 2I_2 = 1 \quad \text{(loop 1)}$$
$$\frac{-60I_2 - 40I_1}{-5\text{ V}} = \frac{-5\text{ V}}{-5\text{ V}} \text{ or } 12I_2 + 8I_1 = 1 \quad \text{(loop 2)}$$

To solve for the currents, isolate the I_2 currents by making them the same value. Multiply the loop 1 equation by 6:

$$42I_1 + 12I_2 = 6 \quad \text{(loop 1)}$$

Note:

$$6 = \frac{12I_2}{2I_2}$$

Note that either loop could be changed. In the case above, only loop 1 was changed (multiplied by 6) so that I_2 would have the same value for both loop equations:

$$42I_1 + 12I_2 = 6 \quad \text{(loop 1)}$$
$$8I_1 + 12I_2 = 1 \quad \text{(loop 2)}$$

Subtracting the two equations, term by term, will eliminate I_2 because $12I_2 - 12I_2 = 0$. Therefore,

$$34I_1 = 5$$
$$I_1 = 147\text{ mA}$$

By using the loop equation method, Kirchhoff's law, the current through the resistance of R_1 is determined. Also, the direction of current flow is correct as assumed because the answer was a positive value.

To calculate I_2, substitute 147 mA for I_1 in either loop equation. Substituting in loop 2,

$$8(147\text{ mA}) + 12I_2 = 1$$
$$1.18\text{ A} + 12I_2 = 1$$
$$12I_2 = 1 - 1.18\text{ A} = -0.18\text{ A}$$
$$I_2 = -0.015\text{ A} = -15\text{ mA}$$

Because the negative sign appears in the solution for the current I_2, it follows that the current through I_2 is opposite in direction from the polarity shown in the circuit of Fig. 10-2, the assumed direction using the loop method.

Now that Kirchhoff's law has been used to determine the currents through the circuit of Fig. 10-2, the circuit can be redrawn to reflect the correct values, as shown in Fig. 10-3.

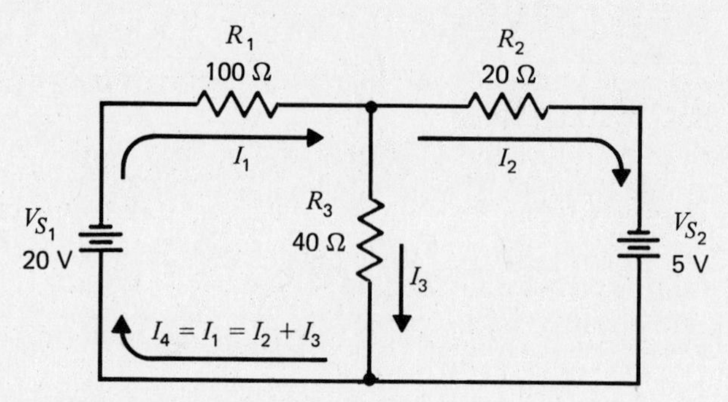

Fig. 10-3. Redrawn version of the circuit in Fig. 10-2.

In the case of circuit Fig. 10-3, the 20-V dc source actually overrides the 5-V dc source. This series opposition results in approximately 15 V dc.

Because R_1 is 100 Ω, it is easy to verify the Kirchhoff's law loop equation solution as follows:

$$\frac{15\text{ V}}{100\ \Omega + (20\ \Omega \parallel 40\ \Omega)} = \frac{15\text{ V}}{113.3\ \Omega} = 132\text{ mA}$$

Loop 1 $I_T = 147$ mA

Loop 2 $I_T = -15$ mA (Subtract negative value)

Total $I_T = 132$ mA

Finally, note that the circuit in Fig. 10-4 is a bridge type of circuit. Kirchhoff's law could be used to calculate the unknown values in this circuit as well as circuits that contain many voltage sources and many current paths.

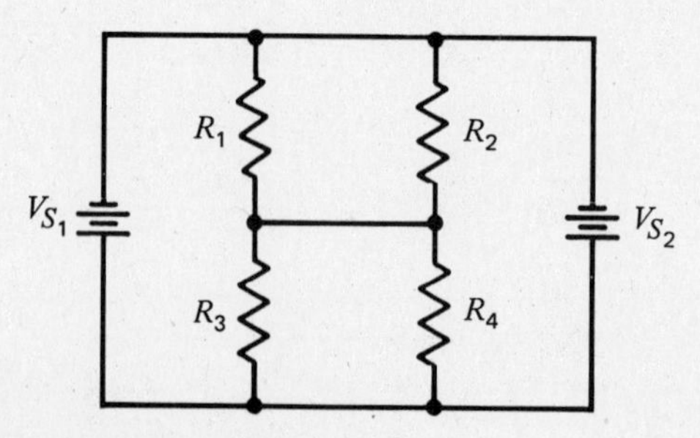

Fig. 10-4. Bridge-type circuit.

EQUIPMENT

DC power supply, 0–10 V
Ammeter (milliamp capability)
Voltmeter or VTVM

Springboard or protoboard
Connecting leads

COMPONENTS

Resistors (all 0.5 W):

(1) 100 Ω (1) 390 Ω
(1) 220 Ω (1) 470 Ω
(1) 330 Ω

(2) Dry cells (flashlight batteries, 1.5 V)

PROCEDURE

1. Measure and record the values of voltage for each dry cell you are using. It may be necessary to solder connecting wires onto the ends of the batteries. Record the results in Table 10-1. Together, series-aiding, the two batteries should be approximately 3.0 V.

2. For each of the three circuits in Figs. 10-5 to 10-7, use batteries for voltages of 1.5 V and use the power supply for the other source voltage. If 3.0 V and 1.5 V are both used, connect the two 1.5-V batteries in series for 3.0 V. Measure and record the values of current in Table 10-2. Also, measure and record the value of the *IR* voltage drop across each resistance. Finally, use arrows to indicate the direction of current flow.

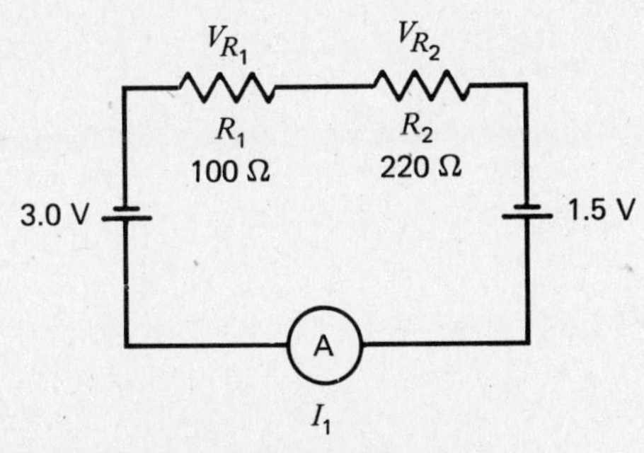

Fig. 10-5. Series-opposing voltage circuit.

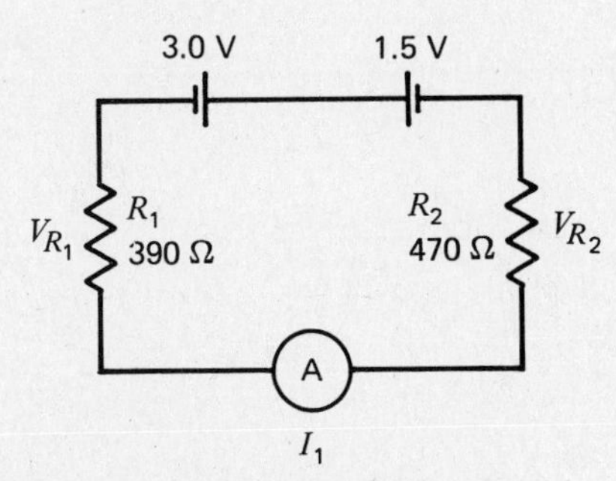

Fig. 10-6. Series-opposing voltage circuit.

3. For the circuit of Fig. 10-8, refer to the procedure outlined in the introduction to this experiment. Preferably, use the loop method of Kirchhoff's law to determine the values and polarities of both current and voltage for the circuit. Then use the lab to verify

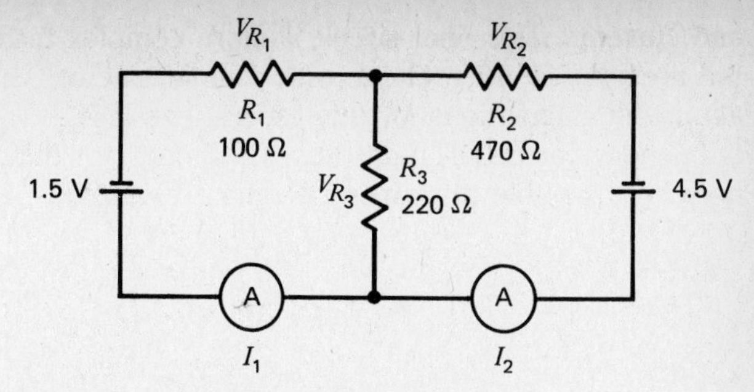

Fig. 10-7. Kirchhoff's circuit for analysis.

the values (within 10 to 15 percent). Record all the calculated values, and then measure and record in Table 10-3 all the values determined in the lab for the circuit. Be sure to include arrows alongside or directly above the meters marked A. Include Fig. 10-8 in your report. Note that Fig. 10-8 will be used to verify Kirchhoff's law. That is, you will compare your calculations to your measurements. Do not forget to include the values of *IR* drops across each resistance.

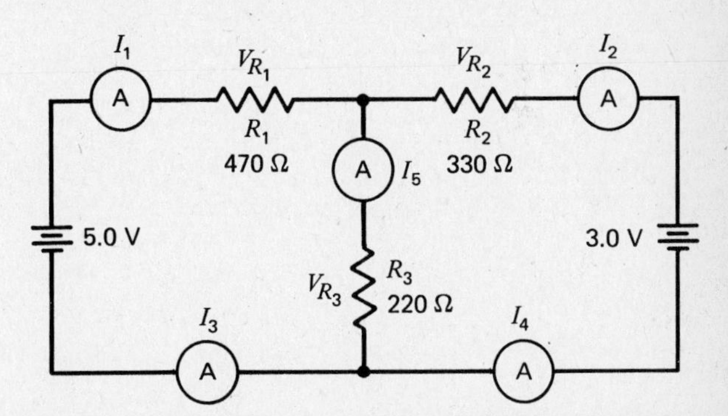

Fig. 10-8. Kirchhoff's circuit. Use loop method to determine the values and polarities of current and voltage.

4. Connect the circuit of Fig. 10-9. Measure and record in Table 10-4 all the values of *IR* voltage drops

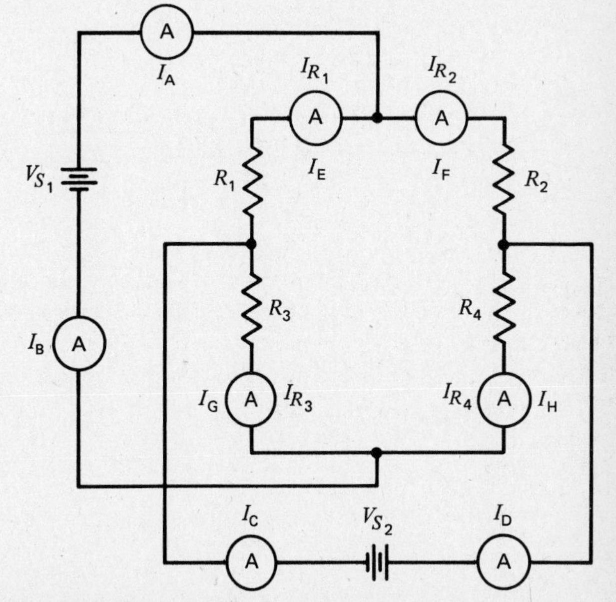

Fig. 10-9. Kirchhoff's law circuit (values to be given by instructor).

and currents. This final circuit is more complex than the previous circuits. Check with your instructor about the calculations for this experiment. It is recommended for extra credit only. The important thing here is to be able to properly connect the circuit and properly measure the unknown values.

Note: Your instructor will assign values for V_{S_1}, V_{S_2}, R_1, R_2, R_3, and R_4. This ensures that the experiment cannot be precalculated.

NAME ______________________ DATE ____________

RESULTS FOR EXPERIMENT 10

QUESTIONS

1. State Kirchhoff's current law.

2. State Kirchhoff's voltage law.

REPORT

Write a complete report. Discuss the measured and calculated results. Discuss the three most significant aspects of the experiment and write a conclusion.

TABLE 10–1. Dry Cell Voltages

Battery	Measured Voltage
1	______
2	______

TABLE 10–2

Circuit	V_{R_1}	V_{R_2}	V_{R_3}	I_1	I_2
Fig. 10-5	______	______		______	
Fig. 10-6	______	______		______	
Fig. 10-7	______	______	______	______	______

TABLE 10–3

	Calculated	Measured
I_1		
I_2		
I_3		
I_4		
I_5		
V_{R_1}		
V_{R_2}		
V_{R_3}		

TABLE 10–4

	Current	Voltage
R_1		
R_2		
R_3		
R_4		
I_A		
I_B		
I_C		
I_D		
I_E		
I_F		
I_G		
I_H		

EXPERIMENT 11

THE WHEATSTONE BRIDGE

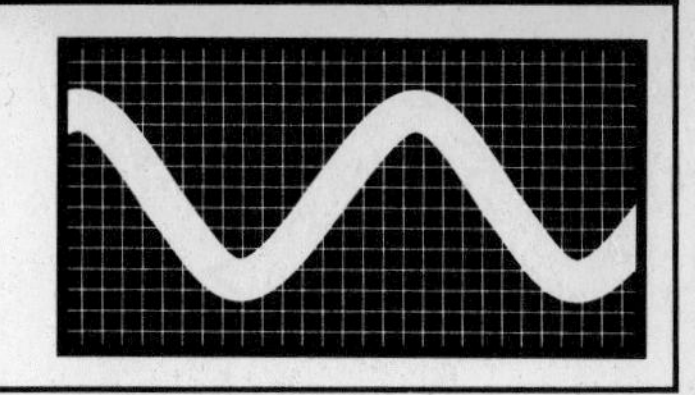

OBJECTIVES

At the completion of this experiment, you will:

- Be able to measure current and voltage in a Wheatstone bridge.
- Know how to use a galvanometer.
- Understand how a Wheatstone bridge can be balanced.

SUGGESTED READING

Chapters 5, 6, and 10, *Basic Electronics,* B. Grob, Seventh Edition

INTRODUCTION

Many types of bridge circuits are used in electronics. The bridge circuit in this experiment can become a functional ohmmeter as well as a balanced circuit. Here, the Wheatstone bridge has two input terminals, where the battery (or power supply) terminals are connected to the ratio arms. The ratio arms are the key to understanding how the Wheatstone bridge is balanced. The bridge is balanced when there is an equal division of voltages across the bridge output. This output is a current path across the ratio arms, like a bridge between two series-parallel paths. It does not matter what the ratios are for the bridge to be in a balanced state, because resistances in parallel (two series resistors in this case) have the same voltage across both branches. However, some ratios may create a more sensitive balance than others, depending upon the total current and total resistance of the ratio arms.

If there is an imbalance in the bridge, current will flow through the output path from one ratio arm to the other. In this experiment, a galvanometer is used. It is a dual directional microammeter with the needle zeroed at top dead center.

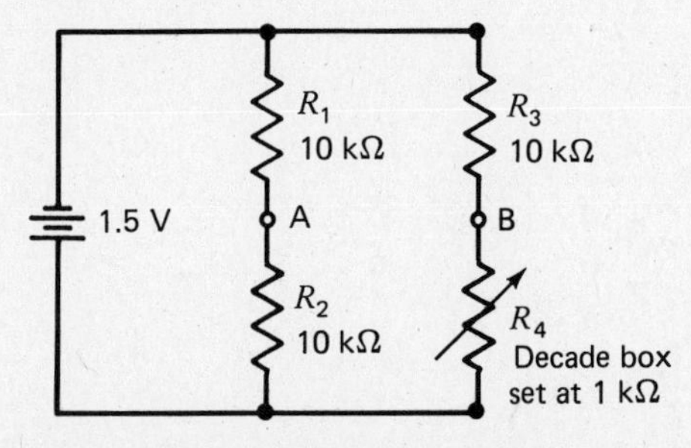

Fig. 11-1. Series-parallel circuit for analysis.

It is possible to use a VOM. However, extreme caution will be needed to prevent damage to the meter.

EQUIPMENT

DC Power supply, 0–10 V
Decade box
Galvanometer
Test leads
VTVM or DVM

COMPONENTS

(3) 10-kΩ 0.5 W resistors
Plus other desired resistors (all 0.5 W):

(1) 5.6 kΩ (1) 2.7 kΩ
(1) 1 kΩ (1) 560 Ω
(1) 2.2 kΩ

PROCEDURE

1. Connect the circuit of Fig. 11-1.
2. Usually, a galvanometer is placed across the output terminals (A and B) to complete the bridge. However, it is better to study the series-parallel aspects of the Wheatstone bridge first to gain a better understanding of how it works. Thus, measure the voltage across points A and B and record the results in Table 11-1. Put the ground side of the voltmeter on point B. Also, determine the direction of current flow.
3. Increase the value of R_4 to 5 kΩ and measure the voltage across points A and B and determine current direction. Record the results in Table 11-1.
4. Increase the value of R_4 in 1-kΩ steps from 5 to 15 kΩ. Measure the voltage across points A and B at each step, and determine current direction and record in Table 11-1. This should lead to an understanding of how the value of the ratio arms determines the voltage across the AB output. When finished, turn the power off.

CAUTION: It will be necessary to reverse the voltmeter leads at some point.

5. Insert a galvanometer across points A and B as shown in Fig. 11-2. Be sure R_4 is 10 kΩ. The circuit is now a Wheatstone bridge, because there is a true current path across points A and B.

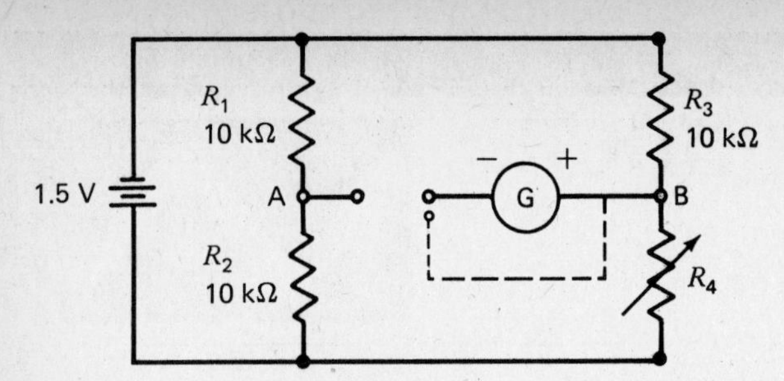

Fig. 11-2. Galvanometer circuit.

Note: Many galvanometers have a switch (button or toggle switch on the front of the meter). Until the switch is closed, current flows through a wire (or short circuit) to protect the meter movement. When the switch is closed, current flows through the meter movement. Be careful not to peg the meter.

6. With equal nominal values of R_1 and R_4, turn the power on and finely adjust R_4 so that no current flows across points A and B. This is the same as 0 V across points A and B. Thus, R_4 (decade box) will be finely adjusted and may be 10.5 kΩ or 11.1 kΩ, etc., unless, of course, you are using precision resistors (1 percent tolerance). The important thing is that both voltage and current at points A and B = 0. If so, the bridge is now balanced. Record the exact value of R_4 (decade box) in Table 11-2.

7. Now adjust R_4 so that the needle deflects to the right, approximately halfway to full scale, as shown in Fig. 11-3. Measure and record in Table 11-2 the voltage across R_4 and R_2. Repeat this procedure by adjusting R_4 with the needle deflecting to the left. Record the value of R_4 and the voltage across R_4 and R_2. Keep in mind the following: the bridge was balanced. Then, the bridge was unbalanced on both ratio arms (left and right). This procedure can be analyzed later to determine how the direction of current flow across the bridge indicates the condition of the bridge. Also record the adjusted value of R_4.

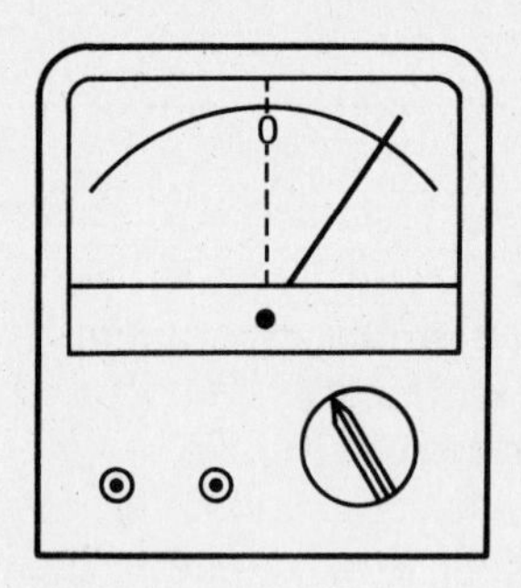

Fig. 11-3. Galvanometer.

8. Bring the bridge back to a balanced condition by readjusting R_4 and monitoring the galvanometer. The needle should be dead center—zero current.

9. Turn the power off. Replace R_2 with an unknown resistor with a value between 2 and 8 kΩ. Simply put electrical tape over a resistor. Or, if this is not possible, use a 5.6-kΩ resistor, and pretend you do not know the value.

10. Turn power on. Adjust R_4 until the bridge is balanced. The value of R_4 (decade box) should be the true value of R_2. Repeat steps 9 and 10 with several resistors (1 kΩ, 2.2 kΩ, 2.7 kΩ, and 560 Ω, for example). R_4 should be equal to R_2 (unknown resistor) each time. Every time you adjust R_4, notice the amount of needle deflection. Record the results in a separate table. Be sure to label all values.

NAME DATE

RESULTS FOR EXPERIMENT 11

QUESTIONS

1. In procedure steps 9 and 10, the Wheatstone bridge was balanced by matching a decade box to the unknown resistor. Thus, the Wheatstone bridge was used as what kind of meter?

2. Explain the difference between a galvanometer and an ohmmeter.

3. In procedure step 5, suppose the schematic of Fig. 11-2 showed R_4 as 50 kΩ, and R_1 to R_3 remained at 10 kΩ. With power on, would the needle deflect to the right or the left?

4. Which circuit in Fig. 11-4 would be more sensitive (greater needle deflection) when attempting to balance the bridge? Why?

5. Explain any differences and/or similarities between the two circuits in Fig. 11-5.

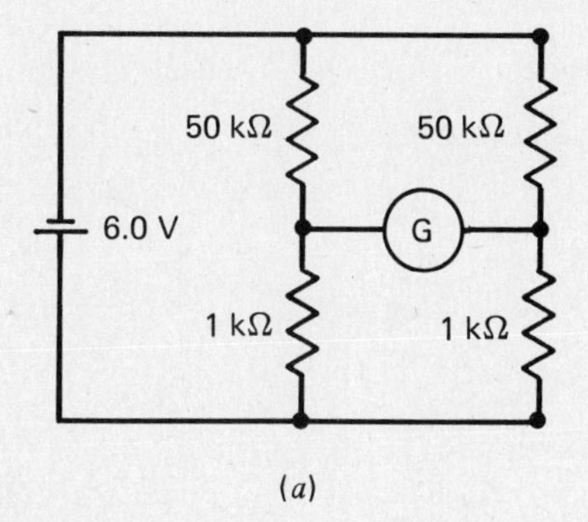

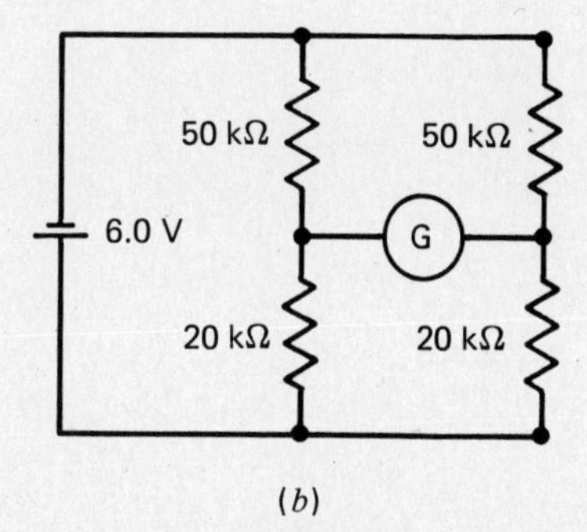

Fig. 11-4. Circuits for question 4.

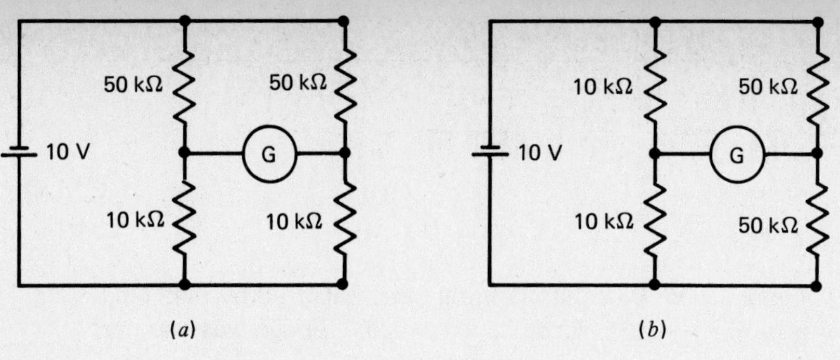

Fig. 11-5. Circuits for question 5.

REPORT

Write a report that describes how the bridge operates, based upon your data.

TABLE 11–1. Data from Fig. 11–1

R_4, kΩ	Voltage AB	Current Direction
1		
5		
6		
7		
8		
9		
10		
11		
12		
13		
14		
15		

TABLE 11–2. R_1 and R_3 = 10 kΩ, R_4 = Decade Box

	Voltage and Current at A and B = 0	Needle Deflection to Right			Needle Deflection to Left		
	R_4, Ω	R_4, Ω	V_{R_4}	V_{R_2}	R_4, Ω	V_{R_4}	V_{R_2}
R_2 = 10 kΩ							
R_2 = 5.6 kΩ							
R_2 = 1 kΩ							
R_2 = 2.2 kΩ							
R_2 = 2.7 kΩ							
R_2 = 560 Ω							

EXPERIMENT 12

VOLTAGE DIVIDERS WITH LOADS

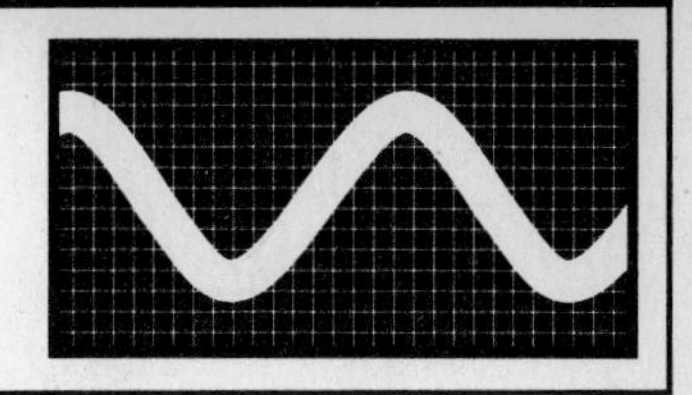

OBJECTIVES

At the completion of this experiment, you will be able to:

- Identify a voltage divider circuit.
- Define the purpose for voltage divider circuits.
- Describe voltage divider loading effects.

SUGGESTED READING

Chapter 7, *Basic Electronics,* B. Grob, Seventh Edition

INTRODUCTION

A series circuit is also a voltage divider. That is, the total voltage applied to a series circuit is divided among the series resistors. Because the current is the same value in all parts of a series circuit, voltage is divided among the series resistances in direct proportion to the value of resistance.

For example, imagine four resistors in series. Each resistor is equal in value. Therefore, any one resistor will receive one-fourth the total applied voltage. In other words, the total voltage is divided by 4 across each resistor.

Another way to determine the voltage across any resistor in series, without using a meter, is to add the total resistance, divide that value into any single resistor in series, and multiply by the total voltage. This is the proportional method. For example, see Fig. 12-1. There,

$$V_{R_2} = \frac{R_2}{R_T} \times V_T = \frac{2\text{ k}\Omega}{6\text{ k}\Omega} \times 6\text{ V} = 2\text{ V}$$

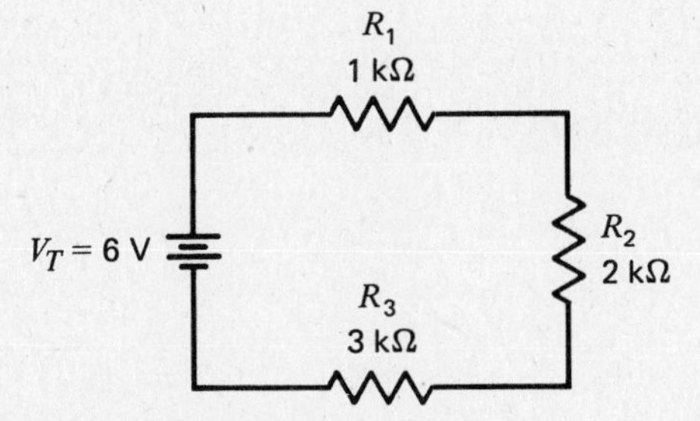

Fig. 12-1. Proportional method circuit.

Although these simple voltage dividers are limited in their use, when a load is placed across any series resistance, the voltage divider is then extremely useful as a voltage tap. For example, the circuit in Fig. 12-2 shows a parallel load current through R_{load}.

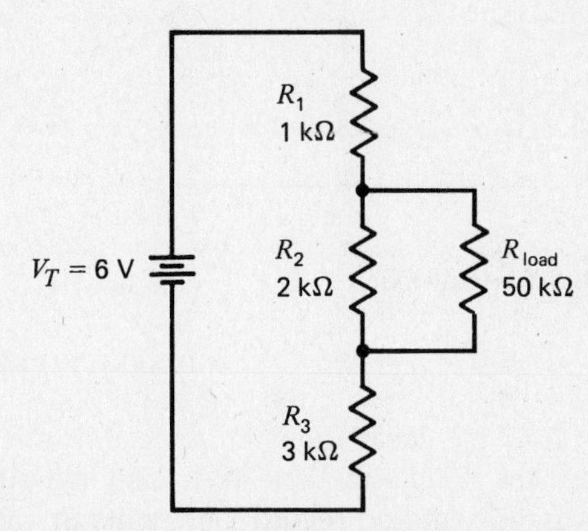

Fig. 12-2. Simple voltage divider with parallel load.

This parallel load, if its resistance was more than 10 times the value of R_2, would actually be sharing the *IR* voltage drops across R_2 without changing the division of voltage considerably. For example, if $R_{\text{load}} = 50\text{ k}\Omega$,

$$\frac{R_2 \times R_{\text{load}}}{R_2 + R_{\text{load}}} = \frac{100 \times 10^6}{52 \times 10^3} = 2\text{ k}\Omega \text{ (approx.)}$$

Thus, R_{load} could share the same approximate voltage as R_2, but it would have its own current.

While voltage dividers are mainly used to tap off part of a total voltage, it is necessary to remember that the addition of a load will always have some effect upon the circuit current and, many times, the proportional *IR* voltage drops.

EQUIPMENT

DC power supply, 0–10 V
Voltmeter (VTVM, VOM)
Test leads

COMPONENTS

Resistors (all 0.5 W):

(3) 1 kΩ	(1) 47 kΩ
(3) 10 kΩ	(1) 4.7 kΩ
(1) 470 Ω	(1) 100 kΩ

PROCEDURE

1. Connect the circuit of Fig. 12-3. Do not connect R_{load}.

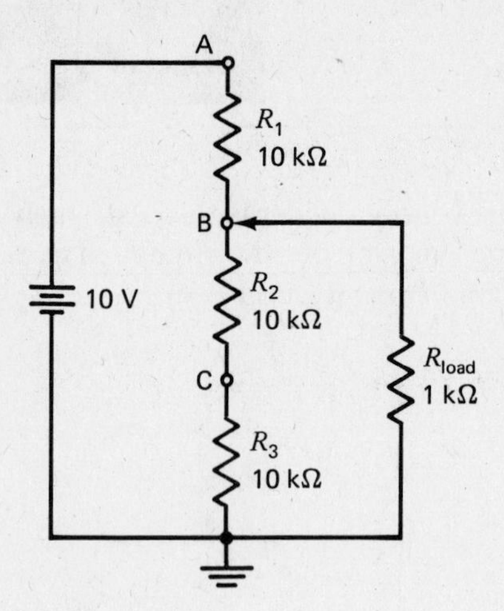

Fig. 12-3. Voltage divider.

2. Calculate the *IR* voltage drop across each resistor without the load in the circuit. Use the proportional method. Show the calculations on a separate sheet and record the results in Table 12-1. Also, measure total circuit current and record the results in Table 12-1.

3. With the load connected to point B, measure the total circuit current, load current, and bleeder current. The bleeder current is the steady drain on the source, the current through R_3. Record the values in Table 12-2.

4. Measure the voltages across R_1, R_2, R_3, and R_{load}. Record these values in Table 12-2.

5. Connect R_{load} to point A and repeat steps 3 and 4. Record the results.

6. Connect R_{load} to point C and repeat steps 3 and 4. Record the results.

7. Change the value of R_{load} to 100 kΩ and repeat steps 3 to 6.

8. Connect the circuit of Fig. 12-4. This is a voltage divider with loads.

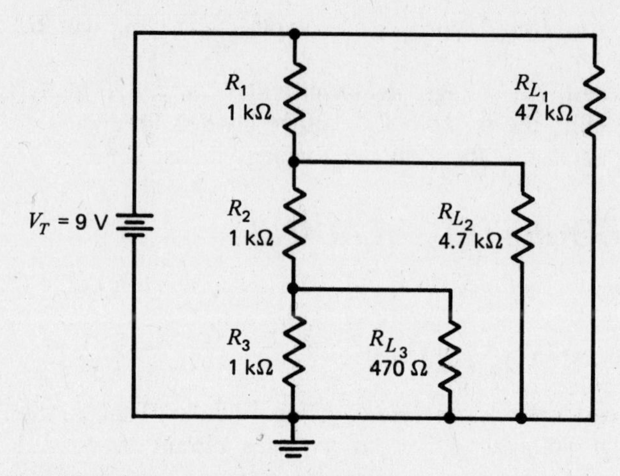

Fig. 12-4. Complex voltage divider.

9. Measure and record the *IR* voltage drops across R_1, R_2, R_3, R_{L_1}, R_{L_2}, and R_{L_3}.

10. Calculate the current through each resistor and the total circuit current. Record the results in Table 12-3.

NAME ____________________ DATE __________

RESULTS FOR EXPERIMENT 12

QUESTIONS

1. Explain what is meant by the term *voltage divider.*

2. Refer to the circuit of Fig. 12-4 (loaded voltage divider). Explain what would happen to the total circuit current, voltage, and resistance if R_{L_2} and R_{L_3} were removed.

3. If Fig. 12-4 were used to tap the total voltage of 9 V into three equal parts (without the loads), explain why V_{R_1}, V_{R_2}, and V_{R_3} are no longer 3 V each. Does the value of any one load greatly affect the original unloaded divider?

4. Redraw Fig. 12-4 and show the path of current flow by drawing arrows where necessary.

5. What effect would reversing the battery polarity have on the circuit of Fig. 12-4?

REPORT

Write a complete report. Discuss the measured and calculated results. Discuss the three most significant aspects of the experiment and write a conclusion.

TABLE 12–1. No Load Values for Fig. 12-3

	Calculated *IR* Drop	Calculated Current
R_1	______	
R_2	______	
R_3	______	
I_T		______

TABLE 12–2. Measured Values for Fig. 12-3, Circuit under Load

R_L = 1 kΩ	Load at Point A *IR* Drops	Load at Point A Current	Load at Point B *IR* Drops	Load at Point B Current	Load at Point C *IR* Drops	Load at Point C Current
R_1	______		______		______	
R_2	______		______		______	
R_3	______		______		______	
R_{load}	______	______	______	______	______	______
I_T		______		______		______
I_B		______		______		______
R_L = 100 kΩ						
R_1	______		______		______	
R_2	______		______		______	
R_3	______		______		______	
R_{load}	______	______	______	______	______	______
I_T		______		______		______
I_B		______		______		______

TABLE 12–3. Circuit Values for Fig. 12-4

	IR Drop, Measured	Current, Calculated*
R_1	______	______
R_2	______	______
R_3	______	______
R_{L_1}	______	______
R_{L_2}	______	______
R_{L_3}	______	______
R_T		______

*V (*IR* drop measured)/R nominal = calculated current.

EXPERIMENT 13

CURRENT DIVIDERS

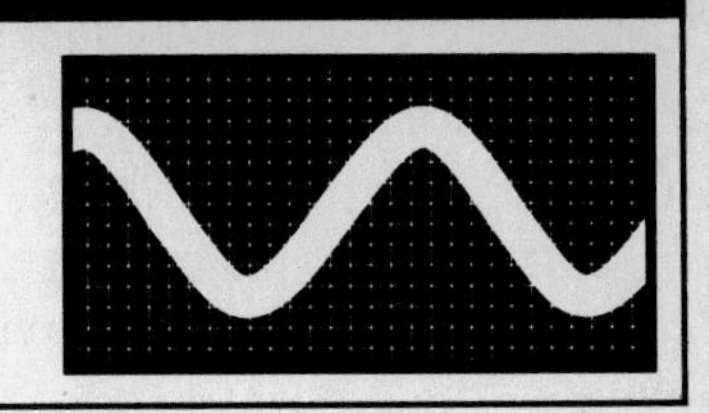

OBJECTIVES

At the completion of this experiment, you will be able to:

- Understand how parallel circuits act as current dividers.
- Be familiar with the proportional method for solving branch currents.
- Be familiar with the parallel conductance method for solving branch currents.

SUGGESTED READING

Chapter 7, *Basic Electronics*, B. Grob, Seventh Edition

INTRODUCTION

In the same way that series circuits are also voltage dividers, parallel circuits are also current dividers. That is, the total current is divided among the parallel branches in inverse proportion to the resistance in any branch. Therefore, main-line current increases as branches are added.

To find the branch currents without knowing the total voltage across the bank, a formula can be used. This formula is a proportional method for solving unknown branch currents. For example, the branch currents for Fig. 13-1 can be found by using this proportional method. It is based on the fact that currents divide in inverse proportion to their resistances.

Notice two things in Fig. 13-1. First, the numerator for each branch resistance is the value of the opposite branch. Second, it is only necessary to calculate one branch current and subtract it from the total current. The remainder will be the other branch current.

Another method for determining branch current division is the parallel conductance method. Remember that conductance $G = 1/R$. Note that conductance and current are directly proportional. This is true because the greater the resistance, the less will be the current. With any number of parallel branches, each branch current can be calculated without knowing the voltage across the bank. The formula is

$$I_x = \frac{G_x}{G_T} \times I_T$$

For example, the branch current I_{R_1} in Fig. 13-2 can be found as follows:

$$I_{R_1} = \frac{G_1}{G_T} \times I_T$$

$$G_1 = \frac{1}{R_1} = \frac{1}{10\ \Omega} = 0.1\ \text{S}$$

Note: Siemens (S) is the reciprocal of ohms.

$$R_T = \frac{1}{1/R_1 + 1/R_2 + 1/R_3} = 6.25\ \Omega$$

$$G_T = \frac{1}{R_T} = \frac{1}{6.25\ \Omega} = 0.16\ \text{S}$$

$$I_{R_1} = \frac{0.1}{0.16} \times 40\ \text{mA} = 25\ \text{mA}$$

$I_T = 30$ A

$I_{R_1} = \frac{40}{60} \times I_T = 20$ A

$V_{unknown}$; R_1 20 Ω; R_2 40 Ω; I_T

$I_{R_2} = \frac{20}{60} \times I_T = 10$ A

Note: $I_{R_x} = \frac{R_{\text{opposite } x}}{R_1 + R_2} \times I_T$

Fig. 13-1. Two-branch current divider.

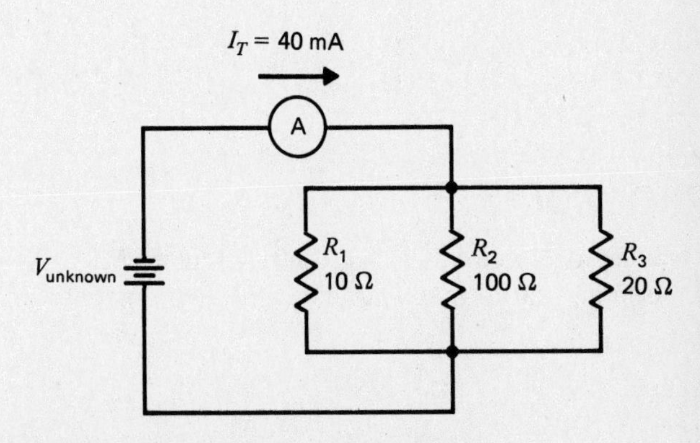

Fig. 13-2. Three-branch current divider.

Because $I_{R_1} = 25$ mA, it is easy to see that the remaining currents will be

$$I_{R_2} + I_{R_3} = +40 \text{ mA} - 25 \text{ mA}$$
$$= 15 \text{ mA}$$

Thus, the previous formula (proportional method) can be used to solve for the remaining two branch currents.

Another method is to find the total conductance and use the following formula to solve for each branch current in Fig. 13-2:

$$G_1 = \frac{1}{R_1} = \frac{1}{10\ \Omega} = 0.10 \text{ S}$$

$$G_2 = \frac{1}{R_2} = \frac{1}{100\ \Omega} = 0.01 \text{ S}$$

$$G_3 = \frac{1}{R_3} = \frac{1}{20\ \Omega} = 0.05 \text{ S}$$

Therefore, the total conductance is

$$G_1 + G_2 + G_3 = 0.1 + 0.01 + 0.05 = 0.16 \text{ S}$$

To calculate branch currents, use the following formula:

$$I_{R_x} = \frac{G_{R_x}}{G_T} \times I_T$$

In the case of Fig. 13-2,

$$I_{R_1} = 10/16 \times 40 \text{ mA} = +25.0 \text{ mA}$$
$$I_{R_2} = 1/16 \times 40 \text{ mA} = +2.5 \text{ mA}$$
$$I_{R_3} = 5/16 \times 40 \text{ mA} = +12.5 \text{ mA}$$
$$I_T = +40.0 \text{ mA}$$

EQUIPMENT

DC power supply
Protoboard or springboard
Ammeter
VTVM/VOM
Leads

COMPONENTS

Resistors (all 0.5 W):

(1) 100 Ω (1) 560 Ω
(1) 150 Ω (1) 820 Ω
(1) 390 Ω

PROCEDURE

1. Refer to the circuit of Fig. 13-3. Calculate R_T.
2. Assume that $I_T = 100$ mA. Calculate the current through R_1 and R_2 by using the proportional method for two branches in parallel. Show your calculations and record the results in Table 13-1.

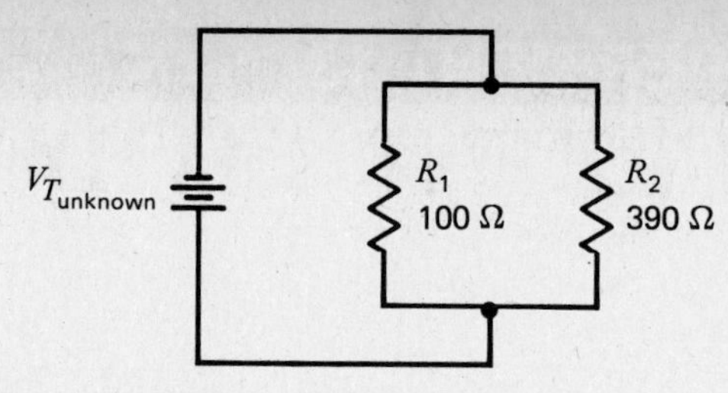

Fig. 13-3. Two-branch current divider.

3. Using the values you calculated for I_{R_1} and I_{R_2}, calculate the total voltage across the bank (use Ohm's law). Show your calculations on a separate sheet, and record the value of V_T in Table 13-1.
4. Connect the circuit of Fig. 13-3. Calculate R_T.
5. Add another resistor, $R_3 = 820\ \Omega$, to the bank and apply the total voltage you calculated in step 3 to the circuit.
6. Measure the total current and the current through each branch. Record the results in Table 13-2.
7. Connect the circuit of Fig. 13-4.

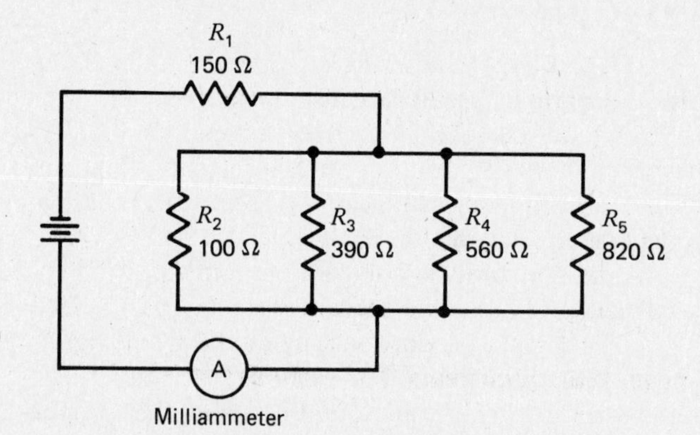

Fig. 13-4. Four-branch current divider.

8. Adjust the voltage so that total current = 30 mA.
9. Use the parallel conductance formula to determine the branch conductances G_{R_2} to G_{R_5}. Also, measure the current through each branch and record the results in Table 13-3. Show all your calculations and record the results in Table 13-3.

Note: Do not measure *any* voltages. Doing so will defeat the purpose of this experiment.

10. Using the values you have determined for the circuit of Fig. 13-4, calculate the voltages across R_1 and the bank. Remember, do not measure these voltages. How does the calculated voltage compare to the applied voltage? (Answer in your report.)

NAME _______________ DATE _______________

RESULTS FOR EXPERIMENT 13

QUESTIONS

Answer TRUE (T) or FALSE (F) to the following:

_____ **1.** Series circuits divide current, and parallel circuits divide voltages.

_____ **2.** Conductance G is the reciprocal of branch current.

_____ **3.** Refer to Fig. 13-4. If another resistor, $R_6 = 100\ \Omega$, were added in parallel to the bank, the voltage across the bank would increase.

_____ **4.** Refer to Fig. 13-4. If the series resistor R_1 were short-circuited, the total current would decrease.

_____ **5.** Refer to Fig. 13-4. If R_3 and R_4 were opened, the voltage across the series resistor R_1 would decrease.

_____ **6.** Refer to Fig. 13-4. If the total voltage were halved and the total circuit resistance doubled, there would be no effect upon total current.

_____ **7.** Refer to Fig. 13-4. If R_1 were opened, the total circuit current would increase.

_____ **8.** The voltage across any parallel bank is increased as the total conductance of the bank is increased.

_____ **9.** The total current in a parallel bank is inversely proportional to its conductance.

_____ **10.** The total current in a parallel bank is directly proportional to its total resistance.

REPORT

Write a complete report.

TABLE 13-1

	Nominal R, Ω	Calculated I, mA	Total Calculated V, V
R_1	100		
R_2	390		
R_T			

TABLE 13-2

	Nominal R, Ω	Measured I, mA	From Table 13-1 V, V
R_1	100		
R_2	390		
R_3	820		
R_T			

TABLE 13-3

	Nominal R, Ω	Calculated G, S	Calculated I, mA	Calculated V, V*	Measured I, mA
R_1	150		30.0		30.0
R_2	100				
R_3	390				
R_4	560				
R_5	820				
R_T			30.0		3.0

Note: Calculate R_T.
* $V_x = I$ calculated × R nominal

EXPERIMENT 14

VOLTAGE DIVIDER DESIGN

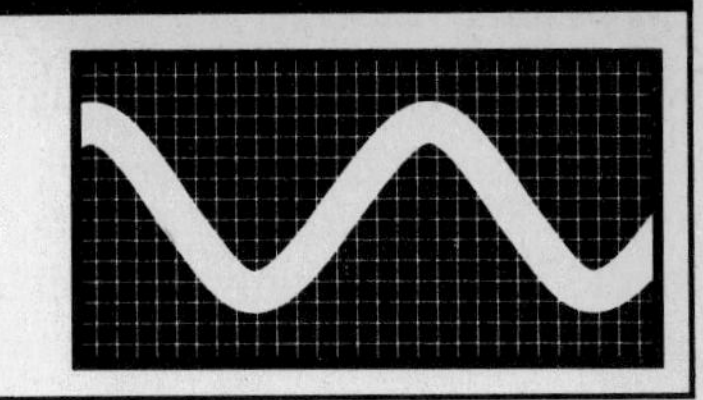

OBJECTIVES

At the completion of this experiment, you will be able to:

- Design a voltage divider for given load requirements.
- Understand the concept of negative voltage.
- Understand the concepts of common ground and bleeder current.

SUGGESTED READING

Chapter 7, *Basic Electronics,* B. Grob, Seventh Edition

INTRODUCTION

This experiment requires that you design and test a loaded voltage divider circuit. You can imagine that you are actually designing this divider for the output of a power supply in any piece of electronic equipment. Often, designing a circuit teaches you more than simply measuring and analyzing data.

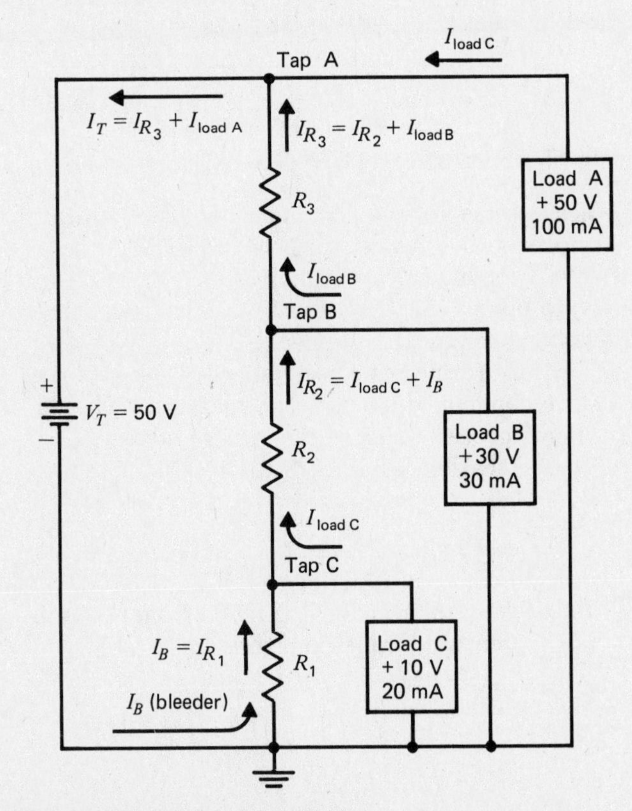

Fig. 14-1. Loaded voltage divider.

In previous experiments, you became familiar with the concepts of voltage and current dividers. You may recall that the loads on a voltage divider are also current dividers because they are parallel circuits.

To review, consider the circuit of Fig. 14-1.

1. *Loads*: The loads are resistances that require specific voltages and, often, specific currents. Here, imagine each load is a separate circuit board that requires certain voltages and currents in order to properly perform a function such as amplification.
2. I_B: This is known as *bleeder current.* It is a steady drain on the source and has no load currents passing through it. Typically, bleeder currents are calculated to be 10 percent of the total current.
3. V_T: Total voltage is fixed. It is like a budget that the designer is allowed to work with.
4. *Taps*: These are the places where the loads are connected in order to tap off their specified voltages.
5. R_1, R_2, R_3: These are the voltage divider resistances that you, the designer, will be determining by your calculations. In this case, Fig. 14-1 has a given budget of 50 V (total). Your calculations will divide this voltage into the necessary values to supply the loads.
6. *Ground*: This symbol shows a common or earth ground that is connected to one side of the power supply. Remember, ground is a reference point, like zero.

How to Calculate the Voltage Divider

Use a table, as shown in Table 14-1, in order to keep your values organized. Find the current in each *R*.

$$I_{R_1} = I_B = \text{approx 10 percent of } I_T \text{ loads}$$

Thus

$$I_B = 150 \text{ mA} \times 0.1 = 15 \text{ mA}$$

$$\begin{aligned} I_{\text{load A}} &= 100 \text{ mA} \\ I_{\text{load B}} &= 30 \text{ mA} \\ I_{\text{load C}} &= \underline{20 \text{ mA}} \\ I_T &= 150 \text{ mA} \end{aligned}$$

Knowing the value of $I_B = I_{R_1}$, I_{R_2} and I_{R_3} are calculated as follows (compare to Fig. 14-1):

$$I_{R_2} = I_B + I_{\text{load C}} = 15 \text{ mA} + 20 \text{ mA} = 35 \text{ mA}$$
$$I_{R_3} = I_{R_2} + I_{\text{load B}} = 35 \text{ mA} + 30 \text{ mA} = 65 \text{ mA}$$

How to Calculate the Voltage Across Each Resistor

The voltages at the taps are also the voltages across the loads with respect to ground. However, the voltages across R_2 and R_3 are not in parallel only with these loads. Therefore, only the voltage across R_1 is the same as $V_{\text{load C}} = 10$ V. V_{R_2} and V_{R_3} are calculated as follows (compare to Fig. 14-1):

$$V_{R_1} = V_{\text{tap C}} = 10\text{ V}$$
$$V_{R_2} = V_{\text{tap B}} - V_{\text{tap C}} = 30\text{ V} - 10\text{ V} = 20\text{ V}$$
$$V_{R_3} = V_{\text{tap A}} - V_{\text{tap B}} = 50\text{ V} - 30\text{ V} = 20\text{ V}$$

How to Calculate Each Resistor

Now that the voltages and currents for each resistor have been determined, it is easy to use Ohm's law to calculate the value of each resistor:

$$R_1 = \frac{V_{R_1}}{I_{R_1}} = \frac{10\text{ V}}{15\text{ mA}} = 666.7\ \Omega$$
$$R_2 = \frac{V_{R_2}}{I_{R_2}} = \frac{20\text{ V}}{35\text{ mA}} = 571.4\ \Omega$$
$$R_3 = \frac{V_{R_3}}{I_{R_3}} = \frac{20\text{ V}}{65\text{ mA}} = 307.7\ \Omega$$

These values of voltage divider resistance should provide the specified load requirements.

EQUIPMENT

DC power supply
Ohmmeter
Ammeter (VOM) optional
Voltmeter (VOM/VTVM)
Protoboard or springboard
Leads

COMPONENTS

Resistors (all 0.5 W):

(1) 150 Ω (3) 470 Ω

PROCEDURE

1. Design a voltage divider, similar to Fig. 14-1, with the following specifications:

$V_T = 20$ V
Load A = 20 V, 40 mA
Load B = 15 V, 25 mA
Load C = 5 V, 10 mA

2. On a separate sheet, draw the circuit similar to Fig. 14-1 and fill in Table 14-1 (columns 1, 2 and 4).

3. Verify that your circuit works, by connecting the circuit and measuring the voltages across R_1, R_2, R_3, and each load. Use resistors within 10 percent of the load values. Record the voltages (measured) in Table 14-1 (column 3). If any value of voltage falls outside plus or minus 20 percent of your calculated values, redesign the circuit.

4. With your circuit connected, remove earth ground (if attached).

5. Using the circuit of Fig. 14-2 as an example, you will move the ground point (common: without an earth ground connection) to point C. Although no actual connection is yet made, point C now becomes the zero reference point.

Note: $V_{R_3} = V_{\text{point A}} - V_{R_2}$.

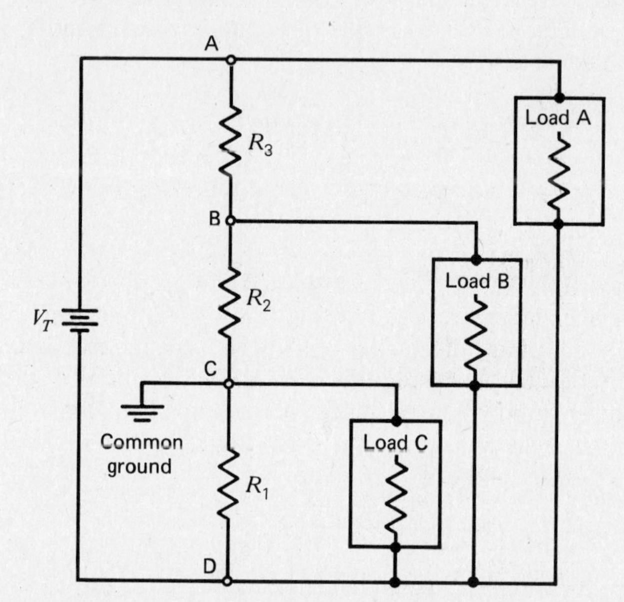

Fig. 14-2. Loaded voltage divider. Note that the common ground is shown as an earth (symbol) ground here, but it is not actually a true earth ground.

6. Measure the voltage across each resistor and load with respect to common. This means connect the negative (common) lead of the voltmeter (VTVM or VOM) to point C and measure from there. Note that the voltage across R_1 will now become a negative voltage. Record the results in Table 14-2.

7. Move common to point B and measure each voltage. Remember to subtract the voltage across R_2 when you measure V_{R_1}. Record the results in Table 14-2.

Note: The loads (A, B, C) are always measured from the original common (point D) in all three circuits. But the divider voltages are measured from points B, C, or D, depending on the configurations.

NAME DATE

RESULTS FOR EXPERIMENT 14

QUESTIONS

1. Explain the difference between earth ground and a common reference point ground.

2. Explain how a negative voltage can be obtained from a voltage divider. Explain the effects upon *I, V,* and *R,* if any.

3. Explain, in your own words, what is meant by bleeder current.

4. Explain the difference, if any, between a loaded voltage divider and a series-parallel circuit.

5. Explain how the circuit of Fig. 14-1 would be affected if:
   A. R_2 were short-circuited **B.** R_2 were opened

REPORT

Write a complete report. Discuss how load changes affect the design and the current values.

TABLE 14–1

Divider	(1) Calculated R, Ω	(2) Measured R Value, Ω	(3) Measured V, V	(4) Calculated I, mA
Load A	________	________	________	________
Load B	________	________	________	________
Load C	________	________	________	________
R_1	________	________	________	________
R_2	________	________	________	________
R_3	________	________	________	________

TABLE 14–2

Divider	Measured V, V Common = Point C	Measured V, V Common = Point B
Load A	________	________
Load B	________	________
Load C	________	________
R_1	________	________
R_2	________	________
R_3	________	________

EXPERIMENT 15

POSITIVE AND NEGATIVE VOLTAGES TO GROUND

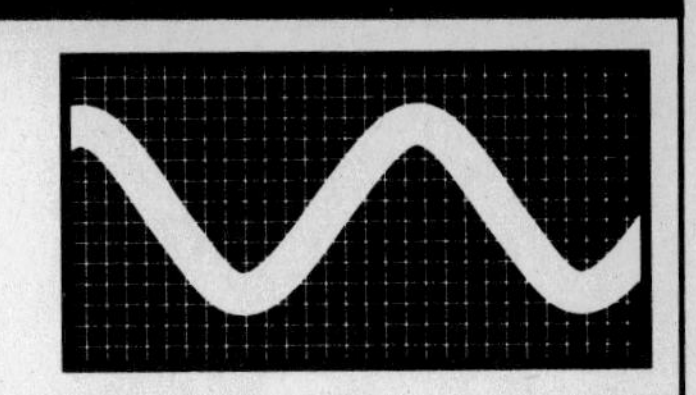

OBJECTIVES

At the completion of this experiment, you will be able to:

- Calculate circuit current and voltage drops found in a voltage-divider circuit.
- Determine the polarity of voltages found in a voltage divider circuit with a common ground.

SUGGESTED READING

Chapters 6 and 7, *Basic Electronics,* B. Grob, Seventh Edition

INTRODUCTION

In the wiring of practical circuits, one side of the voltage source is usually grounded. In electronic equipment the ground often indicates a metal chassis, which is used as a common return for connections to the voltage source. Where printed-circuit boards are used, usually a common ground path is run around the outside perimeter of the circuit board. In other cases the entire back side of a two-sided board may be used as a common-return path. Note that the chassis ground may or may not be connected to earth ground.

When a circuit has a chassis as a common return, measure the voltages with respect to chassis ground. Consider the voltage divider in Fig. 15-1. In Fig. 15-1, the circuit shows no ground system. It is instead a closed series circuit. This circuit has an applied power supply voltage V_A of 20 V. To determine the total circuit current, the total circuit resistance R_T must be determined. And R_T can be calculated as

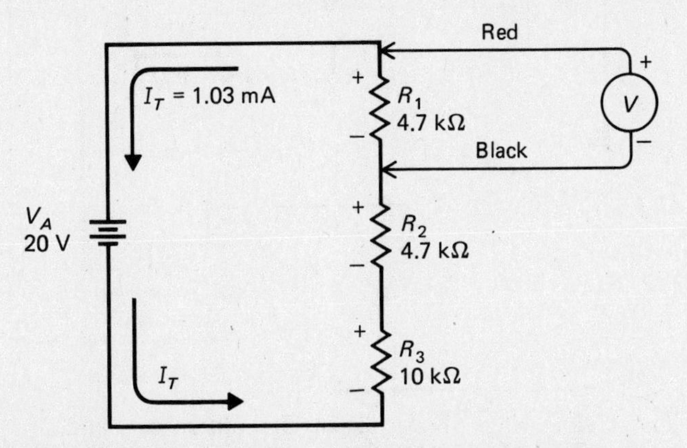

Fig. 15-1. Voltage divider circuit without a common ground.

$$\begin{aligned} R_T &= R_1 + R_2 + R_3 \\ &= 4.7\ \text{k}\Omega + 4.7\ \text{k}\Omega + 10\ \text{k}\Omega \\ &= 19.4\ \text{k}\Omega \end{aligned}$$

The circuit current can now be calculated from the Ohm's law relationship:

$$\begin{aligned} I_T &= \frac{V_A}{R_T} \\ &= \frac{20\ \text{V}}{19.4\ \text{k}\Omega} \\ &= 1.03\ \text{mA} \end{aligned}$$

After the current has been calculated, the individual voltage drops V_1, V_2, and V_3 of the voltage divider can be found from the Ohm's law relationship of

$$V = I \times R$$

For V_1:

$$\begin{aligned} V_1 &= I_T \times R_1 \\ &= 1.03\ \text{mA} \times 4.7\ \text{k}\Omega \\ &= 4.84\ \text{V} \quad (\text{or } +4.84\ \text{V}) \end{aligned}$$

For V_2:

$$\begin{aligned} V_2 &= I_T \times R_2 \\ &= 1.03\ \text{mA} \times 4.7\ \text{k}\Omega \\ &= 4.84\ \text{V} \quad (\text{or } +4.84\ \text{V}) \end{aligned}$$

For V_3:

$$\begin{aligned} V_3 &= I_T \times R_3 \\ &= 1.03\ \text{mA} \times 10\ \text{k}\Omega \\ &= 10.31\ \text{V} \quad (\text{or } +10.31\ \text{V}) \end{aligned}$$

Also the sum of the voltage drops will equal the applied voltage:

$$\begin{aligned} V_A &= V_1 + V_2 + V_3 \\ &= 4.84\ \text{V} + 4.84\ \text{V} + 10.31\ \text{V} \\ 20\ \text{V} &= 19.99\ \text{V} \end{aligned}$$

Refer to Fig. 15-1 and note that the polarities are included on this schematic. The polarity is determined by how the circuit current and the individual voltmeters are connected. The polarity of the resistors, which indicates the direction of current flow, and the color of the test leads are also indicated in Fig. 15-1.

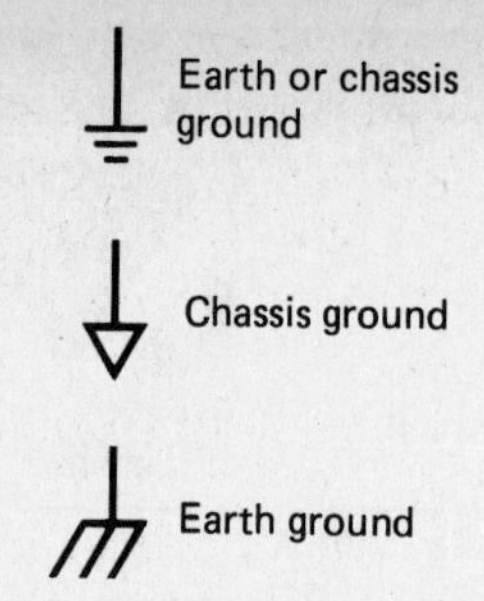

Fig. 15-2. Schematic symbols for ground.

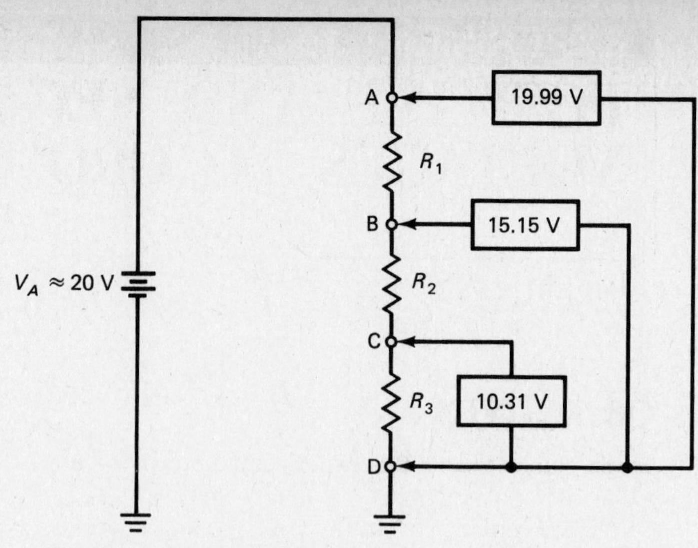

Fig. 15-4. Voltage divider circuit where voltages are measured to common ground.

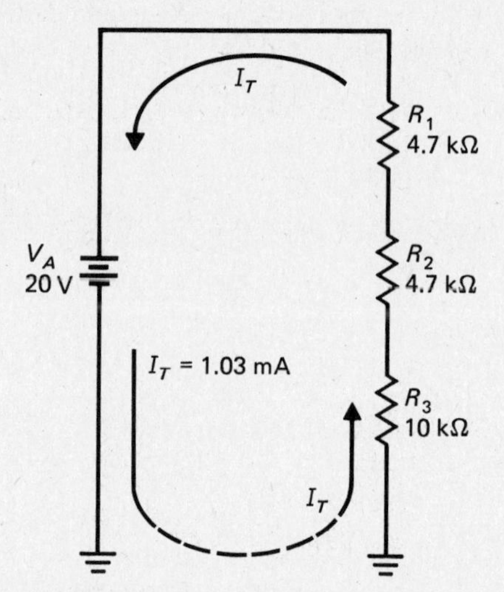

Fig. 15-3. Voltage divider circuit using a ground return.

Figure 15-2 shows the schematic symbols for ground. The ground symbol is used in Fig. 15-3. Here the same circuit with the same component values and applied voltage is shown. The only addition is the "ground return." Also note that the voltage drops of this circuit are equivalent to those shown in Fig. 15-1. The 1.03 mA generated by the battery is pushed out into ground point A and returns to ground point B. This, in effect, means that the ground symbols are connected, perhaps through a cable, foil pattern, or metal chassis.

The circuit shown in Fig. 15-4 details the same circuit of Figs. 15-3 and 15-1. The difference here is that all voltages are taken "with respect to ground." In this case, the voltage from point C to D is +10.31 V. The voltage from point B to D is +15.15 V (where $V_{BD} = VR_2 + VR_3$). The voltage from point A to D is +19.99 V (or 20 V), where $V_{AD} = VR_1 + VR_2 + VR_3$).

The circuit in Fig. 15-5 is similar to the one in Fig. 15-4. The main difference is that the ground has been moved to point C. This then indicates that all measurements should be taken "with reference" to this point. In this case $V_{AC} = +9.68$ V (where $V_{AC} = VR_1 + VR_2$). The voltage $V_{BC} = +4.84$ V (where $V_{BC} = VR_2$). The voltage $V_{DC} = -10.31$ V (where $V_{DC} = VR_3$). The voltage measured from point D to C results in a negative voltage. This circuit is known as a positive and negative voltage divider.

In summary, these dc circuits operate in the same way with or without the ground symbol shown in the schematic. The only factor that changes is the reference point for measuring the voltage.

While this experiment focuses on the use of ground as a reference point, keep in mind that there are several different symbols for ground, as shown in Fig. 15-2. The earth ground symbol usually indicates that one side of the power supply is connected to the earth, usually by a metal pipe in the ground (this is the third prong on the ac wall plug). The other two ground symbols are chassis grounds and may or may not be connected to earth (an automobile or an airplane is a good example).

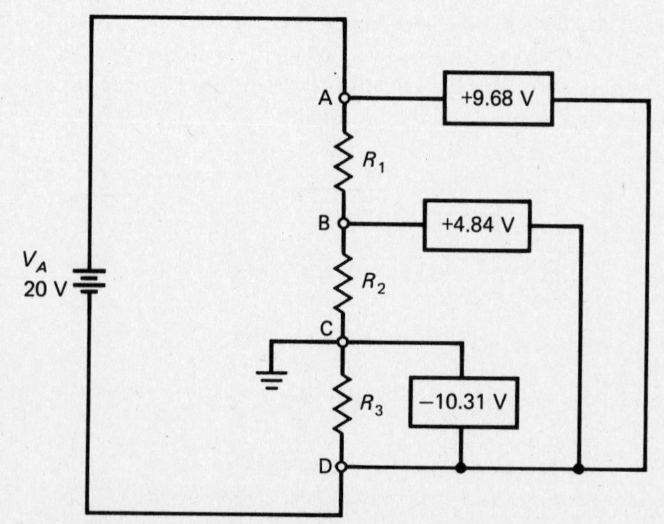

Fig. 15-5. Voltage divider displaying both positive and negative voltages to ground.

EQUIPMENT

DC power supply, 0 to 20 V
Leads
Breadboard
Voltmeter

COMPONENTS

Resistors (all 0.5 W):
(2) 4.7 kΩ
(1) 10 kΩ

PROCEDURE

1. Connect the circuit of Fig. 15-6 to the dc power supply as shown.

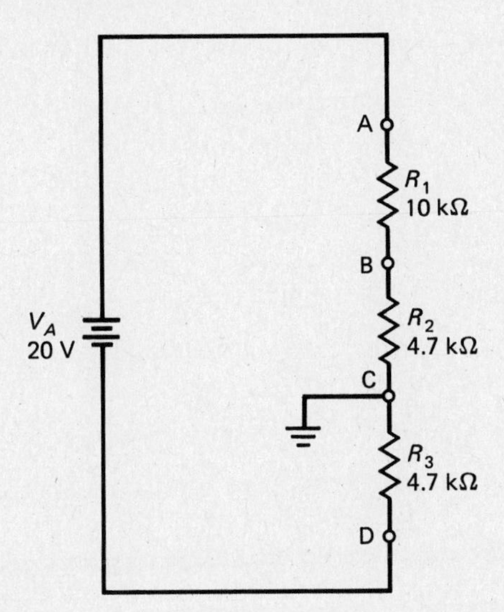

Fig. 15-6. Positive and negative voltage divider circuit with ground.

2. With the power supply turned off and disconnected, measure and record in Table 15-1 the resistance values of R_1, R_2, and R_3.
3. With the power supply turned off and disconnected, measure and record in Table 15-1 the resistive value of R_T.
4. Reconnect the power supply to the circuit shown in Fig. 15-6.
5. Calculate and record in Table 15-1 the values of R_T, I_T, V_{R_1}, V_{R_2}, and V_{R_3} referenced to ground.
6. Turn on the power supply and adjust its voltage value to 20 V dc.
7. Measure and record in Table 15-1: I_T, V_A, V_{R_1}, V_{R_2}, and V_{R_3}. Also note the polarities of V_{R_1}, V_{R_2}, and V_{R_3} in Table 15-1.
8. Turn off the power supply, and reverse its polarity.
9. Reconnect the circuit with $V_A = -20$ V, repeat steps 2–7, and record the results in Table 15-2.

NAME ______________________ DATE ____________

RESULTS FOR EXPERIMENT 15

QUESTIONS

1. Explain the circuit function of a voltage divider.

2. What is the purpose of a circuit ground?

3. Draw an example of a circuit where a voltage is negative with respect to ground.

REPORT

Write a complete report. Discuss the measured and calculated results. Discuss the three most significant aspects of the experiment, and write a conclusion.

TABLE 15–1 (Steps 2–7)

	Measured	Calculated	Polarity
R_1	______		
R_2	______		
R_3	______		
R_T	______	______	
I_T	______	______	
V_{R_1}	______	______	______
V_{R_2}	______	______	______
V_{R_3}	______	______	______
V_A	______		

*Note: Polarity with respect to common ground.

TABLE 15–2 (Steps 8 and 9)

	Measured	Calculated	Polarity
R_1	______		
R_2	______		
R_3	______		
R_T	______	______	
I_T	______	______	
V_{R_1}	______	______	______
V_{R_2}	______	______	______
V_{R_3}	______	______	______
V_A	______		

EXPERIMENT 16

AMMETERS

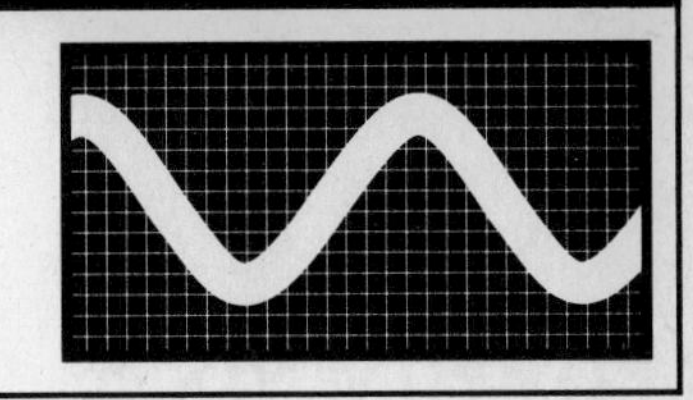

OBJECTIVES

At the completion of this experiment, you will be able to:

- Determine the internal resistance of a basic D'Arsonval meter movement.
- Design an ammeter circuit from this meter movement.
- Use this ammeter for actual ammeter measurements.

SUGGESTED READING

Chapter 8, *Basic Electronics*, B. Grob, Seventh Edition

INTRODUCTION

Range of an Ammeter

The small size of the wire with which an ammeter's movable coil is wound places severe limits on the current that may be passed through the coil. Consequently, the basic D'Arsonval movement may be used to indicate or measure only very small currents—for example, microamperes or milliamperes, depending on meter sensitivity.

To measure a larger current, a shunt must be used with the meter. A shunt is a heavy, low-resistance conductor connected across the meter terminals to carry most of the load current. This shunt has the correct amount of resistance to cause only a small part of the total circuit current to flow through the meter coil. The meter current is proportional to the load current. If the shunt is of such a value that the meter is calibrated in milliamperes, the instrument is called a *milliammeter.* If the shunt is of such a value that the meter is calibrated in amperes, it is called an *ammeter.*

A single type of standard meter movement is generally used in all ammeters, no matter what the range of a particular meter. For example, meters with working ranges of 0 to 10 A, 0 to 5 A, or 0 to 1 A all use the same galvanometer movement. The designer of the ammeter calculates the correct shunt resistance required to extend the range of the meter movement to measure any desired amount of current. This shunt is then connected across the meter terminals. Shunts may be located inside the meter case (internal shunt) or somewhere away from the meter (external shunt), with leads going to the meter.

Extending the Range by Use of Shunts

For limited current ranges (below 50 A), internal shunts are most often employed. In this manner, the range of the meter may be easily changed by selecting the correct internal shunt having the necessary current rating. Before the required resistance of the shunt for each range can be calculated, the resistance of the meter movement must be known.

For example, suppose it is desired to use a 100-μA D'Arsonval meter having a resistance of 100 Ω to measure line currents up to 1 A. The meter deflects full scale when the current through the 100-Ω coil is 100 μA. Therefore, the voltage drop across the meter coil is IR, or

$$0.0001 \times 100 = 0.01 \text{ V}$$

Because the shunt and coil are in parallel, the shunt must also have a voltage drop of 0.01 V. The current that flows through the shunt is the difference between the full-scale meter current and the line current. In this case, the meter current is 0.0001 A. This current is negligible compared with the line (shunt) current, so the shunt current is approximately 1 A. The resistance R_S of the shunt is therefore

$$R_S = \frac{V}{I} = \frac{0.01}{1} = 0.01\ \Omega \text{ (approx.)}$$

and the range of the 100-μA meter has been increased to 1 A by paralleling it with the 0.01-Ω shunt.

The 100-μA instrument may also be converted to a 10-A meter by the use of a proper shunt. For full-scale deflection of the meter, the voltage drop V across the shunt (and across the meter) is still 0.01 V. The meter current is again considered negligible, and the shunt current is now approximately 10 A. The resistance R_S of the shunt is therefore

$$R_S = \frac{V}{I} = \frac{0.01}{10} = 0.001\ \Omega$$

The same instrument may likewise be converted to a 50-A meter by the use of the proper type of shunt. The current I_S through the shunt is approximately 50 A, and the resistance R_S of the shunt is

$$R_S = \frac{V}{I_S} = \frac{0.01}{50} = 0.0002\ \Omega$$

EQUIPMENT

DC power supply
Protoboard or springboard
Leads
VTVM or DVM

COMPONENTS

(1) 0- to 1-mA meter movement

Resistors:

(1) 150-kΩ 0.5-W Other resistors as calculated.
(1) 470-Ω 0.5-W

Potentiometers:

(1) 5 kΩ (1) 100 kΩ

(1) SPST switch
(1) SPDT switch

PROCEDURE

1. Measure the internal resistance of the meter movement by connecting the circuit shown in Fig. 16-1.

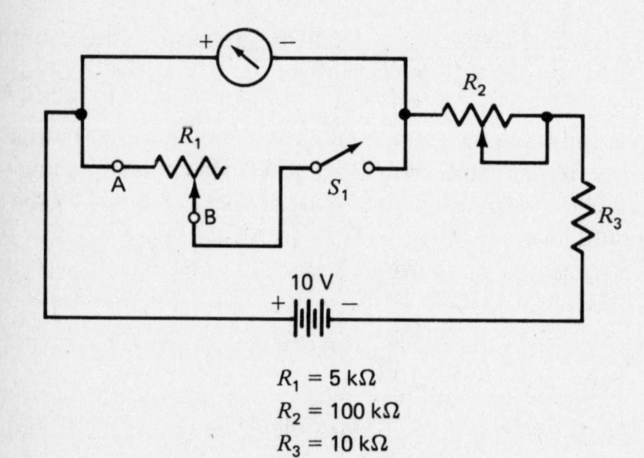

Fig. 16-1. Internal resistance measurement.

2. With S_1 open, turn on the power supply and adjust it for 10 V.
3. Adjust R_2 so that the scale upon the meter movement reads at full-scale deflection.
4. Close S_1 and adjust R_1 so that the scale upon the meter movement reads at half-scale deflection. The currents will evenly divide between R_1 and the internal resistance r_m of the meter movement when $R_I = r_m$.
5. Measure and record in Table 16-1 the voltage dropped across V_m.
6. Measure and record r_m in Table 16-1 by turning the power supply off, disconnecting R_1 from the circuit, and measuring from point A to B. At this point, $r_m = R_1$.
7. Calculate and record I_m in Table 16-1, where

$$I_m = \frac{V_m}{r_m}$$

8. Record in Table 16-2 the value of I_m for full-scale deflection, the r_m (meter movement's internal resistance) for the meter movement, and the value V_m needed for full-scale deflection.
9. Construct the following dual-range ammeter in Fig. 16-2. Range 1 will measure 30 mA full-scale, and range 2 will measure 100 mA full-scale.

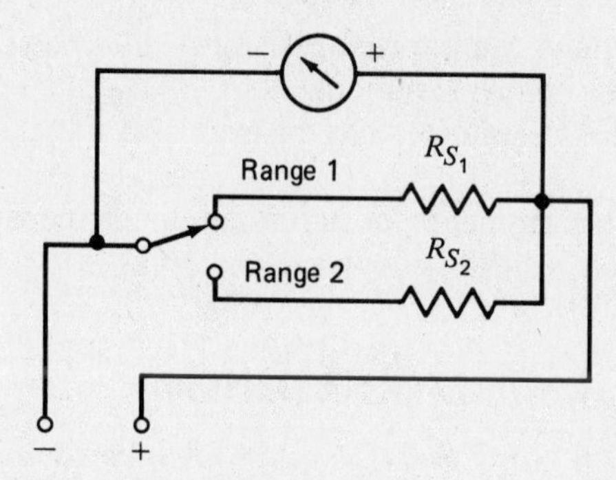

Fig. 16-2. Ammeter circuit.

10. Using the following formula determines the multiplier resistors R_{S_1} and R_{S_2}:

$$R_S = (I_m \times r_m)I_S$$

For a 30-mA full-scale deflection,

$$R_{S_1} = (I_m \times r_m)/30\text{ mA}$$

Record this value in Table 16-2.
For a 100-mA full-scale deflection,

$$R_{S_2} = (I_m \times r_m)/100\text{ mA}$$

Record this value in Table 16-2.

11. Connect your ammeter into the circuit configuration shown in Fig. 16-3, where I_2 is an ammeter of known accuracy and I_1 is your ammeter design. Complete Table 16-3 by turning on and adjusting the power supply in accordance with Table 16-3. Record the values of I_1 and I_2 for each power supply setting.

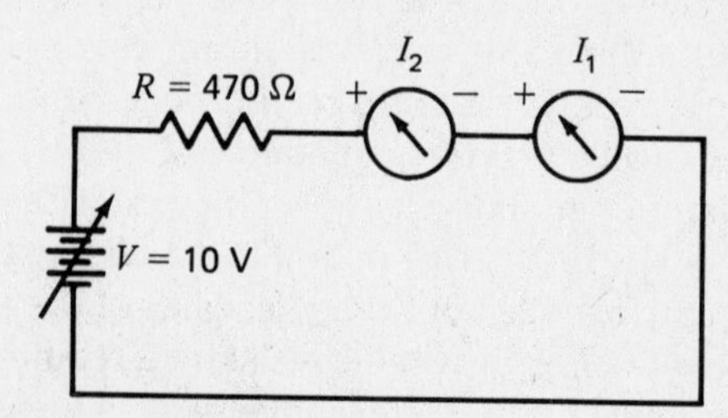

Fig. 16-3. Ammeter test setup.

12. Determine the percentage of accuracy for Table 16-3.

NAME DATE

RESULTS FOR EXPERIMENT 16

QUESTIONS

1. Describe how ammeters are connected in a circuit to measure current.

2. Design an ammeter circuit that will measure 1.5 A with a 0- to 100-mA full-scale deflection meter movement.

REPORT

Write a complete report. Discuss the measured and calculated results. Discuss the three most significant aspects of the experiment and write a conclusion.

TABLE 16–1

r_m	I_m	V_m

TABLE 16–2. Meter Movement

I_m, A	r_m, Ω	V_m, V
Shunt R_{S_1}	**Shunt R_{S_2}**	

TABLE 16–3

Range, mA	Voltage Setting	I_1	I_2	% Accuracy
0–30				
0–100				

EXPERIMENT 17

VOLTMETERS

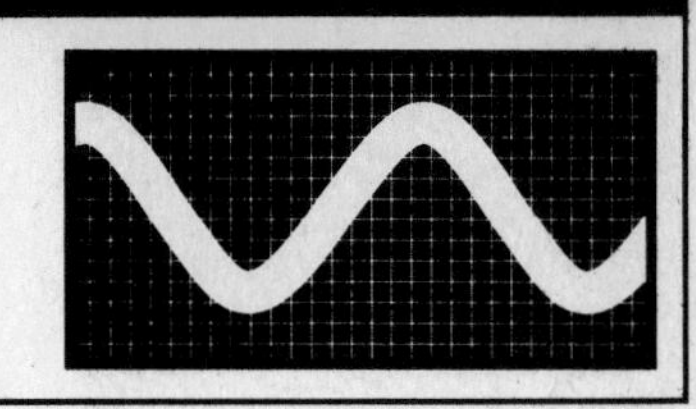

OBJECTIVES

At the completion of this experiment, you will be able to:

- Determine the internal resistance of a basic D'Arsonval movement.
- Design a voltmeter from this meter movement.
- Use this voltmeter for actual voltage measurements.

SUGGESTED READING

Chapter 8, *Basic Electronics,* B. Grob, Seventh Edition

INTRODUCTION

D'Arsonval Meter

The stationary permanent-magnet moving-coil meter is the basic movement used in most measuring instruments for servicing electric equipment. This type of movement is commonly called the D'Arsonval movement because it was first employed by the Frenchman D'Arsonval in making electrical measurements.

The basic D'Arsonval movement consists of a stationary permanent magnet and a movable coil. When current flows through the coil, the resulting magnetic field reacts with the magnetic field of the permanent magnet and causes the coil to rotate. The greater the amount of current flow through the coil, the stronger the magnetic field produced; the stronger this field, the greater the rotation of the coil. To determine the amount of current flow, a means must be provided to indicate the amount of coil rotation.

Voltmeter

The 100-μA D'Arsonval meter used as the basic meter for the ammeter may also be used to measure voltage if a high resistance is placed in series with the moving coil of the meter. When this is done, the unit containing the resistance is commonly called a *multiplier.* A simplified diagram of a voltmeter is shown in Fig. 17-1.

Extending the Range

The value of the necessary series resistance is determined by the current required for full-scale deflection of the meter and by the range of voltage to be measured. Because the current through the meter circuit is directly proportional to the applied voltage, the meter

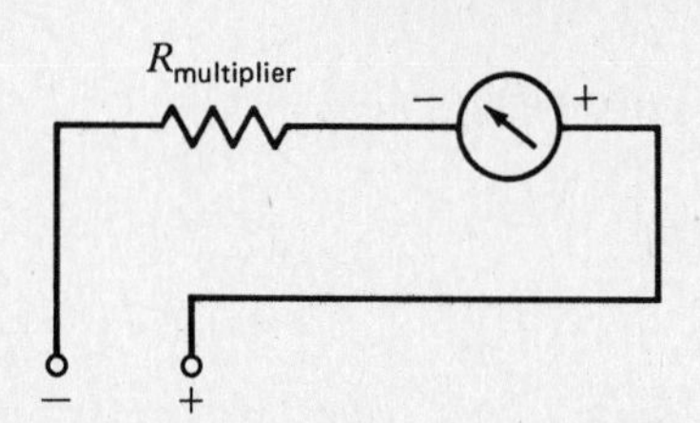

Fig. 17-1. Basic voltmeter circuit with $R_{multiplier}$.

scale can be calibrated directly in volts for a fixed series resistance.

For example, assume that the basic meter (microammeter) is to be made into a voltmeter with a full-scale reading of 1 V. The coil resistance of the basic meter is 100 Ω, and 0.0001 A causes a full-scale deflection. The total resistance R of the meter coil and the series resistance is

$$r_m = \frac{V_m}{I_m} = \frac{1}{100\ \mu A} = 10{,}000\ \Omega$$

and the series resistance alone is

$$R_1 = 10{,}000 - 100 = 9900\ \Omega$$

Multirange voltmeters utilize one meter movement with a convenient switching arrangement. A multirange voltmeter with three ranges is shown in Fig. 17-2. The total circuit resistance for each of the three ranges, beginning with the 1-V range, is

$$R_1 = \frac{V_m}{I_m} = \frac{1}{100} = 0.01\ \text{M}\Omega$$

$$R_2 = \frac{V_m}{I_m} = \frac{100}{100} = 1\ \text{M}\Omega$$

$$R_3 = \frac{V_m}{I_m} = \frac{1000}{100} = 10\ \text{M}\Omega$$

Voltage-measuring instruments are always connected across (in parallel with) a circuit. If the approximate value of the voltage to be measured is not known, it is best to start with the highest range of the voltmeter and progressively lower the range until a suitable middle third reading is obtained.

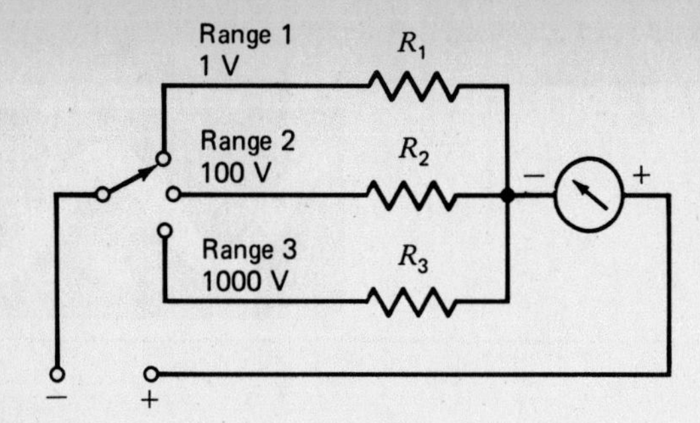

Fig. 17-2. Three-range voltmeter circuit.

EQUIPMENT

DC power supply, 10 V
Voltmeter of known accuracy
Protoboard or springboard
Test leads
VTVM or DVM

COMPONENTS

Resistors:

(1) 150-kΩ 0.5-W Other resistors as calculated.

Potentiometers (linear taper):

(1) 100 kΩ (1) 5 kΩ

(1) SPST switch
(1) SPDT switch
(1) 50-μA meter movement of unknown resistance

PROCEDURE

1. Measure the internal resistance of the meter movement by connecting the circuit shown in Fig. 17-3.

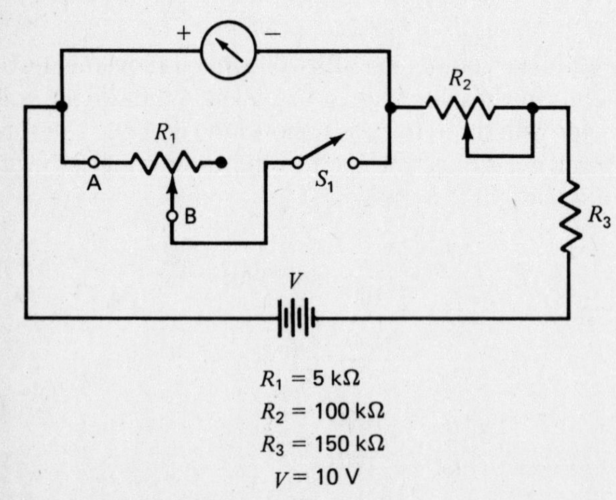

Fig. 17-3. Measuring internal resistance.

2. With S_1 open, turn on the power supply and adjust it for 10 V.
3. Adjust R_2 so that the scale upon the meter movement reads at full-scale deflection.
4. Close S_1 and adjust R_1 so that the scale upon the movement reads at half-scale deflection. The currents will evenly divide between R_1 and the internal resistance r_m of the meter movement when $R_1 = r_m$.
5. Measure and record in Table 17-1 the voltage dropped across V_m.
6. Measure and record r_m in Table 17-1 by turning the power supply off, disconnecting R_1 from the circuit, and measuring from points A to B. At this point, $r_m = R_1$.
7. Calculate and record I_m in Table 17-1, where

$$I_m = \frac{V_m}{r_m}$$

8. Record in Table 17-2 the I_m for full-scale-deflection, the r_m (meter movement's internal resistance) for the meter movement, and the value of V_m for full-scale deflection.
9. Construct the dual-range voltmeter in Fig. 17-4 so that range 1 will measure 5 V full scale and range 2 will measure 10 V full scale. Use the following formula to determine the multiplier resistors R_1 and R_2:

$$R_{\text{multiplier}} = \frac{V_{\text{FS}}}{I_{\text{FS}}} - r_m = \frac{V_{\text{intended}}}{I_m} - r_m$$

Note: FS means "full scale."

For a 5-V full-scale deflection,

$$R_1 = \frac{5\ \text{V}}{I_m} - r_m$$

For a 10-V full-scale deflection,

$$R_2 = \frac{10\ \text{V}}{I_m} - r_m$$

Record these calculated values in Table 17-2.

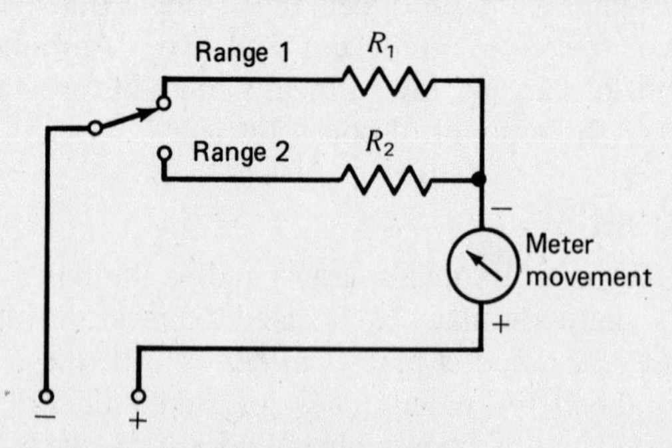

Fig. 17-4. Voltmeter dual-range circuit.

10. Connect your voltmeter into the circuit configuration of Fig. 17-5, where V_2 is a voltmeter of known accuracy and V_1 is your voltmeter design.
11. Complete Table 17-3 by turning on and adjusting the power supply in accordance with Table 17-3. Record the values of V_1 and V_2 for each power supply setting.

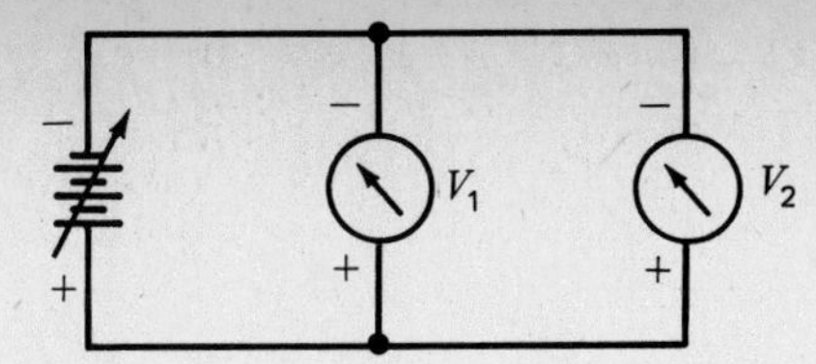

Fig. 17-5. Measuring voltages.

12. Determine the percentage of accuracy for Table 17-3.

NAME DATE

RESULTS FOR EXPERIMENT 17

QUESTION

1. Design a voltmeter that will measure 0 to 30 V dc by using a 100-mA meter movement.

REPORT

Write a complete report. Discuss the measured and calculated results. Discuss the three most significant aspects of this experiment and write a conclusion.

TABLE 17–1

0–50 μA
V_m = ________
r_m = ________
I_m = ________

TABLE 17–2. Meter Movement

I_m = ________
r_m = ________
V_m = ________
R_1 = ________
R_2 = ________

TABLE 17–3

Power Supply Voltages	V_1	V_2	% Accuracy
Range 1: 1–5 V			
1 V	______	______	______
2 V	______	______	______
3 V	______	______	______
4 V	______	______	______
5 V	______	______	______
Range 2: 1–10 V			
1 V	______	______	______
2 V	______	______	______
3 V	______	______	______
4 V	______	______	______
5 V	______	______	______
6 V	______	______	______
7 V	______	______	______
8 V	______	______	______
9 V	______	______	______
10 V	______	______	______

EXPERIMENT 18

OHMMETERS

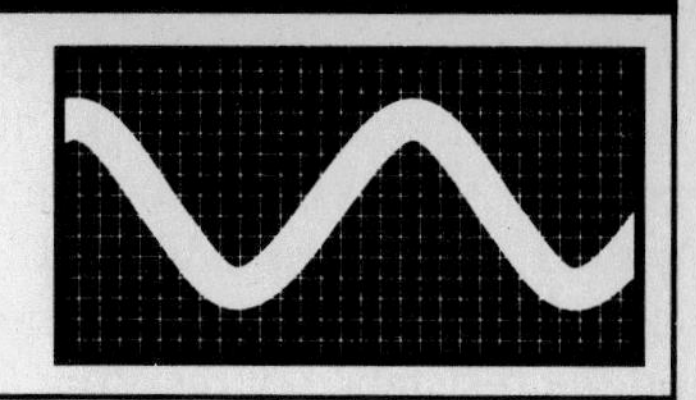

OBJECTIVES

At the completion of this experiment, you will be able to:

- Determine the internal resistance of a basic D'Arsonval meter movement.
- Design an ohmmeter from this meter movement.
- Use this ohmmeter for actual ohm measurements.

SUGGESTED READING

Chapter 8, *Basic Electronics*, B. Grob, Seventh Edition

INTRODUCTION

The ohmmeter consists of a dc milliammeter, with a few added features. The added features are:

1. A dc source of potential
2. One or more resistors (one of which is variable).

A simple ohmmeter circuit is shown in Fig. 18-1.

The ohmmeter's pointer deflection is controlled by the amount of battery current passing through the moving coil. Before measuring the resistance of an unknown resistor or electric circuit, the test leads of the ohmmeter are first short-circuited together, as shown in Fig. 18-1. With the leads short-circuited, the meter is calibrated for proper operation on the selected range. (While the leads are short-circuited, meter current is maximum and the pointer deflects a maximum amount, somewhere near the zero position on the ohms scale.) When the variable resistor is adjusted properly, with the leads short-circuited, the meter pointer will come to rest exactly on the zero graduation. This indicates *zero resistance* between the test leads, which in fact are short-circuited together. The zero readings of series-type ohmmeters are sometimes on the right-hand side of the scale, whereas the zero reading for ammeters and voltmeters is generally to the left-hand side of the scale. When the test leads of an ohmmeter are separated, the meter pointer will return to the left side of the scale, due to the interruption of current and the spring tension acting on the movable-coil assembly.

After the ohmmeter is adjusted for zero reading, it is ready to be connected in a circuit to measure resistance. A typical circuit and ohmmeter arrangement is shown in Fig. 18-2.

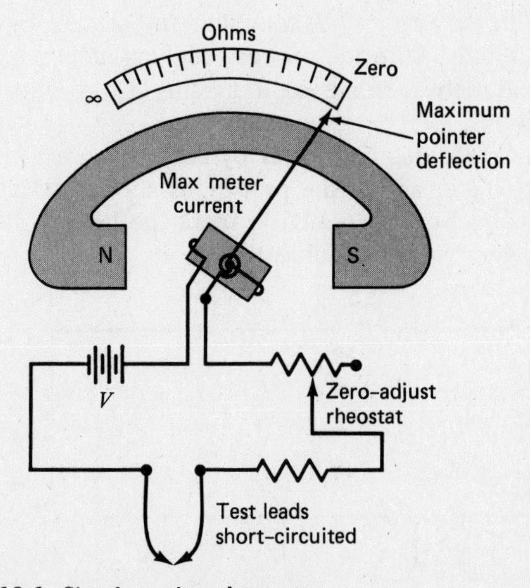

Fig. 18-1. Simple series ohmmeter.

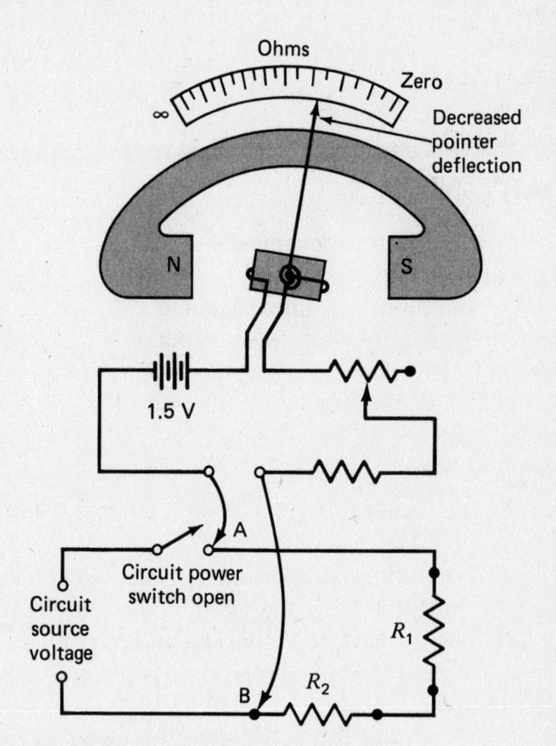

Fig. 18-2. Typical ohmmeter arrangement.

The power switch of the circuit to be measured should always be in the off position. This prevents the circuit's source voltage from being applied across the meter, which could cause damage to the meter movement.

The test leads of the ohmmeter are connected across (in parallel with) the circuit to be measured (see Fig. 18-2). This causes the current produced by the meter's internal battery to flow through the circuit being tested. Assume that the meter test leads are connected at points A and B of Fig. 18-2. The amount of current that flows through the meter coils will depend on the resistance of resistors R_1 and R_2, plus the resistance of the meter. Since the meter has been preadjusted (zeroed), the amount of coil movement now depends solely upon the resistance of R_1 and R_2. The inclusion of R_1 and R_2 raised the total series resistance, decreased the current, and thus decreased the pointer deflection. The pointer will now come to rest at a scale figure indicating the combined resistance of R_1 and R_2. If R_1 and R_2, or both, were replaced with a resistor(s) having a larger ohmic value, the current flow in the moving coil of the meter would be decreased still more. The deflection would also be further decreased, and the scale indication would read a still higher circuit resistance. Movement of the moving coil is proportional to the amount of current flow. The scale reading of the meter, in ohms, is inversely proportional to current flow in the moving coil.

EQUIPMENT

DC power supply, 0–10 V
Ohmmeter
Voltmeter
Ammeter
Test leads
VTVM or DVM

COMPONENTS

(1) 0- to 1-mA meter movement
(1) 150-kΩ 0.5-W resistor
(1) 5-kΩ potentiometer, linear taper
(1) 100-kΩ potentiometer, linear taper
(1) 1-MΩ potentiometer, linear taper
(1) 10-kΩ 0.5-W resistor
(1) SPST switch
(1) Decade box

PROCEDURE

1. Measure the internal resistance of the meter movement by connecting the circuit shown in Fig. 18-3, where $R_1 = 5\,\text{k}\Omega$, $R_2 = 100\,\text{k}\Omega$, and $R_3 = 10\,\text{k}\Omega$.
2. With S_1 open, turn on the power supply and adjust it for 10 V.
3. Adjust R_2 so that the scale upon the meter movement reads at full-scale deflection.
4. Close S_1 and adjust R_1 so that the scale upon the movement reads at half-scale deflection. The currents will evenly divide between R_1 and the internal r_m of the meter movement when $R_1 = r_m$.
5. Measure and record in Table 18-1 the voltage dropped across V_m.

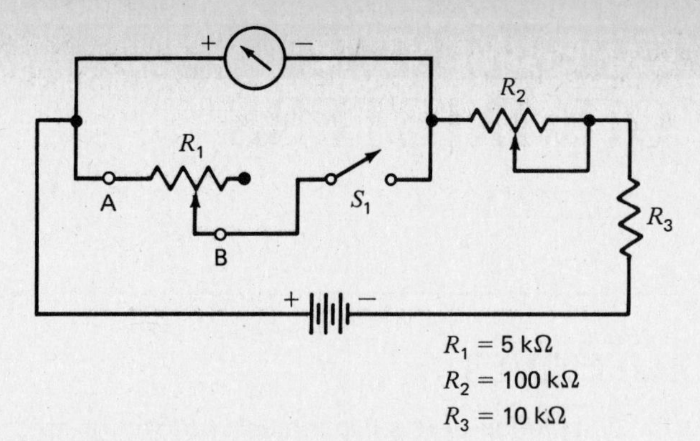

Fig. 18-3. Measuring internal resistance.

6. Measure and record r_m in Table 18-1 by turning the power supply off, disconnecting R_1 from the circuit, and measuring from points A to B. At this point, $r_m = R_1$.
7. Calculate and record I_m in Table 18-1, where

$$I_m = \frac{V_m}{r_m}$$

8. Record in Table 18-1 the r_m for full-scale deflection of a 0- to 1-mA meter movement.
9. Construct the series-type ohmmeter shown in Fig. 18-4, where $R_1 = 1.5\,\text{k}\Omega - r_m$. (The resistance value of R_1 may have to be created by using a resistance decade box). R_2 will be used to set the ohmmeter to zero ohms.

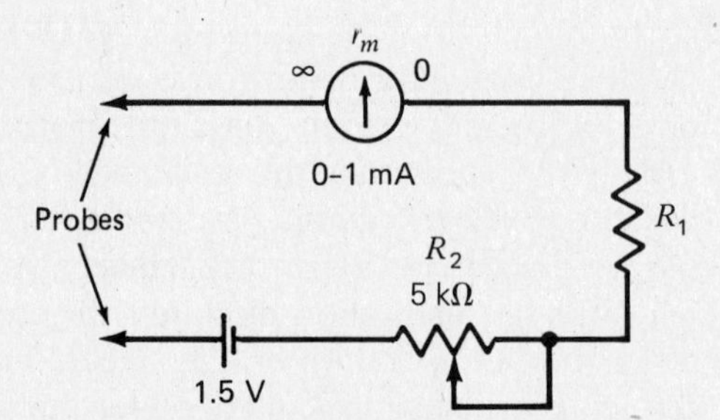

Fig. 18-4. Basic series ohmmeter.

10. With the probes not touching, the ohmmeter reads infinity. With the probes touching, adjust R_2 until the meter reads zero, indicating a zero ohms condition.
11. Calibrate this ohmmeter by connecting a 1-MΩ potentiometer across the probes, as shown in Fig. 18-5, using a grease pencil to mark the face of the ohmmeter. Complete Table 18-2.

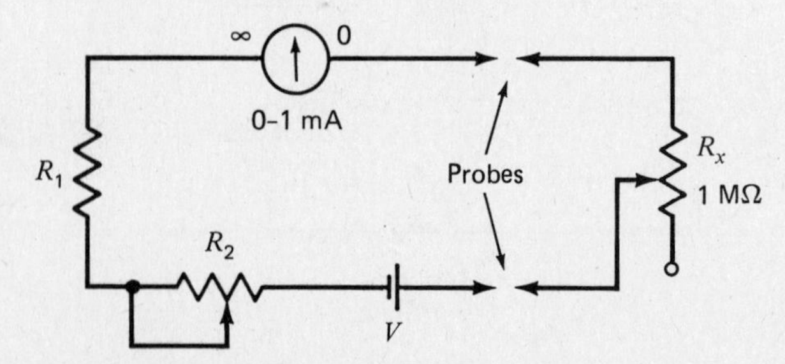

Fig. 18-5. Ohmmeter test circuit.

12. After calibrating the ohmmeter, measure the several resistances shown in Table 18-2 with an ohmmeter of known accuracy and your ohmmeter design. Complete Table 18-3, and determine the percentage of accuracy.

NAME ______________________________ DATE ______________

RESULTS FOR EXPERIMENT 18

QUESTIONS

Answer TRUE (T) or FALSE (F) for each question.

_____ **1.** An ohmmeter is used to measure voltage and current.

_____ **2.** An ohmmeter has an internal battery.

_____ **3.** The infinity symbol (∞) on an ohmmeter indicates a short circuit.

_____ **4.** The ohmmeter's leads are placed across the resistance to be measured.

_____ **5.** When the ohmmeter leads are short-circuited, the needle will probably indicate zero.

_____ **6.** Ohmmeters do not require internal current-limiting resistances or shunt paths.

_____ **7.** The ohms or resistance scale that reads from left to right is called a *back-off scale.*

_____ **8.** The zero-ohms adjustment should not be used when changing ranges.

_____ **9.** For greater values of resistance, a less-sensitive meter is required to read lesser values of current.

_____ **10.** An ohmmeter can be destroyed or have its fuse blown if it is used to measure resistance in a circuit where power is applied.

REPORT

Write a complete report. Discuss the measured and calculated results. Discuss the three most significant aspects of the experiment and write a conclusion.

TABLE 18–1

Steps 5, 6, and 7
V_m = __________
r_m = __________
I_m = __________

Step 8
r_m = __________

TABLE 18–2

External R_x, Ω	Deflection	Scale Reading
0		
750		
1,500		
3,000		
150,000		
500,000		

TABLE 18–3

R	Known Meter	Design Meter	% Accuracy
100 Ω			
1 kΩ			
4.7 kΩ			
22 kΩ			
100 kΩ			
1 MΩ			

EXPERIMENT 19
NETWORK THEOREMS

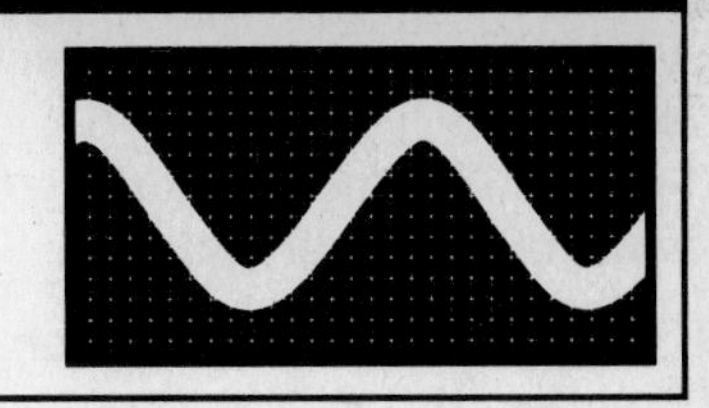

OBJECTIVES

At the completion of this experiment, you will be able to:

- Thevenize a circuit.
- Nortonize a circuit.

SUGGESTED READING

Chapter 10, *Basic Electronics*, B. Grob, Seventh Edition

INTRODUCTION

The analysis of Ohm's law and Kirchhoff's laws have been of primary use in the solution of relatively simple solutions of dc circuits. In the analysis of relatively complex circuits, a more powerful method is required. In the case of simplifying complex circuits, Thevenin and Norton theorems are used. This technique involves reducing a complex network to a simple circuit, which acts like the original circuit. In general, any circuit with many voltage sources and components, with no regard made to interconnection, can be represented by an equivalent circuit with respect to a pair of terminals in the equivalent circuit.

Thevenin's theorem states that a circuit can be replaced by a single voltage source V_{Th} in series with a single resistance R_{Th} connected to two terminals. This is shown in Fig. 19-1.

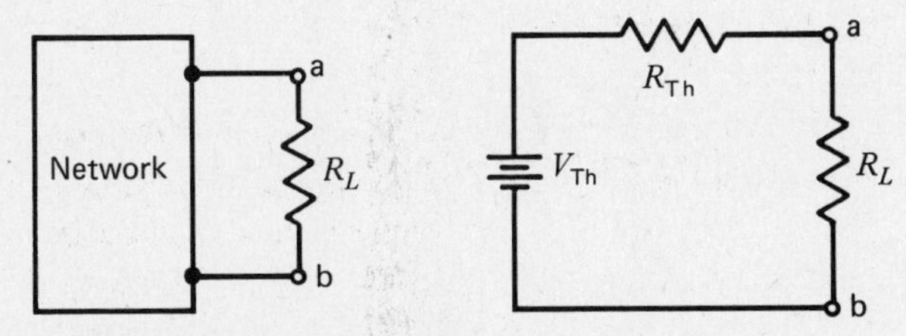

Fig. 19-1. Thevenin's circuit.

Norton's analysis is used to simplify a circuit in terms of currents rather than voltage, as is done in Thevenin circuits. Norton's circuit can be used to reduce a complex network into a simple parallel circuit that consists of a current source I_N and a parallel resistance to R_N. An example of this is shown in Fig. 19-2.

In the procedure that follows, the techniques of thevenizing and nortonizing simple voltage source–resistor networks will be developed as the procedure is completed.

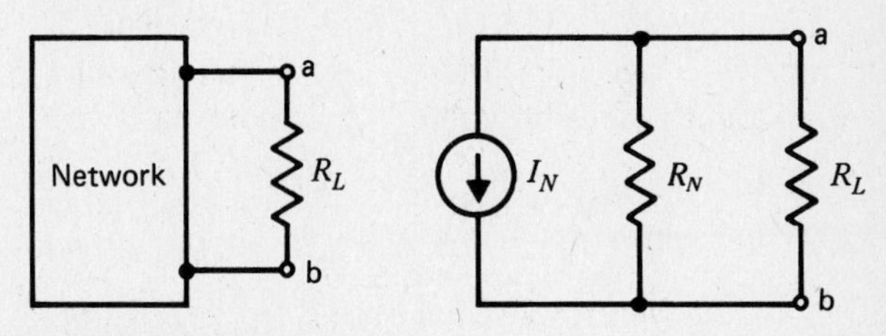

Fig. 19-2. Norton's circuit.

EQUIPMENT

DC power supply
Ammeter
Voltmeter
Protoboard or springboard
Test leads

COMPONENTS

Resistors:

(1) 100-Ω 1-W (1) 270-Ω 0.5-W
(1) 220-Ω 0.5-W

PROCEDURE

Thevenizing

1. Construct the circuit shown in Fig. 19-3, where $R_1 = 100\ \Omega$, $R_2 = 270\ \Omega$, $R_L = 220\ \Omega$, and V is adjusted to 10 V.

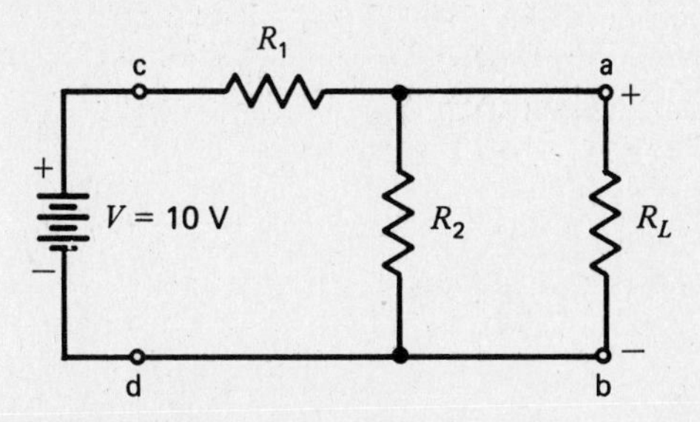

Fig. 19-3. Thevenizing a circuit.

2. Open the circuit at points a and b by disconnecting R_L from the circuit. The remainder of the circuit connected to a and b will be thevenized. Calculate, measure, and record in Table 19-1 the voltage across ab. Note that $V_{ab} = V_{Th}$.

3. Turn off the power supply and completely remove it from the circuit.
4. With the power supply removed, connect points c and d.
5. Calculate, measure, and record in Table 19-1 the value of R_{ab}. Note that $R_{ab} = R_{Th}$.

V_{Th} has now been determined and is found to be in series with R_{Th}. Since R_L was disconnected, this Thevenin equivalent can be applied to any value of R_L.

6. Reconnect the circuit shown in Fig. 19-3. Calculate (using the voltage divider formula), measure, and record V_L and I_L in Table 19-1 by reconnecting R_L. V_L is defined by the voltage divider formula as

$$V_L = \frac{R_L}{R_L + R_{Th}} \times V_{applied}$$

and I_L can be determined as

$$I_L = \frac{V_L}{R_L}$$

The same answers could be determined by using Ohm's law. The advantage of thevenizing the circuit is that the effect of R_L can be calculated easily for different values.

7. Complete Table 19-1 for the percentage of accuracy.

Nortonizing

8. Construct the circuit shown in Fig. 19-4, where $R_1 = 100\ \Omega$, $R_2 = 270\ \Omega$, $R_L = 220\ \Omega$, and V is adjusted to 10 V.

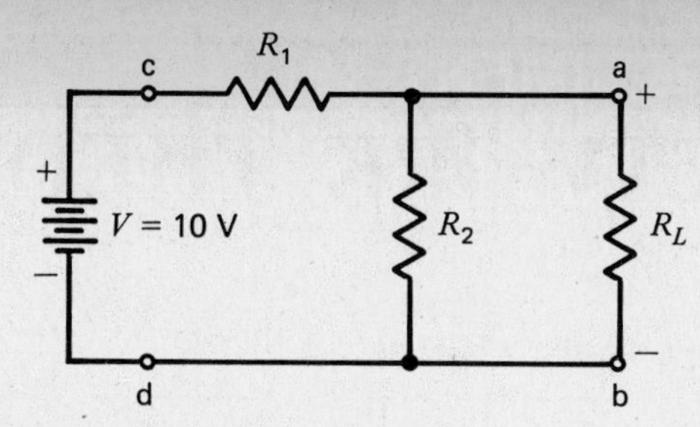

Fig. 19-4. Nortonizing a circuit.

9. Short-circuit points a and b together. This will also short-circuit R_2, and this will create a circuit condition in which resistor R_1 is in series with the power supply. Calculate, measure, and record in Table 19-2 the current flowing through R_1.

Note: $I_{R_1} = I_N$ (Norton)

10. Determine R_N by removing the short circuit, and remove R_L. This will leave points a and b unconnected to R_L.
11. Turn off the power supply and completely remove it from the circuit.
12. With the power supply removed, connect point c to d.
13. Calculate, measure, and record in Table 19-2 the value R_{ab}. Note that $R_{ab} = R_N$.
14. Reconnect the circuit shown in Fig. 19-4. Calculate, measure, and record I_L and V_L in Table 19-2 by reconnecting R_L.
15. Complete Table 19-2 for the percentage of accuracy.

NAME ______ DATE ______

RESULTS FOR EXPERIMENT 19

QUESTIONS

1. What is the primary use and importance of thevenizing a circuit?

2. What is the primary use and importance of nortonizing a circuit?

3. Draw a Thevenin equivalent of the circuit shown in Fig. 19-3.

4. Draw a Norton equivalent of Fig. 19-4.

5. Is the statement made at the end of procedure step 6, which reads, "the advantage of thevenizing the circuit is that the effect of R_L can be calculated easily for different values," valid? Explain and prove by example.

REPORT

Write a complete report. Discuss the measured and calculated results. Discuss the three most significant aspects of the experiment and write a conclusion.

TABLE 19–1

	Calculated	Measured	% Accuracy
V_{Th}			
R_{Th}			
V_L			
I_L			

TABLE 19–2

	Calculated	Measured	% Accuracy
I_N			
R_N			
I_L			
V_L			

EXPERIMENT 20

POTENTIOMETERS AND RHEOSTATS

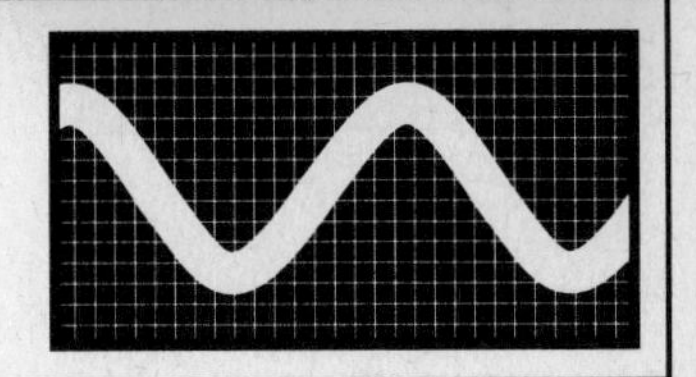

OBJECTIVES

At the completion of this experiment, you will be able to:

- Identify the circuit configuration of a potentiometer.
- Identify the circuit configuration of a rheostat.

SUGGESTED READING

Chapter 2, *Basic Electronics*, B. Grob, Seventh Edition

INTRODUCTION

Potentiometers and rheostats are variable resistances and are used to vary voltage and current in a circuit. A rheostat is a two-terminal device. The potentiometer is a three-terminal device, as shown in Fig. 20-1.

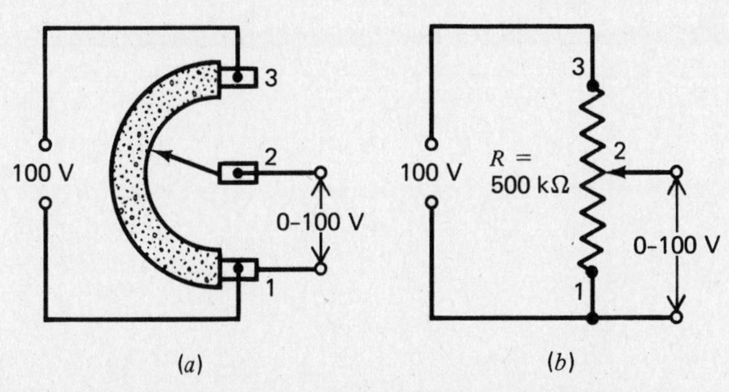

Fig. 20-1. Potentiometer connected across voltage source to function as a voltage divider. (*a*) Wiring diagram. (*b*) Schematic diagram.

The maximum resistance is seen between the two end terminals. The middle terminal mechanically adjusts and taps a proportion of this total resistance. A potentiometer can be used as a rheostat by connecting one end terminal to the other, as shown in Fig. 20-2.

The primary purpose of a potentiometer (pot) is to tap off a variable voltage from a voltage source, as shown in Fig. 20-3.

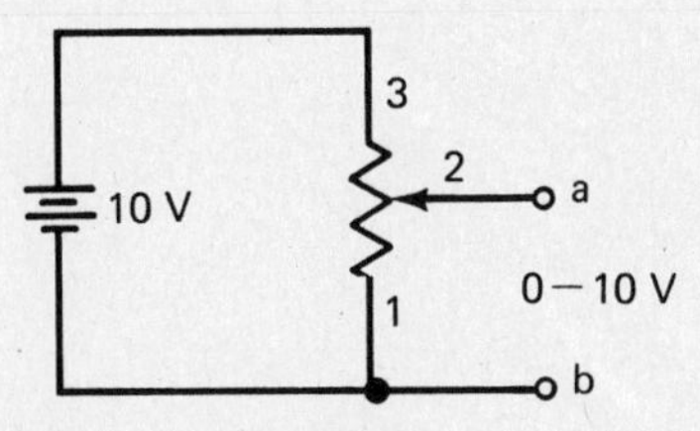

Fig. 20-3. Variable voltage source.

As pin 2 is rotated up toward pin 3, the voltage at ab increases until a 10-V level is achieved. If pin 2 is rotated downward toward pin 1, then the voltage present at ab decreases to zero, or approximately zero. The variance of voltages may appear to be presented in a linear or a nonlinear fashion, depending upon the manufacturer's type of potentiometer.

The primary purpose of a rheostat is to vary current though a load. This is accomplished by locating the rheostat in series with the load and source voltage. In this way, the total resistance R_T can be varied and indirectly vary the total current. This circuit configuration is shown in Fig. 20-4.

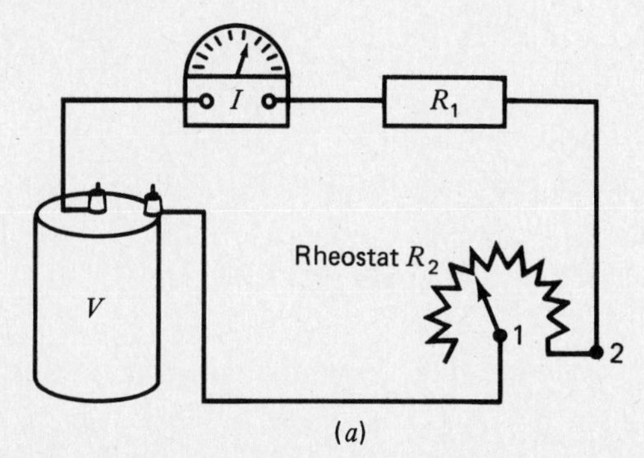

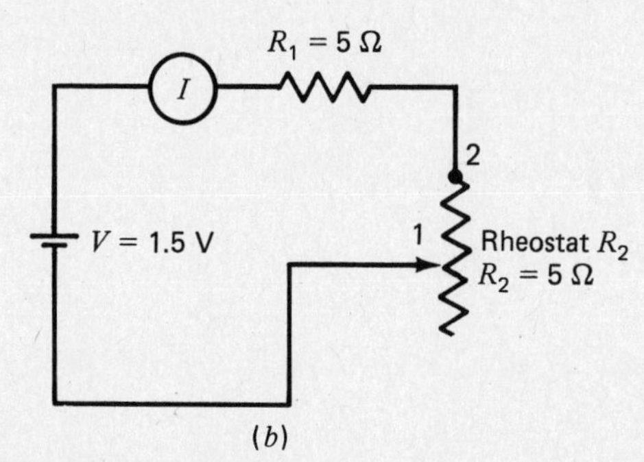

Fig. 20-2. Rheostat connected in series circuit to vary the current. (*a*) Wiring diagram with ammeter to measure *I*. (*b*) Schematic diagram.

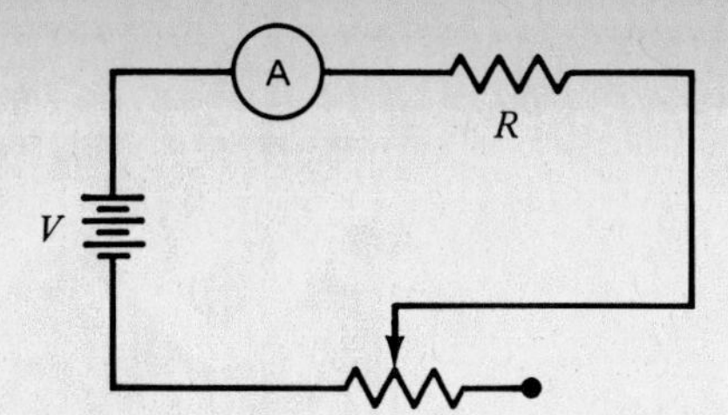

Fig. 20-4. Rheostat circuit to vary current.

In summary, rheostats are:

Two-terminal devices
Found in series with loads and voltage sources
Used to vary total current

Potentiometers are:

Three-terminal devices
Found to have end terminals connected across voltage sources
Used to tap off part of the voltage source

EQUIPMENT

DC power supply, 0–10 V
Voltmeter
Ammeter
Ohmmeter
Protoboard
Test leads

COMPONENTS

(1) 100-Ω, 1-W resistor
(1) 1-kΩ, 1-W potentiometer, linear taper

PROCEDURE

1. Connect the circuit shown in Fig. 20-5, where V = 10 V, R_1 = 100 Ω, and R_2 (pot) = 1 kΩ.

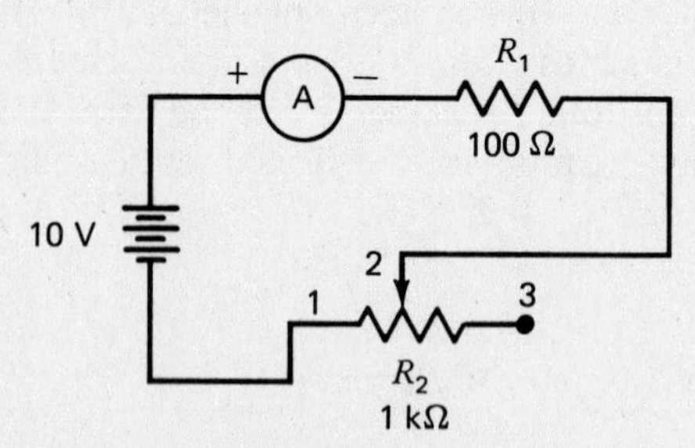

Fig. 20-5. Rheostat controlling current.

2. Turn on the power supply and adjust R_2 for a minimum resistance value (maximum I).
3. Remove R_2 from the circuit and connect it to an ohmmeter (points 1 and 2). Set R_2 to 100 Ω and reconnect R_2 to the circuit. Measure the current flow and record in Table 20-1.
4. Repeat step 3 in 100-Ω increments (100, 200, 300 Ω, etc.) up to 1000 Ω.
5. Make a graph of resistance (horizontal axis) versus current (vertical axis) from Table 20-1.
6. Connect the circuit shown in Fig. 20-6, where V = 10 V and R_2 (pot) = 1 kΩ.

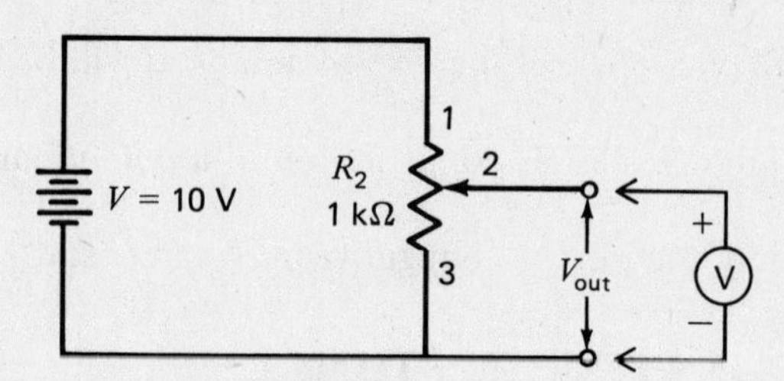

Fig. 20-6. Potentiometer voltage divider.

7. Turn on the power supply and adjust R_2 for a minimum resistance value (minimum V).
8. Remove R_2 from the circuit and connect it to an ohmmeter (points 2 and 3). Set R_2 to 100 Ω, and reconnect R_2 to the circuit. Measure the voltage drop across R_2 (pins 2 and 3). Record the results in Table 20-2.
9. Repeat step 8 in 100-Ω increments (100, 200, 300 Ω, etc.) up to 1000 Ω.
10. Make a graph of resistance (horizontal axis) versus voltage (vertical axis) from Table 20-2.

NAME DATE

RESULTS FOR EXPERIMENT 20

QUESTIONS

1. How many circuit connections to a potentiometer are needed?

2. How many circuit connections to a rheostat are needed?

3. Determine maximum power consumption from the graphs. What are the actual necessary wattages of R_1 and R_2?

REPORT

Write a complete report. Discuss the measured and calculated results. Discuss the three most significant aspects of the experiment and write a conclusion.

TABLE 20–1

Resistance, Ω	Measured Current
100	________
200	________
300	________
400	________
500	________
600	________
700	________
800	________
900	________
1000	________

TABLE 20–2

Resistance, Ω	Measured Voltage
100	______
200	______
300	______
400	______
500	______
600	______
700	______
800	______
900	______
1000	______

EXPERIMENT 21

INTERNAL RESISTANCE

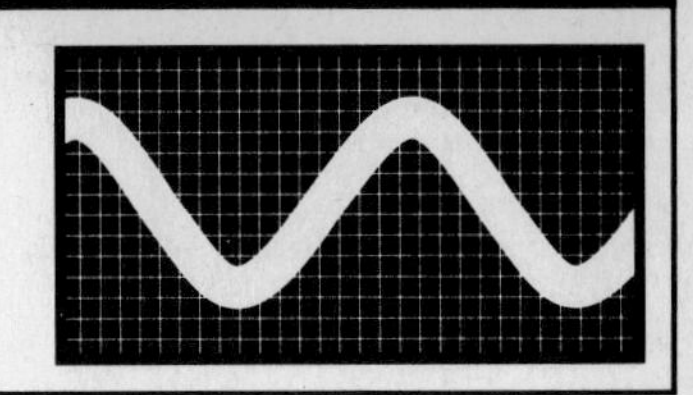

OBJECTIVES

At the completion of this experiment, you will be able to:

- Validate the concept of internal resistance in a power source.
- Determine the internal resistance of a dry cell battery and a dc power supply (generator).
- Graph or plot decreasing terminal voltage versus load current.

SUGGESTED READING

Chapter 12, *Basic Electronics,* B. Grob, Seventh Edition

INTRODUCTION

Any source of electric power that produces a continuous output voltage can be called a *generator.* All generators have some internal resistance, labeled r_i. This internal resistance has its own *IR* voltage drop, because it is in series with any load connected to the generator. In other words, the internal resistance of a source subtracts from the generated voltage, resulting in a decreased voltage across the output terminals. In a battery, r_i is due to the chemical makeup inside; in a power supply, r_i is due to the internal circuitry of the supply.

For example, Fig. 21-1 is a schematic representation of a 9-V battery with 100 Ω of internal resistance.

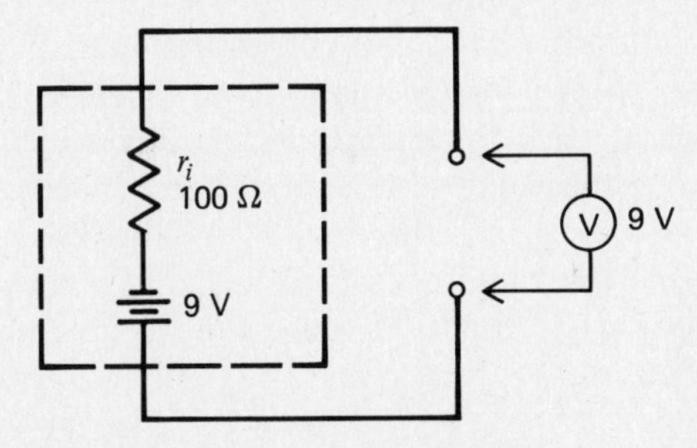

Fig. 21-1. A 9-V battery with r_i = 100 Ω.

Notice that the dotted line indicates that r_i is actually inside the battery. This battery has 9 V across its output terminals when it is measured with a voltmeter. If r_i were equal to 100 kΩ, the voltmeter would still measure 9 V across the output terminals. Thus, the value of r_i does not affect the output voltage. However, if a load is connected across the output terminals, then the value of r_i becomes significant. In any case, Fig. 21-2 shows that the battery's internal resistance now becomes a resistance in series with the load.

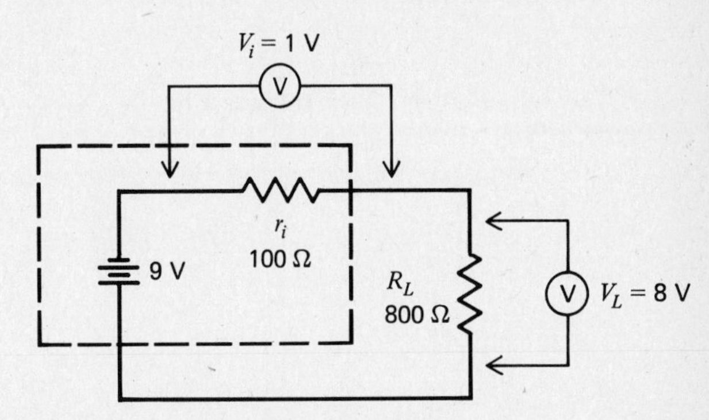

Fig. 21-2. A 9-V battery under load.

With a load resistance of 800 Ω connected across the output terminals, the voltmeter will now measure 8 V instead of 9 V. The other 1 V is now across the internal resistance of the battery. If r_i were equal to 100 kΩ, for example, the voltage across the output terminals would be almost 0 V due to the excessive value of r_i. In that case, the battery would be worn out or depleted.

Most bench power supplies have a fixed value of internal resistance that does not vary, regardless of the load value. Remember, without the load connected, the circuit is an open load. Therefore, the voltage drop across r_i equals zero. In this case, the total voltage is still available across the output, and it is called *open-circuit voltage,* or *no-load voltage.*

In the example of Fig. 21-2, the total circuit current I_T is equal to

$$I_T = \frac{V_L}{R_L} = \frac{8\ \text{V}}{800\ \Omega} = 0.01\ \text{A}$$

As the load resistance decreases, more circuit current will flow. If R_L decreases to 350 Ω, the current will increase and the load will require more current. Also, the voltage drop across the load will decrease and the voltage drop across V_i will increase.

Notice that as the load resistance decreased, the circuit current increased. Thus, the terminal voltage (the same thing as the load *IR* voltage) decreased. Therefore, the terminal voltage drops with more load current.

There is a method for determining the internal resistance of a source (generator) based on the examples given. Simply put, it is this:

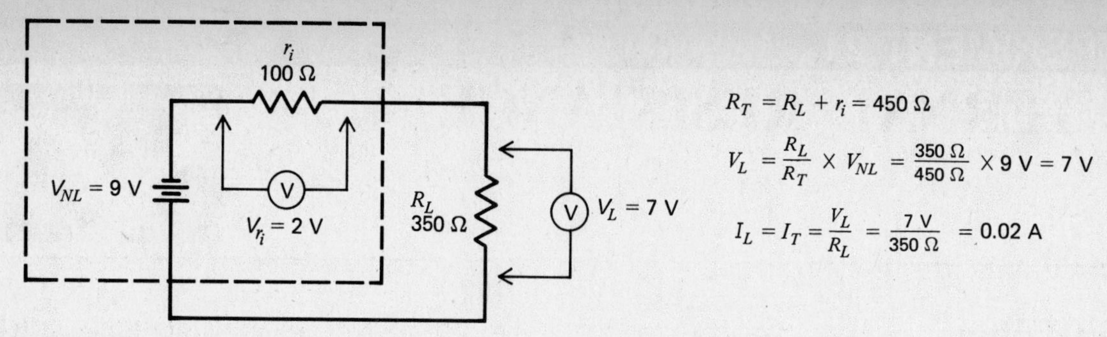

Fig. 21-3. Determining r_i.

1. Measure the no-load voltage.
2. Connect a load, and measure the voltage across the load and the circuit current.
3. Use the following formula to determine r_i:

$$r_i = \frac{V_{\text{no load}} - V_{\text{load}}}{I_{\text{load}}}$$

In Fig. 21-3, this would be

$$r_i = \frac{V_{NL} - V_L}{I_L} = \frac{9\ V - 7\ V}{0.02\ A} = \frac{2\ V}{0.02\ A} = 100\ \Omega$$

Finally, in general, if a generator has a very low internal resistance in relation to load resistance, it is considered a constant voltage source, because the voltage across r_i will subtract very little from the load voltage. If the value of r_i is very great in relation to load resistance, the generator is considered a constant current source, because the load resistance will have little effect upon the total resistance ($r_i + R_L$) and the total circuit current.

EQUIPMENT

DC power supply
Voltmeter
Ammeter
VOM
1.5-V battery
Protoboard or springboard
Test leads

COMPONENTS

Resistors (all 0.5 W unless indicated otherwise):

(1) 220 Ω (1) 2.2 kΩ
(2) 560 Ω (1) 5.6 kΩ
(2) 1 kΩ (1) 10 kΩ

PROCEDURE

1. Connect the circuit of Fig. 21-4. Do not connect the load yet. Measure and record in Table 21-1 the no-load voltage (V_{NL}) across the output terminals.

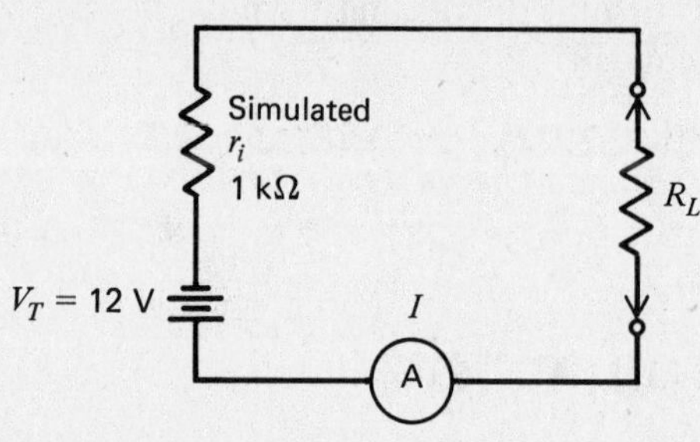

Fig. 21-4. Power supply with simulated r_i. Note that r_i is a 1-kΩ resistor in series.

2. Connect the following loads to the circuit of Fig. 21-4. Measure and record in Table 21-1 the load voltage and current for each load resistance.

R_L = 10 kΩ
= 5.6 kΩ
= 2.2 kΩ
= 1 kΩ
= 560 Ω
= 220 Ω

3. Calculate the value of r_i by using the measured values of load voltage and current for each load resistance. Show your calculations on a separate sheet of paper, and record the results in Table 21-1.
4. Change the value of r_i to 560 Ω and repeat steps 1 to 3 above. Record the results in Table 21-2.
5. Change the value of V_T to 6 V (using r_i = 560 Ω), and repeat steps 1 to 3. Record the results in Table 21-3.
6. Measure the voltage across a 1.5-V battery and record the value in Table 21-4.
7. Measure the short-circuit current of a 1.5-V battery by placing an ammeter across the output terminals of the battery for no longer than approximately 5 s, as shown in Fig. 21-5. Record the value in Table 21-4.

CAUTION: Use a VOM on its 10- or 12-A scale range. Some VOMs have special input jacks for this purpose.

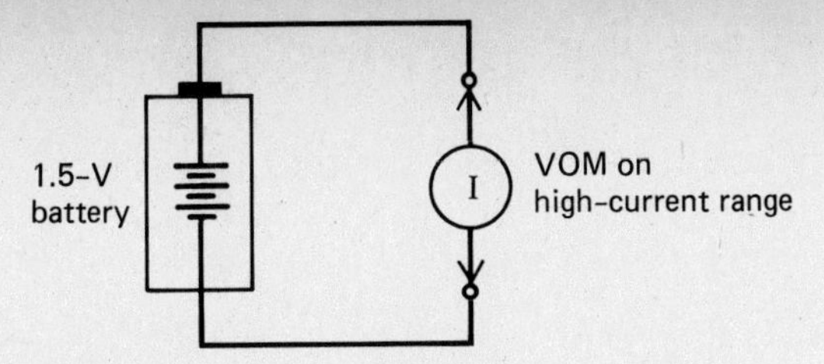

Fig. 21-5. Short-circuit method.

8. Calculate r_i by using Ohm's law. If available, repeat with a 22-V or any other size battery that will not damage the meter.

Note: You can do this for any value of battery, provided a meter of large enough current capacity is used.

9. Using the method for determining r_i (steps 1 to 3), use an unknown value of r_i (three times) and use a 1-kΩ load resistor. This can be done by disguising the value of a resistance with black electrical tape or by placing the resistance inside a chassis, as illustrated in Fig. 21-6. Use the circuit in Fig. 21-7 for this step. Record the results in Table 21-5.

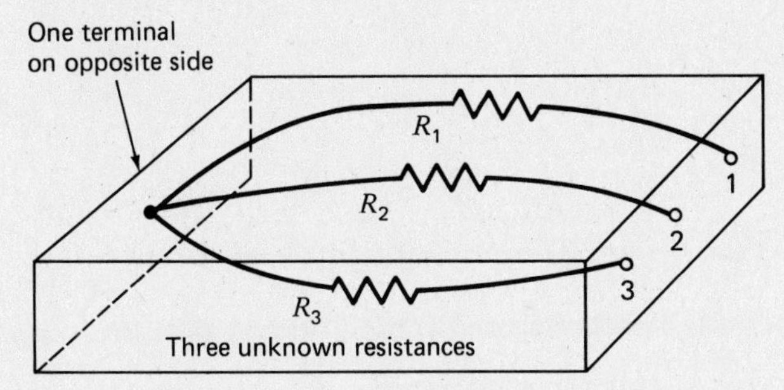

Fig. 21-6. Resistance box containing three unknown resistors.

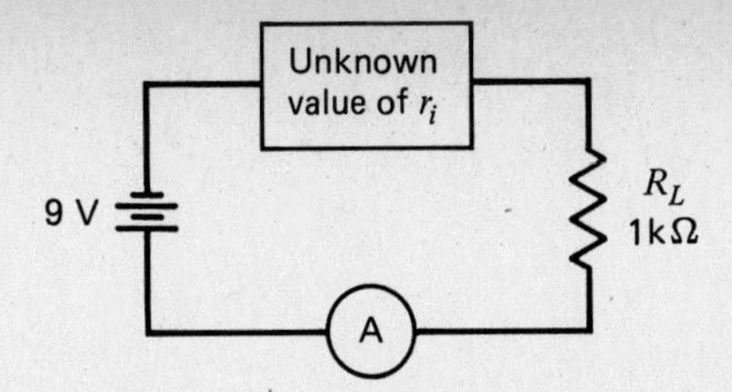

Fig. 21-7. Circuit for determining the internal resistance.

10. *OPTIONAL:* Plot the results of steps 1 to 3 (Table 21-1). For example, V_L versus I_L. Use regular graph paper, *not* semilog graph paper. See Fig. 21-8. See the Appendix for suggestions on how to make graphs.

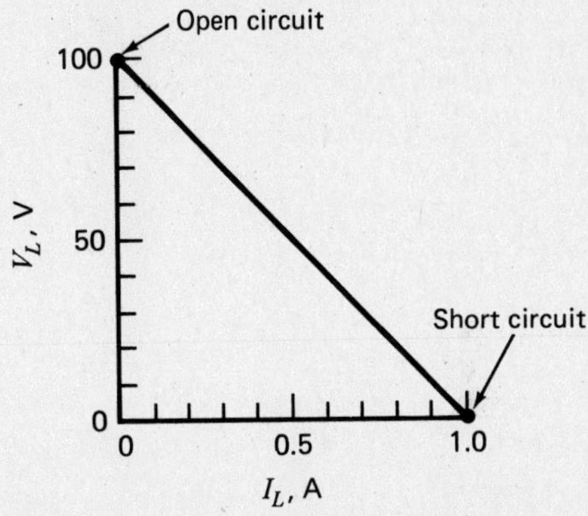

Fig. 21-8. How terminal voltage V_L drops with more load current I_L.

NAME ______________________ DATE ______________

RESULTS FOR EXPERIMENT 21

QUESTIONS

Answer TRUE (T) or FALSE (F) to the following.

_____ **1.** Batteries have internal resistance, but dc power supplies do not.

_____ **2.** As load resistance increases, the terminal voltage decreases.

_____ **3.** As load current increases, terminal voltage increases.

_____ **4.** Connecting four batteries in parallel, each with $r_i = 100\ \Omega$, would increase the total r_i four times.

_____ **5.** Connecting four batteries in series, each with $r_i = 100\ \Omega$, would increase the total r_i four times.

_____ **6.** The internal resistance of a generator is always in parallel with a load.

_____ **7.** Subtracting the load voltage from the no-load voltage gives a remainder that is equal to the *IR* voltage drop in the internal resistance of the source.

_____ **8.** Internal resistance is in series with the load resistance.

_____ **9.** Short-circuiting a battery will not drain the battery.

_____ **10.** A 1.5-V dry cell battery with 1 Ω of internal resistance is probably a depleted battery.

REPORT

Write a complete report. Discuss the measured and calculated results. Discuss the most significant aspects of the experiment and write a conclusion.

TABLE 21–1. $r_i = 1\ k\Omega$ (Steps 1–3)

R_L	Measured V_L, V	Measured I_L, A	Calculated r_i, Ω
10 kΩ	______	______	______
5.6 kΩ	______	______	______
2.2 kΩ	______	______	______
1 kΩ	______	______	______
560 Ω	______	______	______
220 Ω	______	______	______

V_{NL} = ______

TABLE 21–2. $r_i = 560\ \Omega$ (Step 4)

R_L	Measured V_L, V	Measured I_L, A	Calculated r_i, Ω
10 kΩ	________	________	________
5.6 kΩ	________	________	________
2.2 kΩ	________	________	________
1 kΩ	________	________	________
560 Ω	________	________	________
220 Ω	________	________	________

TABLE 21–3. $V_T = 6$ V; $r_i = 560\ \Omega$ (Step 5)

R_L	Measured V_L, V	Measured I_L, A	Calculated r_i, Ω
10 kΩ	________	________	________
5.6 kΩ	________	________	________
2.2 kΩ	________	________	________
1 kΩ	________	________	________
560 Ω	________	________	________
220 Ω	________	________	________

TABLE 21–4. Steps 6–8

	V_{NL}, V	Short-circuit I, A	Calculated r_i, Ω
1.5-V battery	________	________	________
Additional ____-V battery	________	________	________

Note: Only the instructor will know the value of the three unknown values of r_i. In this way, your lab techniques and your ability to follow procedures will be tested.

TABLE 21–5. r_i Unknown; $r_L = 1$ kΩ (Step 9)

	Measured V_L, V	Measured I_L, A	Calculated r_i, Ω
Measured V_{NL} = __________			
r_i No. 1	__________	__________	__________
r_i No. 2	__________	__________	__________
r_i No. 3	__________	__________	__________

Chassis or box number (if applicable): __________

EXPERIMENT 22

LOAD MATCH AND MAXIMUM POWER

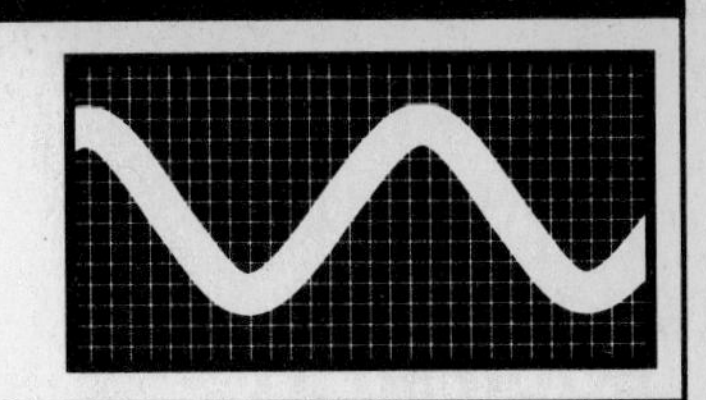

OBJECTIVES

At the completion of this experiment, you will be able to:

- Validate that maximum source power is transferred to a load when the value of source $r_i = R_L$.
- Plot a graph of load power for differing values of load resistance.
- Understand the concept of maximum efficiency versus maximum power.

SUGGESTED READING

Chapter 12, *Basic Electronics,* B. Grob, Seventh Edition

INTRODUCTION

When the internal resistance of a generator is equal to the load resistance, the load is considered matched to the source. The matching of load to source resistance is significant because the source can then transfer maximum power to the load.

Whenever $R_L = r_i$, maximum power is transferred to the load. When load resistance is more than r_i, the output voltage is more but the circuit current is less. When the load resistance is less than r_i, the output voltage is less but the circuit current is more. This experiment will provide data that you can analyze and thus prove that these concepts are valid.

The circuit of Fig. 22-1 and the accompanying graph of Fig. 22-2 illustrate the concept of matching a load to an internal source resistance to obtain maximum power transfer.

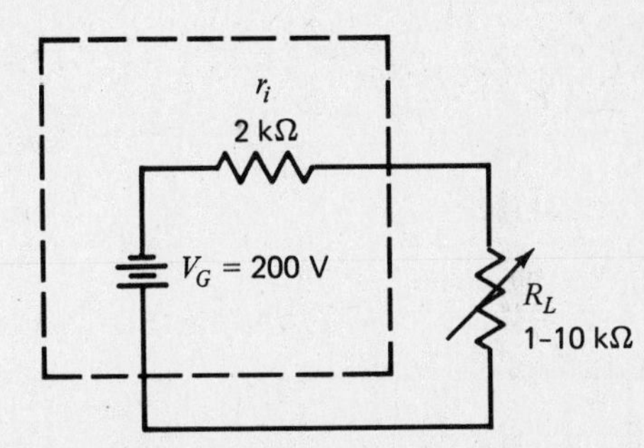

Fig. 22-1. Maximum power transfer circuit for analysis.

Because of the voltage divider formed by r_i and R_L, there is an equal voltage division: half of V_G is across r_i and half of V_G is across R_L. Under these circumstances, the load develops the maximum power that is possible using the particular source.

Referring to Fig. 22-1, as R_L increases, current decreases, resulting in less power dissipated in r_i. This results in more circuit efficiency because less power is lost across r_i. However, when $r_i = R_L$, the circuit efficiency is 50%.

$$\frac{P_L}{P_T} \times 100 = \text{Circuit Efficiency}$$

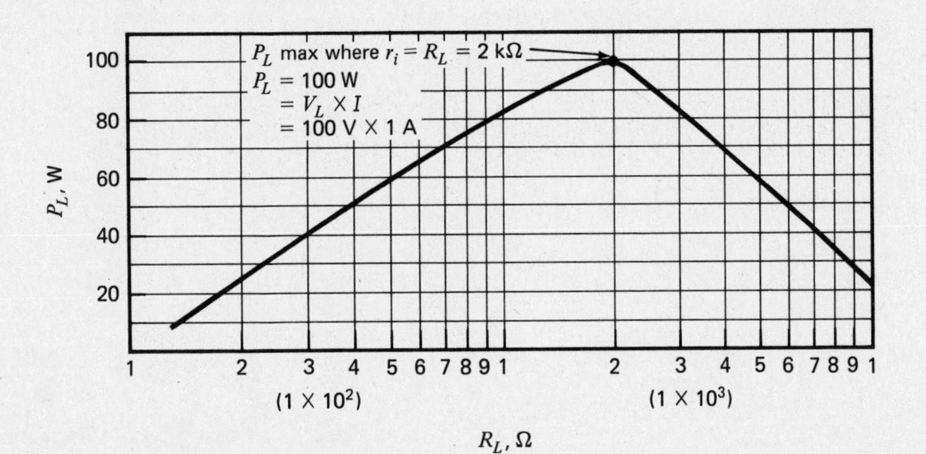

Fig. 22-2. Semilog graph of P_L versus R_L.

where P_T is the total power dissipated by the circuit, or

$$P_T = P_L + P_{r_i}$$

By this definition, 100% circuit efficiency means that absolutely no power is being dissipated.

EQUIPMENT

Ammeter
Voltmeter
DC power supply
Decade box
Protoboard or springboard
Leads

COMPONENTS

(1) 820-Ω, 0.5-W resistor

PROCEDURE

1. Connect the circuit of Fig. 22-3.

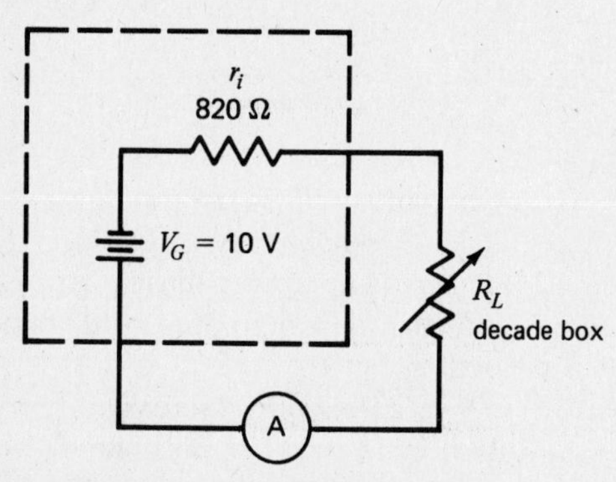

Fig. 22-3. Maximum power transfer circuit.

2. Increase the load resistance in 100-Ω steps from 100 Ω to 1 kΩ. Measure the voltage across R_L and the current at each step. Record the results in Table 22-1 for each step.

3. Increase the load resistance from 1 kΩ to 10 kΩ in 1-kΩ steps. Measure the voltage across R_L and the current at each step. Record the results in Table 22-1 for each step.

4. Calculate the *IR* voltage drop across r_i at each step, as $V_{r_i} = V_G - V_{R_l}$ and record the results in Table 22-1.

5. Calculate the load power dissipated at each step of R_L, as

$$P_L = V_L \times I$$

Record in Table 22-1.

6. Calculate the power dissipated across r_i at each step of R_L as

$$P_{r_i} = V_{r_i} \times I$$

Record in Table 22-1.

7. Calculate the total power dissipated in the circuit for each step of R_L as

$$P_T = V_G \times I$$

Note that

$$V_G = V_L + V_{r_i}$$

Record in Table 22-1.

8. Calculate circuit efficiency for each step of R_L as

$$\frac{P_L}{P_T} \times 100$$

expressed as a percentage.

9. Plot a graph of load resistance versus load power, using your data. Use two-cycle semilog paper. Prepare this graph as if it were to be used in a professional situation. It should be neat, well-organized, include a title and all possible values, and critical parameters should be labeled.

NAME DATE

RESULTS FOR EXPERIMENT 22

QUESTIONS

1. Explain the difference between circuit efficiencies of 1, 50, and 100 percent. In other words, explain what is meant by circuit efficiency as it relates to transfer of maximum power.

2. Explain what would happen if the circuit of Fig. 22-3 had an internal resistance of 100 kΩ.

3. Explain what would happen if the circuit of Fig. 22-3 had an internal resistance of 0.001 Ω.

4. Explain why semilog paper is used to graph the data.

5. Explain how you could get maximum power transferred to a 15-kΩ load if the internal resistance of your source were 10 kΩ.

REPORT

Write a complete report. Discuss the measured and calculated results. Discuss the three most significant aspects of the experiment and write a conclusion.

TABLE 22-1. Data for Circuit Fig. 22-3

R_L	Measured V_L, V	Measured I, A	Calculated V_{r_i}, V	Calculated P_L, W	Calculated P_{r_i}, W	Calculated P_T, W	% Efficiency
100 Ω							
200 Ω							
300 Ω							
400 Ω							
500 Ω							
600 Ω							
700 Ω							
800 Ω							
900 Ω							
1 kΩ							
2 kΩ							
3 kΩ							
4 kΩ							
5 kΩ							
6 kΩ							
7 kΩ							
8 kΩ							
9 kΩ							
10 kΩ							

EXPERIMENT 23

MAGNETISM

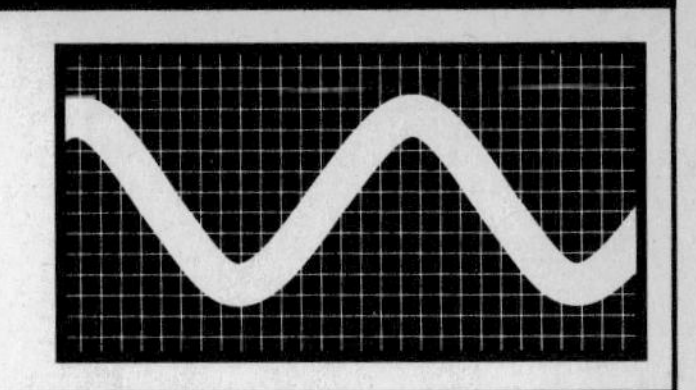

OBJECTIVES

At the completion of this experiment, you will be able to:

- Validate that current in a conductor has an associated magnetic field.
- Understand the concept of shielding.
- Examine the left-hand rule to determine magnetic polarity.

SUGGESTED READING

Chapters 13, 14, and 15, *Basic Electronics*, B. Grob, Seventh Edition

INTRODUCTION

Any electric current has an associated magnetic field that can do the work of attraction or repulsion. Not only is the magnetic field useful for doing work, it is also the cause of unwanted attraction and repulsion. Thus, it is often necessary to shield particular circuits to prevent one component from affecting another.

The most common example of magnetic force is that produced by a magnet. The magnet, with its north and south poles, acts as a generator that produces an external magnetic field provided by the opposite magnetic poles of the magnet. The idea is like the two opposite terminals of a battery that have opposite charges. Also, the earth itself is a huge natural magnet, having both north and south poles. Thus, the needle of a compass (also a magnet) is attracted to the north pole, because the atoms that make up the needle have been aligned in such a way that their magnetic field is attracted to the magnetic field of the earth's north pole.

It is these magnetic fields that are the subject of electromagnetism. These fields are thought of as lines of force, called *magnetic flux*, as shown in Fig. 23-1.

Glass Iron filings Magnet

Lines of force

(*a*) (*b*)

Fig. 23-1. Magnetic field of force around a bar magnet. (*a*) Field outlined by iron filings. (*b*) Field indicated by lines of force.

If current is flowing in a conductor, there is a similar magnetic field that can be used in conjunction with the fields of a magnet. For example, PM (permanent magnet) loudspeakers found in most radios, televisions, and public address systems all use the principles of magnetism to produce the audible sound we listen to.

Finally, the opposite effect of current moving through a conductor is a magnetic field in motion, forcing electrons to move. This action is called *induction*. Inductance is produced by the motion of magnetic lines of flux cutting across a conductor, thus forcing free electrons in the conductor to move, as shown in Fig. 23-2.

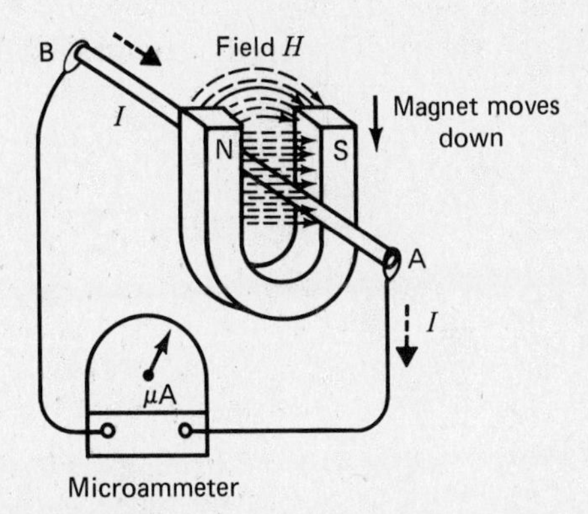

Fig. 23-2. Magnetically induced current in a conductor. Current I is electron flow.

EQUIPMENT

DC power supply
Galvanometer or microammeter
Heavy-duty horseshoe magnet (>20 lb pull)
Magnetic compass
Shield (6 × 6 in.) conductance sheet metal

COMPONENTS

2 to 3 ft of thin insulated wire
(1) No. 18 iron nail
Iron filings

PROCEDURE

1. Connect the circuit of Fig. 23-3. Have the compass

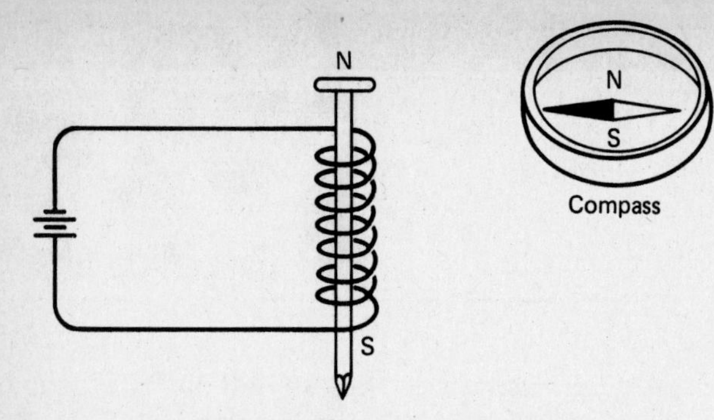

Fig. 23-3. Electromagnetic circuit for step 1.

and the iron filings nearby. Wrap the insulated wire evenly around the nail (about 10 to 15 turns).

2. The left-hand rule states that if a coil is grasped with the fingers of the left hand curled around the coil in the direction of electron flow, the thumb (extended) points to the north pole of the coil. Imagine your left hand grasped around the coil of wire wound around the nail. Determine which end is the north pole.

3. Turn the power on, and slowly move the compass close to both ends of the needle. Determine which end of the nail is north and which is south. Compare the results to step 2 above.

4. Place the shield between the compass and the nail in the circuit and repeat step 3. Note the results.

5. Turn the power off. Remove the nail and pass it through the iron filings. Note the results. Disconnect the circuit.

6. Connect the circuit of Fig. 23-4. Using the same wire as in steps 1 to 4 above, loop the wire around the magnet many times, making sure that the galvanometer is on the lowest range.

7. Move the horseshoe magnet up and down, as necessary, and note the amount of current produced. Try moving the magnet more rapidly, then slowly.

8. Disconnect the circuit.

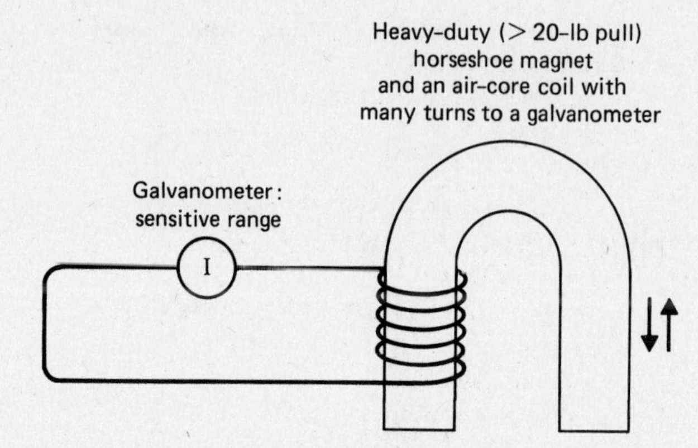

Fig. 23-4. Electromagnetic generator circuit for step 6.

NAME _______________ DATE _______________

RESULTS FOR EXPERIMENT 23

QUESTIONS

1. Explain what is meant by shielding.

2. Discuss the results of moving the magnet (in step 7) faster or slower.

3. Explain which end of the nail attracted the iron filings, and why.

4. Discuss any differences between the results of steps 1 to 4 and steps 6 to 8.

5. Explain the left-hand rule as it was applied in this experiment.

REPORT

Write a complete report. Discuss the results. Discuss the three most significant aspects of the experiment and write a conclusion.

EXPERIMENT 24

AC VOLTAGE AND OHM'S LAW

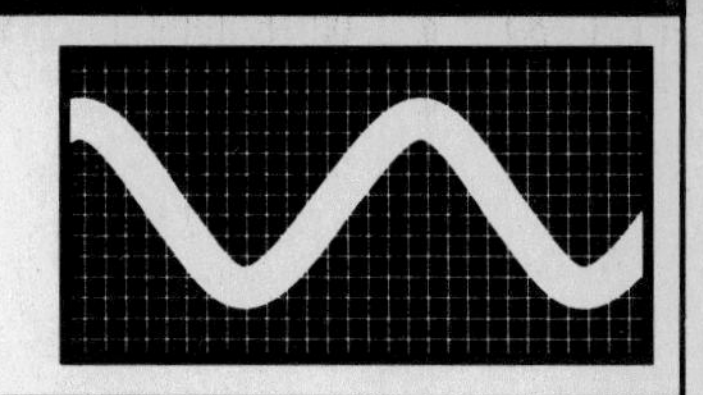

OBJECTIVES

At the completion of this experiment, you will be able to:

- Validate the Ohm's law expression for alternating current, where

$$V = I \times R$$

$$I = \frac{V}{R}$$

$$R = \frac{V}{I}$$

- Operate an ac oscillator or signal generator.

SUGGESTED READING

Chapter 16, *Basic Electronics,* B. Grob, Seventh Edition

INTRODUCTION

This experiment is designed to introduce you to ac voltages and to validate that Ohm's law can still be used to determine current, voltage, or resistance if two of the three terms of an ac circuit are known. Also the amount of electric power, measured in watts, can be determined for ac circuits by using Ohm's law in the same way as for dc circuits.

Alternating current is electron flow in two directions: positive and negative. As shown in Fig. 24-1, two batteries can be switched on and off alternately to achieve the effect of ac voltage. Notice that, depending on which battery is switched on, the direction of current flow through the resistor will alternate. Now, imagine that you could move the rocker switch back and forth 100 times in 1 s. If you could do it, you would create an alternating current of 100 cycles/s, expressed as 100 hertz (Hz). Note that 1 Hz = 1 cycle/s, named after the 19th-century scientist Hertz.

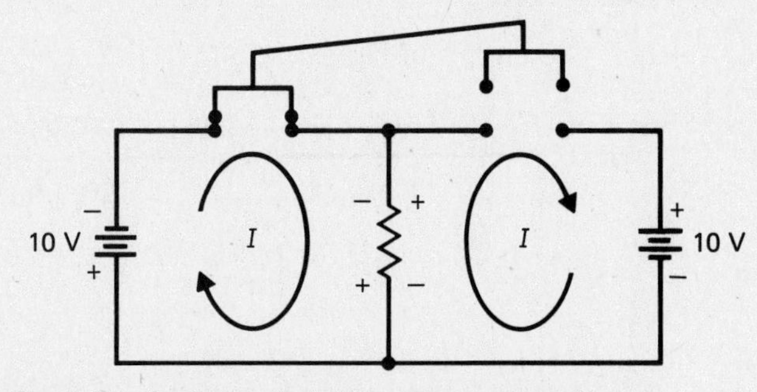

Fig. 24-1. Rocker switch alternates current flow.

The concept of alternating current is described not only by the frequency (in hertz) at which it alternates, but also by the amount of voltage that is being alternated. In the circuit of Fig. 24-1, the two batteries supply a voltage to the resistor at 10 V in each polarity or direction. Therefore, it is valid to say that the voltage across the resistor has +10 or −10 V across it at a particular time. In fact, this is a total potential of 20 V from one peak value (+10 or −10 V) to another.

Because it takes some amount of time for the voltage to reach the resistor, depending on the frequency of the alternating current, each potential voltage rises to its peak value (+10 or −10 V) and returns to zero before rising to the alternate peak value. And because this alternation of current flow occurs in a back-and-forth or oscillating manner, it is represented by a sine-wave symbol. Notice that Fig. 24-3 shows the number line of Fig. 24-2 combined with a sine wave to illustrate how ac voltage is represented.

Fig. 24-2. Peak values on a number line.

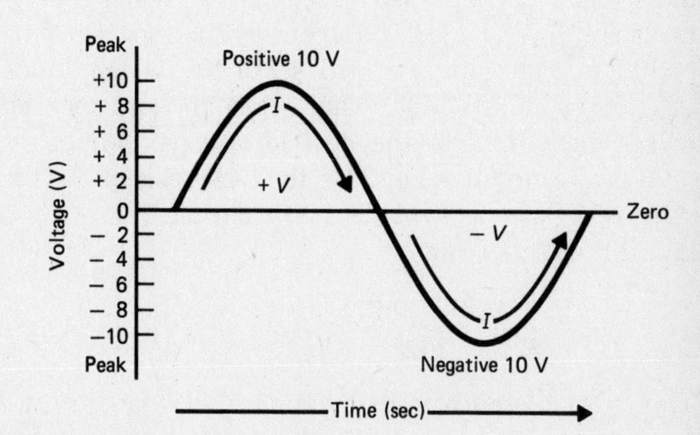

Fig. 24-3. Sine-wave illustration of ac voltage: peak to peak.

The most common form of applying ac voltage to a circuit is by using an oscillator or signal generator. This instrument can be simple or complex, depending on its

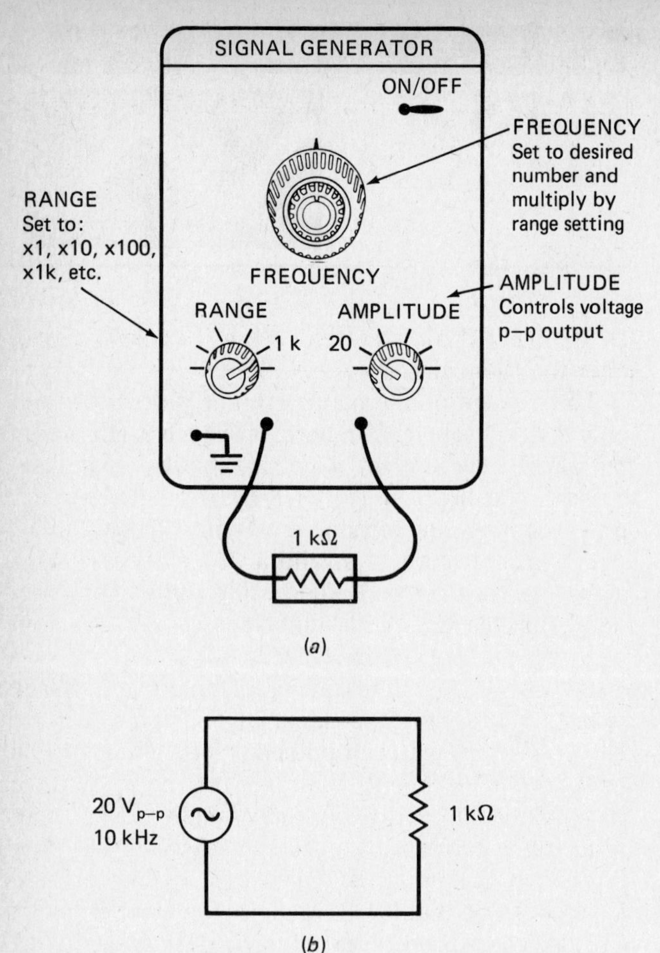

Fig. 24-4. (*a*) Simplified signal generator or oscillator connected to a 1-kΩ resistor. (*b*) Schematic.

style and manufacture. However, Fig. 24-4*a* shows a simplified version of such an instrument.

Study the signal generator in your lab. If necessary, refer to its operating manual for details or instructions.

Notice that Fig. 24-4*b* shows the schematic version of the drawing where the sine-wave symbol represents the signal generator. Also notice that the signal generator is set to 10 kHz (10,000 Hz) using the ranges switch (×100) and the tuning control set to 10. The amplitude adjustment control is a voltage adjust knob that is used to set the output to the desired voltage (p-p) level.

In the ac circuit of Fig. 24-4, the 1-kΩ resistor has the same resistance as it would in a dc circuit. Thus, Ohm's law still dictates that

$$I = \frac{V}{R} = \frac{20\text{ V p-p}}{1\text{ k}\Omega} = 20\text{ mA}$$

The only difference is that the current is labeled as I (p-p) to show that it is the result of an ac voltage (p-p).

In ac circuits, an ammeter is not used to measure current because it is only built to measure current flow in one direction (direct current). Do not put a dc ammeter in an ac circuit.

Finally, power in an ac circuit is calculated slightly differently from that in a dc circuit. To calculate the power dissipation, the p-p voltage must be converted to its dc equivalent value. This is also called the *rms* (root-mean-square) *value* and is equal to 70.7 percent of the peak voltage. For example, a 20-V p-p voltage will produce the same heating or lighting power (in watts) as a 7-V dc battery. Thus, ac power is calculated as

$$\frac{V_{p\text{-}p}}{2} \times 0.707 = V_{dc} \text{ or } V_{rms} \text{ for use in the power equation}$$

$$\text{Power (watts)} = \frac{V^2}{R}$$

$$\text{where } V = \frac{V_{p\text{-}p}}{2} \times 0.707$$

In summary, ac voltage can be used in any Ohm's law equation where I or V is labeled as a p-p value and converted to its dc equivalent (rms) for power calculations. Although ac voltages are usually measured on an oscilloscope, the following procedure will use a voltmeter (VOM or DVM) to validate Ohm's law.

EQUIPMENT

AC voltmeter (DVM or VOM) using the ac scale
Signal generator or oscillator
DC power supply
Springboard
Leads

COMPONENTS

Resistors (all 0.5 W):
(1) 100 Ω
(1) 1 kΩ

PROCEDURE

1. Connect the circuit of Fig. 24-5.

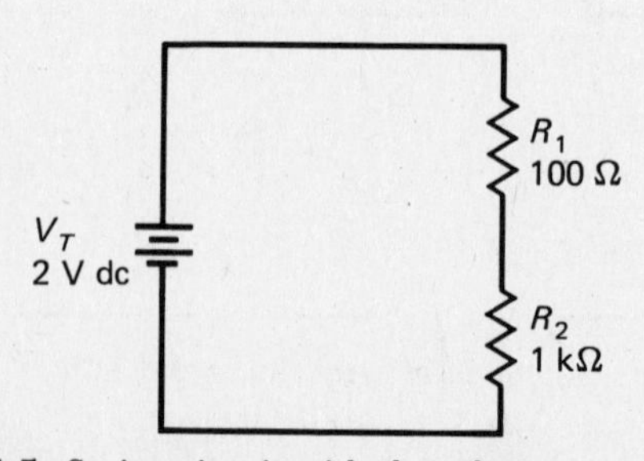

Fig. 24-5. Series circuit with dc voltage.

2. Measure and record in Table 24-1 the total voltage V_{R_1} and V_{R_2}. Compute the total current I_T and I_{R_1} and I_{R_2} by using Ohm's law.

3. Connect the circuit of Fig. 24-6.

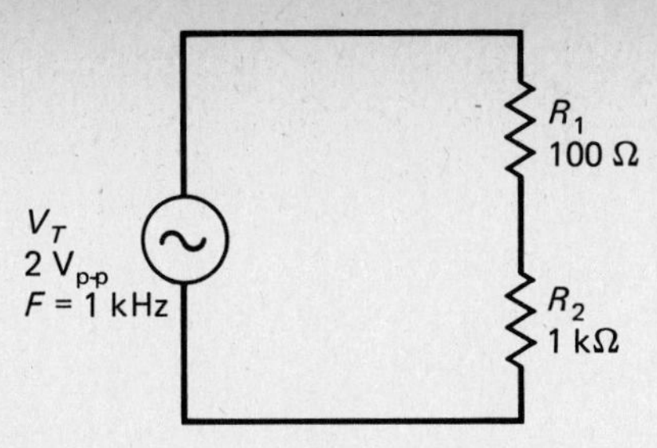

Fig. 24-6. Series circuit with ac voltage.

Note: Connect the entire circuit and then measure V_T to be sure it is 2 V p-p.

4. Repeat step 2 for the circuit of Fig. 24-6. Current should be calculated as:

$$I_{p\text{-}p} = \frac{V_{p\text{-}p}}{R}$$

5. Change the frequency to 5 kHz and repeat step 2 for Fig. 24-6.

6. Calculate the power dissipated by each circuit as shown in Table 24-1.

NAME ______ DATE ______

RESULTS FOR EXPERIMENT 24

QUESTIONS

Answer TRUE (T) or FALSE (F) to the following:

_____ **1.** Resistors do not function the same in ac circuits as in dc circuits.

_____ **2.** A 10-V battery and a 7-V (p-p) signal will produce the same amount of power across a 2-kΩ resistor.

_____ **3.** Ohm's law can only be used to find current in dc circuits.

_____ **4.** Current in an ac circuit can be measured with an ammeter just as a dc circuit can.

_____ **5.** A signal generator does not produce any current in an ac circuit.

REPORT

No formal report is required for this experiment. Hand in the data table and the answers to questions 1 to 5.

TABLE 24–1

DC CIRCUIT—Fig. 24–5				
Resistance	**Nominal Value**	**Volts Measured**	**Current $I = V/R$**	**Power, Watts, V^2/R**
R_1	100 Ω	______	______	______
R_2	1 kΩ	______	______	______
R_T	1.1 kΩ	______	______	______
AC CIRCUIT * Fig. 24–6 * F = 1 kHz				
R_1	100 Ω	______	______	______
R_2	1 kΩ	______	______	______
R_T	1.1 kΩ	______	______	______
AC CIRCUIT * Fig. 24–6 * F = 5 kHz				
R_1	100 Ω	______	______	______
R_2	1 kΩ	______	______	______
R_T	1.1 kΩ	______	______	______

*Note: Use p-p values. For example, 10 V p-p ÷ 100 Ω = 100 mA, I p-p.

Also, power = $(\frac{V_{p\text{-}p}}{2} \times 0.707)^2 \div R$ for ac circuits.

EXPERIMENT 25

ALTERNATING VOLTAGE AND CURRENT

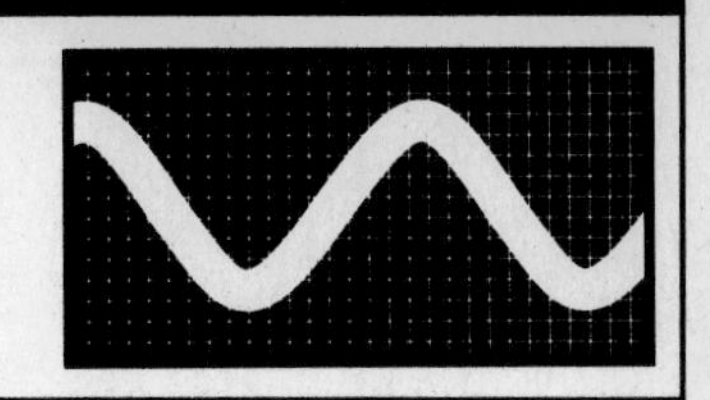

OBJECTIVES

At the completion of this experiment, you will be able to:

- Operate an oscilloscope.
- Measure dc voltages.
- Measure ac voltages.

SUGGESTED READING

Chapter 16, *Basic Electronics,* B. Grob, Seventh Edition, the operating instructions for your oscilloscope, and Appendix H of this lab manual.

INTRODUCTION

This experiment continues the ac portion of your laboratory studies. The same rules for series and parallel circuits and Ohm's law still apply to ac circuits as they did for dc circuits. However, there is a difference. As the name *alternating current* implies, current is alternating its direction. That is, electron flow reverses its direction, and therefore positive and negative polarities will alternate every time electron flow changes direction.

To study ac voltages in the lab, it is necessary to learn how to operate an oscilloscope. Although the oscilloscope is a complex instrument, it is basically a voltmeter. It measures and displays ac voltages as well as dc voltages.

There are some aspects of ac voltage that must be understood before attempting to measure them. AC voltages have frequency; that is, one alternation occurs over some period of time. For example, the ac wall outlet has a frequency of 60 hertz (Hz). This means that 60 complete cycles of electron flow, from negative to positive and back again, occur during 1 s. A frequency of 20 kHz means that 20,000 cycles occur during 1 s.

DC voltage does not have frequency. It is a steady, or direct, current flow that does not change over time. This difference in frequency is a major factor in understanding ac voltages.

The other important difference between ac and dc voltage is the way that magnitude is measured. DC voltage is usually steady and constant. AC voltage, because of its alternating character, takes some time to reach a peak value, then it reverses direction and becomes a

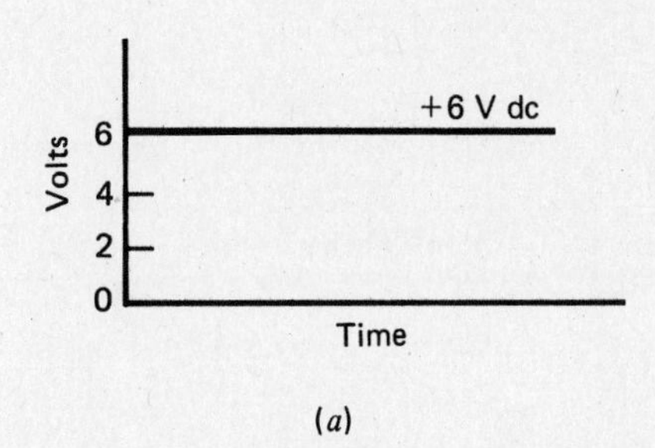

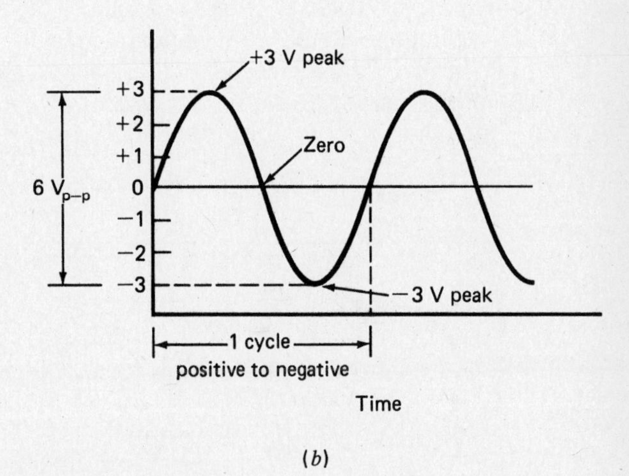

Fig. 25-1. DC versus ac voltages. (*a*) DC voltage plot. (*b*) AC voltage plot.

zero value before it reaches a peak value in the other direction. Figure 25-1 illustrates how this works. Notice how the dc and ac voltages differ. They differ in magnitude and frequency. The +6 V dc is referenced from zero. The 6-V peak to peak (p-p) ac voltage is not referenced from zero. It is referenced from one peak value to another. Zero volts still exists, but the ac voltage goes above it and below it. The other difference is that the dc voltage has no frequency, whereas the ac voltage occurs during some period of time. If, for example, the cycle of Fig. 25-1*b* occurred during 1 ms, the frequency would be

$$f = \frac{1}{t} = \frac{1}{0.001\text{ s}} = 1000\text{ Hz or } 1\text{ kHz}$$

where f = frequency
t = time it takes for a complete alternation

The circuit of Fig. 25-2 shows the ac signal generator

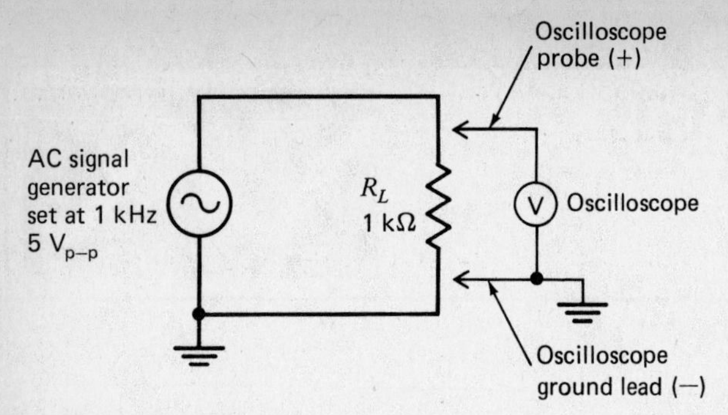

Fig. 25-2. AC circuit.

with its sine-wave symbol. Notice that the ground symbol indicates that one side of the signal generator is grounded. This means that there is a true earth ground on the signal generator. This is where the ground lead of the oscilloscope is also connected. Always remember that the grounds must be connected together. If, for example, the oscilloscope leads in Fig. 25-2 were reversed, the load resistor would be between two earth grounds. Therefore, it would be effectively out of the circuit, and no voltage would be dropped across it. Properly connected, the oscilloscope will display all the applied voltage dropped across R_L, as shown in Fig. 25-3.

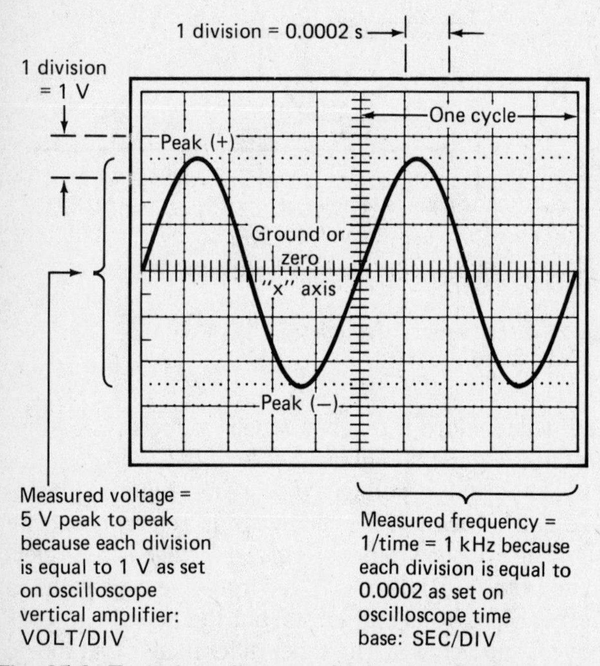

Fig. 25-3. Typical oscilloscope display for circuit shown in Fig. 25-2: 5 $V_{p\text{-}p}$ sine wave at 1 kHz. Note that $f = 1/t$. Here, five divisions $= 5 \times 0.0002$ s $= 0.001$. Thus $f = 1/0.001 = 1$ kHz.

If a dc voltage were measured, the oscilloscope display would be different. Figure 25-4*b* shows how the oscilloscope would display the measured voltage. Notice that the dc measurement shows a trace that is a straight line. No frequency and no peak-to-peak value are displayed because there is no ac voltage, only a steady dc voltage.

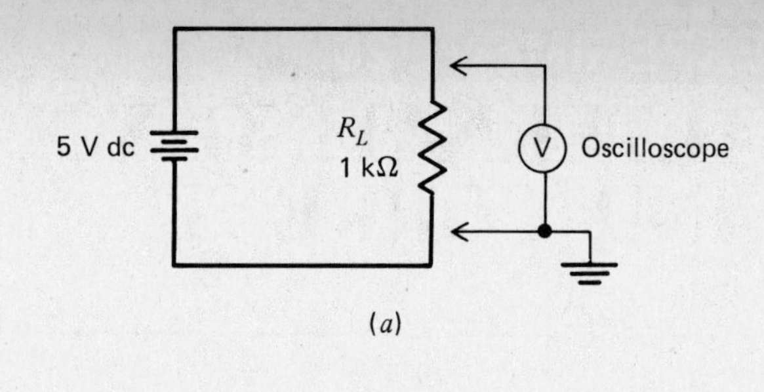

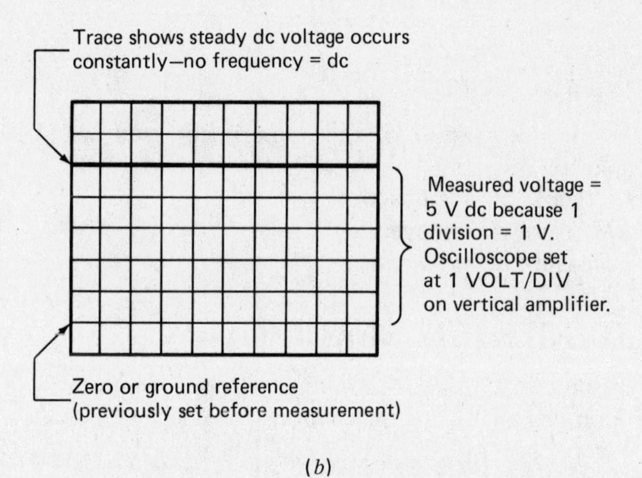

Fig. 25-4. (*a*) Oscilloscope measurement of dc voltage. (*b*) Oscilloscope display of dc measurement.

For both types of measurements, the oscilloscope must be adjusted prior to reading the measurement from the display. In the same way that it may have taken you several attempts before you could adjust an ohmmeter and properly measure a resistor, it may take several attempts before you will be able to measure voltages by using the oscilloscope. The following procedures are meant to instruct you on how to use a scope. However, because there are so many different oscilloscopes manufactured, either you will have to receive some preliminary instructions from your laboratory instructor or you should read the operator's instructions for your oscilloscope. Most oscilloscope manuals have a section that is intended to get you started. It is a good idea to spend some time examining your oscilloscope before you start.

Keep in mind the following main points when you start the following procedures:

1. An oscilloscope is a voltmeter.
2. It measures magnitude and frequency of a signal.
3. The display is like a television display: you adjust it for the best picture. If a sine wave appears larger or smaller, it is due to the operator's adjustment.
4. The magnitude and frequency of the input signal are adjusted by using the signal generator controls.
5. Avoid ground loops by keeping all ground connections at the same point when measuring ac voltages.

EQUIPMENT

Oscilloscope (preferably, a late-model, solid-state, auto-triggering type, including an operator's manual)
Signal generator
VTVM/VOM
DC power supply
Springboard
Leads

COMPONENTS

Resistors (all 0.5 W):

(1) 2.2 kΩ (1) 10 kΩ
(1) 4.7 kΩ

PROCEDURE

Measuring DC Voltages

1. Set up the oscilloscope; locate the oscilloscope probe. It should be connected to the channel 1 (or A) input for dual channel scopes. The probe should have a positive and a negative lead. Be sure the scope is plugged into the ac power line. Do not turn on the power yet. Set the front panel controls as follows:

Set the INTENSITY to midrange.
For dual trace scopes, set the VERTICAL MODE (trigger) to channel 1 or channel A.
Set the VOLTS/DIV (vertical amplifier, channel 1) to 1 V per division.
Set the AC-DC-GROUND switch (input coupling) to the ground position.
Set the SEC/DIV (time base) to 0.2 s.
Set the TRIGGERING for auto, or for normal (manual) if no auto trigger exists.
Also, be sure the trigger control is set to the INTERNAL mode, not to external or line.
Turn the FOCUS control to midrange.
Be sure the vertical amplifier and time base controls are in the CAL position (typically, fully clockwise).

Now, turn the power on.

Note: The remaining front panel controls will vary, depending upon the model. Therefore, be sure to consult your scope manual or your instructor if you have any concerns.

2. The trace should appear as a straight line across the display graticule. Note that the graticule is eight divisions vertically and ten divisions horizontally. If the trace does not appear, push the button marked BEAM, or BEAM FINDER, etc. If some type of trace appears when you push this, it means the scope is probably working. In any case, now use the following controls to adjust the trace so that a sharply defined trace is displayed at the middle of the XY axis, as in Fig. 25-5.

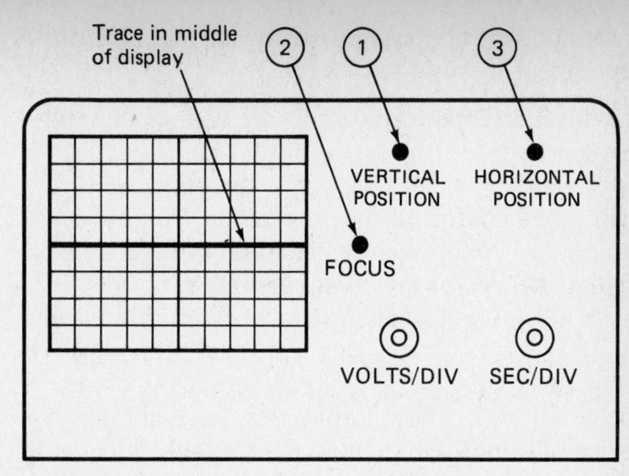

Fig. 25-5. Generic oscilloscope front panel display.

VERTICAL POSITION control ① moves trace up and down.
FOCUS control ② sharpens the trace.
HORIZONTAL POSITION control ③ moves trace right to left.

Here, in Fig. 25-5, the trace is a sharply defined straight line. Because the input coupling switch is set to GROUND, this trace means that ground, or zero volts, is at the division line. Thus, each division line above the ground line equals +1 V, and each division line below equals −1 V. This is similar to zeroing a VTVM.

If your trace looks like the one in Fig. 25-5, then you are doing fine. If not, repeat steps 1 and 2 or ask for assistance.

3. Leave the scope as it is. Connect the circuit of Fig. 25-6.

Note: Use the dc power supply and VTVM or VOM.

4. Using the VTVM, measure and record all the voltages for the circuit of Fig. 25-6. Refer to Table 25-1. These VTVM-measured dc voltages will be compared to the same voltages measured with the scope.

5. Now, use the oscilloscope to measure the same dc voltages. Connect the ×1 (times 1) scope probe across R_1 (2.2 kΩ). The scope probe can be used like the VTVM leads. Move the resistors around to avoid

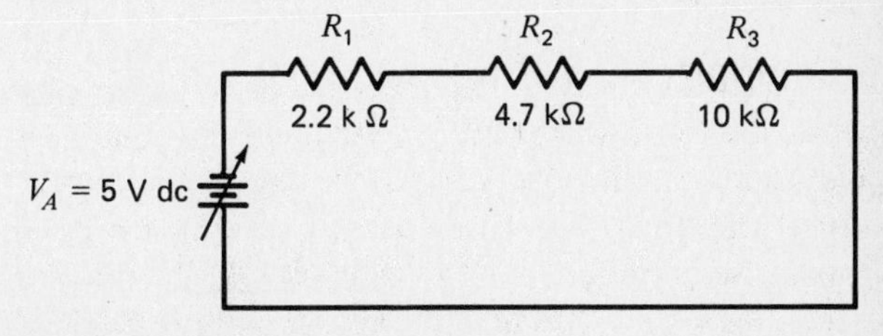

Fig. 25-6. Simple dc series circuit.

group loops if necessary. Usually, for this type of experiment it is not necessary.

With the scope leads across R_1, you should still see no change in the scope display because the input coupling is set to ground. Therefore, now switch this control to the dc position. The trace should now move slightly higher on the graticule display. Here is how you would read it.

Because the channel 1 VOLTS/DIV control is set at 1.0 V per division, each major vertical division on the graticule is equal to 1.0 V. Therefore, if the trace appears at 0.7 divisions away from ground (0 V on the center line), the dc measured voltage is equal to 0.7 divisions multiplied by 1.0 V ($0.7 \times 1.0 = 0.7$ V). Multiply the VOLTS/DIV value by the number of divisions.

Record the voltage in Table 25-1.

6. Go ahead and measure the remaining dc voltages for R_2 and R_3 of Fig. 25-6, using the oscilloscope. These values should be the same as those measured with the VTVM or VOM, plus or minus 15 percent. Do not measure V_T yet.

7. Reverse the power supply leads on the circuit of Fig. 25-6, and remeasure each resistor voltage. Notice that the trace appears below the center zero/ground reference in the same proportions as it appeared in steps 5 and 6 above. These are the same voltages except that, with respect to ground, they are opposite in polarity, and thus, they are read as negative voltages. Do not record these measurements.

8. To measure the total voltage, connect the scope across all three resistors with the ground lead at the negative side. Notice that the trace has disappeared from the graticule display. Now, readjust the vertical amplifier VOLTS/DIV control to 2 V per division. The trace should now appear. This is like changing ranges on the VTVM/VOM. Read the trace and record the value. Because the total voltage is +5 V dc, and the VOLTS/DIV = 2, you should be reading 2.5 divisions positive (above ground reference).

9. With the oscilloscope probe still across V_T, slowly increase and then decrease the applied voltage from the dc power supply. Notice how the trace rises and falls on the display in proportion to the voltage adjustment.

After completing this portion (above) of the experiment, you should be able to understand that the oscilloscope is basically a voltmeter.

Measuring AC Voltages

The oscilloscope will display ac voltages so that their waveshapes can be observed. In this way, the amplitude and the frequency can be measured, referenced to an XY axis: the amplitude is read by analyzing the vertical Y axis; the frequency is read by analyzing the horizontal X axis.

Notice in Fig. 25-3 in the introduction that the sine-wave amplitude is five vertical divisions from peak to peak. Because the VOLTS/DIV control is set at 1 V per division, each peak is about 2.5 divisions with respect to ground, which equals 5 V peak to peak. From the peak-to-peak value, the following values, typical for describing ac voltages, can be determined:

$$V_{peak} \text{ or } V_{max} = \frac{V_{p\text{-}p}}{2}$$

$$V_{av} = \frac{V_{p\text{-}p}}{2} \times 0.636$$

$$V_{rms} = \frac{V_{p\text{-}p}}{2} \times 0.707$$

Frequency is read as the number of cycles that occur during 1 s (cycles per second = hertz). Notice in Fig. 25-3 that one cycle occurs over five horizontal divisions. This is measured from zero to zero on the X axis as long as ground (zero reference) is set at a desired place and noted. Here, time base control (sometimes called the horizontal amplifier) is set for 0.2 ms per division. Thus, one cycle occurs over five divisions, or 5×0.2 ms = 0.001 s (1 ms). Because $f = 1/t$, the frequency equals 1 kHz, or 1000 cycles per second.

1. Readjust the oscilloscope. Set the VOLTS/DIV to 1. Set the input coupling switch back to the ground position, and set the trace so that ground is, once again, at midscale XY, with a sharply defined trace.

2. Connect the circuit of Fig. 25-7.

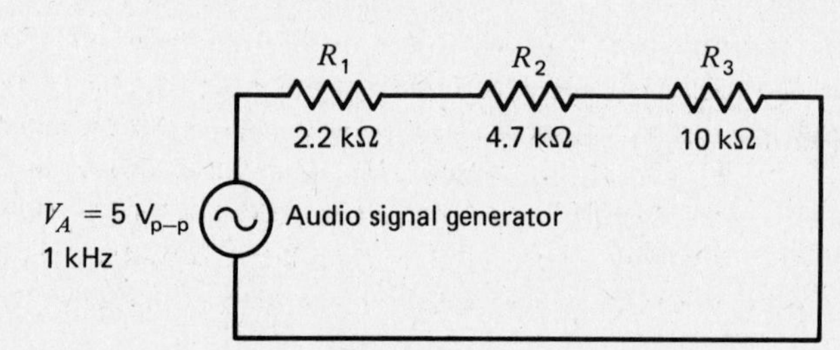

Fig. 25-7. Simple ac series circuit.

Note: The 5-$V_{p\text{-}p}$ 1-kHz sine wave is applied to the circuit by using a signal generator (oscillator). Do not be concerned if you cannot adjust it yet. Simply connect the output of the signal generator as you did the dc power supply. However, there are no positive or negative terminals; they are both the same because it is alternating current. If there is a grounded terminal, avoid using it for this experiment.

3. Connect the scope leads across the entire circuit. Then, set the input coupling for alternating current. The display should be a sine wave. Adjust the amplitude and frequency controls of the signal generator so that the trace is approximately 5 $V_{p\text{-}p}$ at 1 kHz. The display should be very similar to the one shown in Fig. 25-3.

4. Go ahead and measure all the voltages in the circuit—R_1, R_2, R_3—and of course V_T that is now being displayed. Record the values in Table 25-2. Change the VOLTS/DIV, or any other setting, so that you get the biggest display without going off the screen. Record all the values in Table 25-2, and calculate the remaining values.

5. Reconnect the scope leads across the entire circuit and adjust the signal generator so that 4 V_{p-p} at 20 kHz is the applied voltage. Then, adjust the scope so that the biggest amplitude possible (without going off the CRT) is displayed. Also, adjust it so that only two complete cycles appear on the screen. Draw a picture of the trace on the graticule of Fig. 25-8, and record the VOLTS/DIV and SEC/DIV setting you used to get this display. If your lab instructor requires that you hand in Fig. 25-8, you can use one of the graticule sheets in the Appendix, or you can use engineering paper and draw your own.

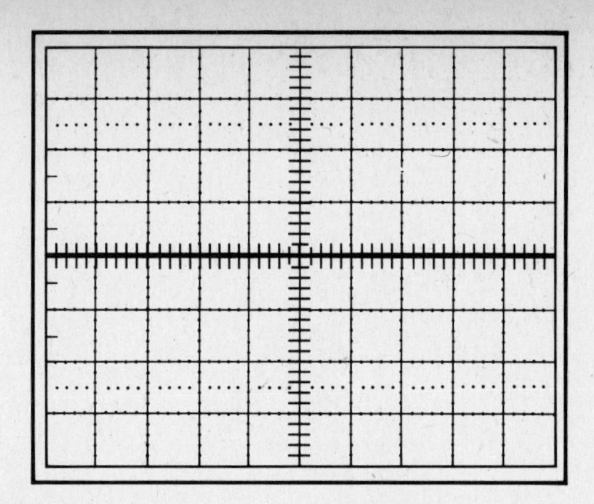

VOLTS/DIV = ____________ SEC/DIV = ____________

Fig. 25-8. Graticule. Displayed waveform = 4 V_{p-p} at 20 kHz.

NAME DATE

RESULTS FOR EXPERIMENT 25

QUESTIONS

Answer TRUE (T) or FALSE (F) to the following:

_____ **1.** An oscilloscope can measure both ac and dc voltages.

_____ **2.** The horizontal amplifier section controls the amplitude of the display on the CRT.

_____ **3.** The vertical amplifier section can be adjusted to increase or decrease the peak-to-peak voltage display.

_____ **4.** The oscilloscope probe does not have a grounded side.

_____ **5.** The oscilloscope input coupling switch is usually set to either GND, AC, or DC, depending upon the desired usage.

REPORT

This concludes the experiment. Hand in Tables 25-1 and 25-2. No report is required for this experiment. The objectives were to learn how to use the oscilloscope. If necessary, repeat this experiment and spend some more time learning to operate the scope. It can be one of the technician's most useful pieces of equipment.

TABLE 25–1. Measured DC Voltages for Fig. 25-6

	VTVM or VOM	Scope
V_{R_1}		
V_{R_2}		
V_{R_3}		
V_T		

TABLE 25–2. AC Voltages for Fig. 25-7

	Measured V_{p-p}	Calculated V_{peak}	Calculated V_{av}	Calculated V_{rms}
$R_1 = 2.2\ k\Omega$				
$R_2 = 4.7\ k\Omega$				
$R_3 = 10\ k\Omega$				
V_T				

EXPERIMENT 26

INDUCTIVE REACTANCE

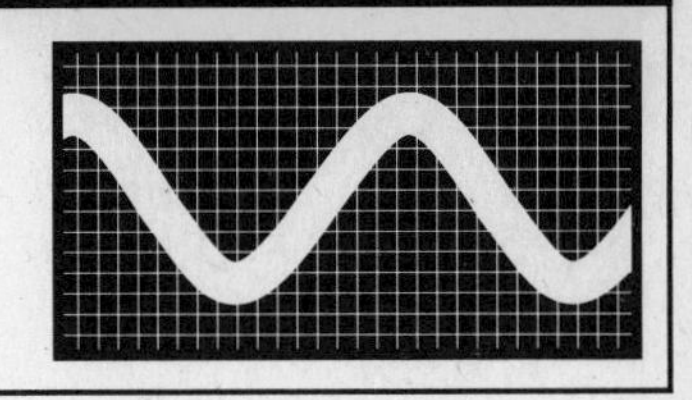

OBJECTIVES

At the completion of this experiment, you will be able to:

- Understand the concept of inductive reactance and validate the formula $X_L = 2\pi fL$.
- Understand phase relationships.
- Plot the frequency response of a series *RL* circuit.

SUGGESTED READING

Chapters 17, 18, and 19, *Basic Electronics*, B. Grob, Seventh Edition

INTRODUCTION

To study inductance, cosine waves, inductive reactance, inductive coupling, and many other aspects of an inductor (coil) in an ac circuit would require weeks of laboratory time. Therefore, this experiment will focus on the basic properties of an inductor that are fundamental to a technician.

As its name implies, an inductor produces its own (self-induced) voltage. In fact, any conductor that is close to another conductor (like two wires in close proximity) can be considered an inductor if current is flowing. Thus, as current flows in any inductor, a voltage is induced. This is accomplished because the inductor (coil) produces a magnetic field, and the magnetic field is greater with more turns of wire and/or more changes of current (frequency).

Remember that a coil's inductance uses the symbol *L* (from linkages of magnetic flux) and is measured in henrys (H). An 8-H coil, for example, is a rather large inductor and is used for 60-Hz ac power lines or other low-frequency (audio range) applications. A 1-μH coil, however, is a rather small inductor and is used at higher frequencies. Also, remember that as the frequency increases in an ac circuit, the inductive reactance X_L increases even though the value of the inductor (in henrys) remains the same. In other words, the coil reacts to, or opposes, a change in current. As frequency increases, X_L increases. Therefore, the formula for inductive reactance is

$$X_L = 2\pi fL$$

where f is the determining factor. Note that 2π is the constant circular motion from which a sine wave is derived (360°).

As X_L increases, the amount of circuit current decreases because the inductive reactance acts like a variable resistance. That is, for any given frequency, the inductive reactance will have a different resistance in ohms. For this reason, two series inductors of the same value will have twice as much reactance (in ohms), behaving like two series resistors. Similarly, two parallel inductors of the same value will have half as much reactance as one. And, according to Ohm's law, the value of circuit current is

$$I_T = \frac{V_T}{X_L}$$

where X_L is in ohms.

When a voltage is induced in a coil, the current lags the induced voltage by 90°. This lag exists in time, as is best represented by Fig. 26-1.

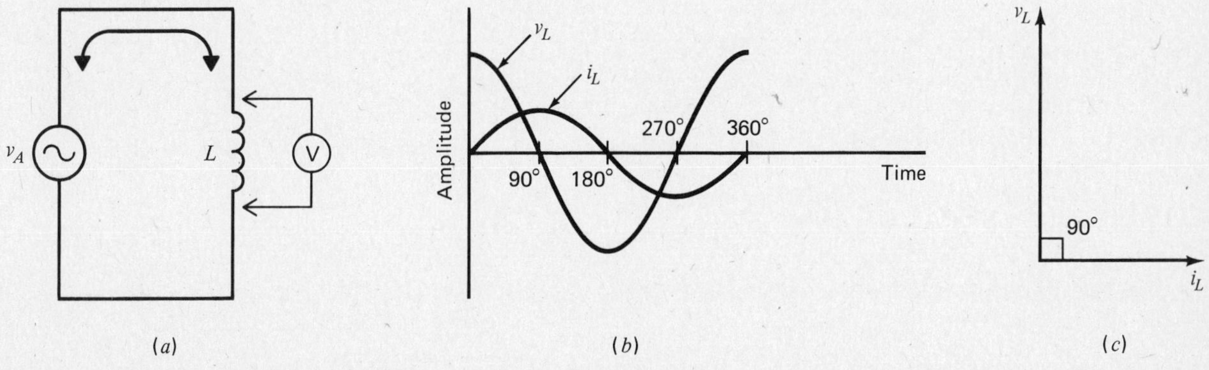

Fig. 26-1. Inductive circuit. (*a*) Purely inductive circuit. (*b*) Sine wave i_L lags v_L. (*c*) Phasor diagram.

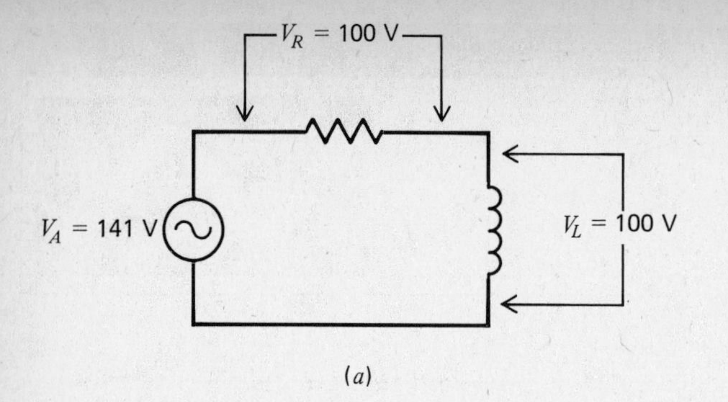

(a)

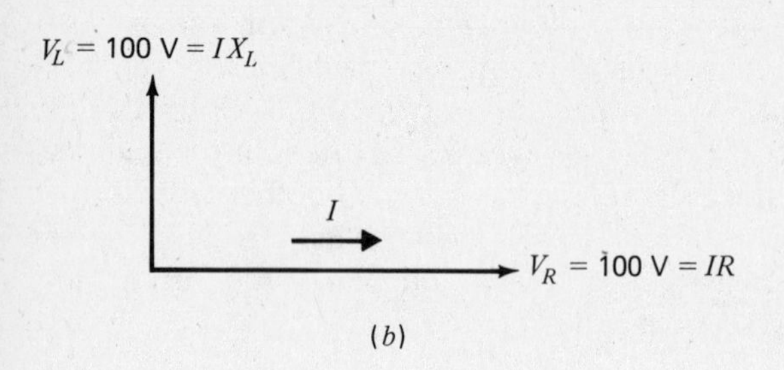

(b)

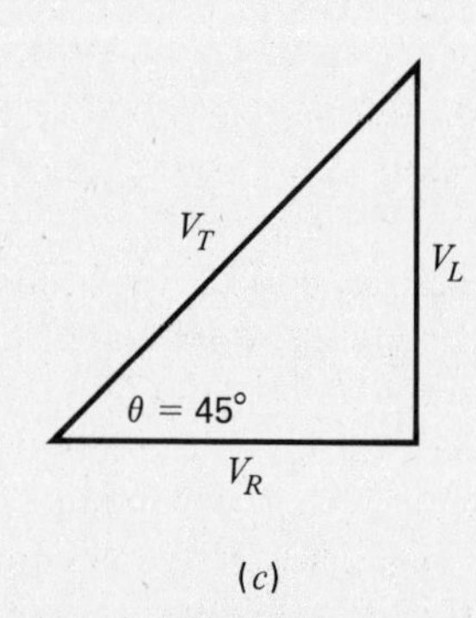

(c)

Fig. 26-2. *RL* series circuit. (*a*) Series *RL* circuit with voltage drops shown. (*b*) Phasor diagram. (*c*) Resultant of two phasors $= V_T = \sqrt{V_R^2 + V_L^2} = 141$ V.

Although the current in the inductor lags by 90°, the current is still the same in all parts of a series circuit. The time lag only exists between the induced voltage and current, and is more theoretical than practical.

When a resistor and an inductor are in series with an ac source, the current is the same in all parts of the circuit, and the inductor and resistor each have their own voltage drop (*IR*) calculated by Ohm's law. However, the voltage across the inductor leads the voltage across the resistor by 90° due to the lagging current. This is often referred to as *ELI*, where *E* (voltage across *L*) leads *I*. Thus, an inductor may have a 100-V ($I \times X_L$) drop and the resistor may also have a 100-V *IR* drop, but the combined total voltage (*V* applied from the generator) will be the applied 141 V, as shown in Fig. 26-2.

Notice that in Fig. 26-2 the pythagorean theorem can easily be used to combine the two *IR* series voltages. The resultant of the phase addition of *R* and X_L (in ohms) is calculated the same way and is called the *impedance*, or total opposition to current, with the symbol *Z*:

$$Z = (R^2 + X_L^2)^{1/2} \text{ or } Z = \sqrt{R^2 + X_L^2}$$

Therefore, if $R = 1\ \text{k}\Omega$ and $X_L = 1\ \text{k}\Omega$,

$$Z = (1000^2 + 1000^2)^{1/2} \text{ or } Z = \sqrt{1000^2 + 1000^2} = 1.414\ \text{k}\Omega$$

The phase angle θ (theta) of the complete circuit is calculated as the inverse tangent of X_L/R. In the case of Fig. 26-2,

$$\frac{X_L}{R_L} = \frac{1000}{1000} = 1 \text{ and } \tan^{-1} = 45°$$

At 45°, notice that X_L and *R* are equal in resistance and also in *IR* voltages.

The procedures that follow will allow you to validate all the concepts discussed above. However, remember that the effects of inductive reactance are sometimes difficult to comprehend in the same way that magnetism produces somewhat mysterious effects.

EQUIPMENT

Oscilloscope
Audio signal generator
Protoboard
Leads
DC power supply

COMPONENTS

(1) 560-Ω resistor
(1) 33-mH inductor

PROCEDURE

Note: Show all calculations on a separate sheet and label which procedure step they apply to.

1. Using an ohmmeter, measure the dc resistance of the inductor and record the value in Table 26-1. Remember, this is the resistance at 0 Hz.
2. Connect the circuit of Fig. 26-3.

Note: Adjust the signal generator to 5 $V_{p\text{-}p}$ by placing the oscilloscope across points A and B only after the entire circuit is connected. This is because the signal generator has its own internal resistance. Therefore, V_A can never be adjusted correctly unless

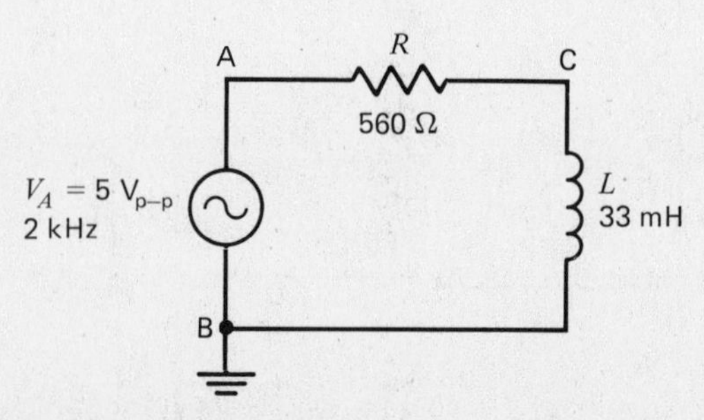

Fig. 26-3. Series *RL* circuit.

it is connected in the circuit. The frequency does not have to be exact.

3. Calculate X_L for this circuit as

$$X_L = 2\pi fL$$

Record the value in Table 26-1.

4. Measure and record the peak-to-peak voltage across the inductor. Be sure the oscilloscope probe (+) is at point C.

5. Measure and record in Table 26-1 the peak-to-peak voltage across the resistor. Be sure to exchange the resistor's place with the coil so that the oscilloscope ground and the signal generator ground are at the same point.

6. Calculate the circuit current and record it in Table 26-1 by using the Ohm's law formula. Do this by using the measured voltages for both components:

$$\text{(a) } I_T = \frac{V_R}{R} \qquad \text{(b) } I_T = \frac{V_L}{X_L}$$

7. Calculate and record in Table 26-1 the total circuit impedance, using two formulas:

$$\text{(a) } Z_T = (R^2 + X_L^2)^{1/2} \qquad \text{(b) } Z_T = \frac{V_G}{I_T}$$

8. Calculate and record in Table 26-1 the frequency as

$$f = \frac{X_L}{2\pi L}$$

9. Calculate and record in Table 26-1 the inductances as

$$L = \frac{X_L}{2\pi f}$$

10. Refer to Fig. 26-3. Readjust the frequency to 100 Hz. Increase the frequency in 100-Hz steps from 100 Hz to 1 kHz. Measure and record V_L at each frequency step in Table 26-2.

11. Beginning at 1 kHz, increase the frequency in 1-kHz steps from 1 kHz to 10 kHz and measure and record V_L at each step in Table 26-2.

12. Calculate X_L for each frequency point in steps 10 and 11 above and record the values in Table 26-2.

13. Plot a graph of frequency versus V_L and X_L for the data of steps 11 and 12, using semilog paper.

OPTIONAL PROCEDURES

Series Inductance

Connect two inductors of equal value in series, and repeat steps 1 to 9. Make your own data table (similar to Table 26-1).

Parallel Inductance

Connect two inductors of equal value in parallel, and repeat steps 1 to 9. Make your own data table (similar to Table 26-1).

NAME ______________________________ DATE ______________

RESULTS FOR EXPERIMENT 26

QUESTIONS

Answer TRUE (T) or FALSE (F) to the following:

_____ **1.** As frequency increases, X_L increases.

_____ **2.** The dc resistance of a coil is always 0 Ω.

_____ **3.** As frequency decreases, X_L increases.

_____ **4.** Two parallel coils of equal value will have twice as much inductance as one coil.

_____ **5.** Two series coils of equal value will have half as much inductance.

_____ **6.** If, in a series *RL* circuit, $V_R = 2\ V_{p\text{-}p}$ and $V_L = 2\ V_{p\text{-}p}$, the applied voltage would equal $4\ V_{p\text{-}p}$.

_____ **7.** In a series *RL* circuit, the voltage across the resistor always lags the voltage across the inductor by 90°.

_____ **8.** The current in an *RL* series circuit is the same in all parts of the circuit.

_____ **9.** There is a phase relationship in a *RL* circuit because the resistor opposes a change in current.

_____ **10.** The phase angle of an *RL* series circuit, where *R* and X_L are equal, will always be 45°.

REPORT

Write a complete report. Discuss the results. Discuss the three most significant aspects of the experiment and write a conclusion.

TABLE 26–1

Procedure Step	Circuit Component	Value
1	R_L Measured	________ Ω
3	X_L Calculated	________ Ω
4	V_L Measured	________ $V_{p\text{-}p}$
5	V_R Measured	________ $V_{p\text{-}p}$
6a	I_T Calculated	________ mA
6b	I_T Calculated	________ mA
7a	Z_T Calculated	________ Ω
7b	Z_T Calculated	________ Ω
8	f Calculated	________ kHz
9	L Calculated	________ H

TABLE 26–2

f	V_L	X_L
100 Hz		
200 Hz		
300 Hz		
400 Hz		
500 Hz		
600 Hz		
700 Hz		
800 Hz		
900 Hz		
1 kHz		
2 kHz		
3 kHz		
4 kHz		
5 kHz		
6 kHz		
7 kHz		
8 kHz		
9 kHz		
10 kHz		

EXPERIMENT 27

CAPACITIVE REACTANCE

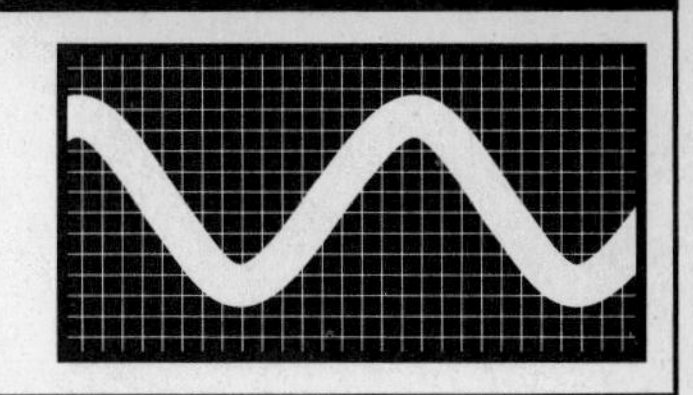

OBJECTIVES

At the completion of this experiment, you will be able to:

- Understand how a capacitor reacts in a series *RC* circuit by calculation and measurement.
- Understand the effects of changing frequency upon an *RC* series circuit.
- Plot a graph of frequency versus V_C and V_R for an *RC* series circuit.

SUGGESTED READING

Chapters 20, 21, and 22, *Basic Electronics*, B. Grob, Seventh Edition

INTRODUCTION

Capacitors have the ability to store an electric charge, much like a battery. The charge exists between the two plates made of conductive material. These plates are separated by an insulator, or dielectric. In a dc circuit, a capacitor will charge up to the potential difference across it and act as an open circuit, blocking any dc current flow. In an ac circuit, the capacitor will charge and discharge in proportion to the frequency of the alternating current, acting like a resistor or, at high frequencies, a short circuit to the alternating current. In fact, one definition of a capacitor is any two conductors separated by an insulator.

A charged capacitor is easily discharged by connecting a conducting path across the dielectric. Placing your fingers across a charged capacitor will discharge the capacitor through your fingers. With large capacitors, this will result in electric shock. *Never* assume that a capacitor is discharged.

Capacitors are manufactured and rated in units called farads (F), named after Michael Faraday. When 1 C (coulomb) is stored in the dielectric, with a potential difference of 1 V, the capacitance is equal to 1 F. Typically, capacitors are most often found in picofarad (pF) and microfarad (μF) values. The voltage rating of a capacitor is a maximum voltage that can be placed across the capacitor without damage. For example, a 10-μF capacitor, rated at 10 V, could be easily ruptured (exploded) if 20 V was across the capacitor.

When two capacitors of equal value are placed in series, the value of capacitance is reduced by one-half. This occurs because the dielectric thickness is increased and the plates are farther apart. Similar to resistors in parallel, the formula for capacitors in series is calculated by using the reciprocal method:

$$\frac{1}{C_T} = \frac{1}{C_1} + \frac{1}{C_2} + \cdots$$

When two capacitors of equal value are placed in parallel, the value of capacitance is increased by twice as much. Similar to resistors in series, the formula for capacitors in parallel is calculated by using simple addition:

$$C_T = C_1 + C_2 + \cdots$$

In an ac circuit, a capacitor will allow current to flow. The current does not actually flow through the capacitor; it flows because of the charging and discharging action of the capacitor. Therefore, the capacitive current varies with the frequency of applied ac voltage. As the frequency increases, the capacitor's opposition (in ohms) to current flow decreases. As frequency decreases, the opposition increases. That is, the capacitor reacts to changes in frequency. This is what is meant by capacitive reactance X_C. The formula for the capacitive reactance (in ohms) of any capacitor is

$$X_C = \frac{1}{2\pi f C}$$

where 2π is the sine-wave rotation, f is the frequency in hertz, and C is the value of capacitance in farads. Also, because $1/2\pi = 0.159$,

$$X_c = \frac{0.159}{fC}$$

$$f = \frac{0.159}{CX_C}$$

$$C = \frac{0.159}{fX_C}$$

Similar to an inductor, a capacitor will have a phase difference between the voltage across it and its current. Capacitive current i_C leads v_C by 90°. The term *ICE* is an easy way to remember that *I* current leads *C* capacitive voltage *E*. Figure 27-1 illustrates this.

In an *RC* series circuit, the value of current is the same in all parts of the circuit. However, each has its own series voltage drop:

$$V_R = I_T \times R \qquad \text{and} \qquad V_C = I_T \times X_C$$

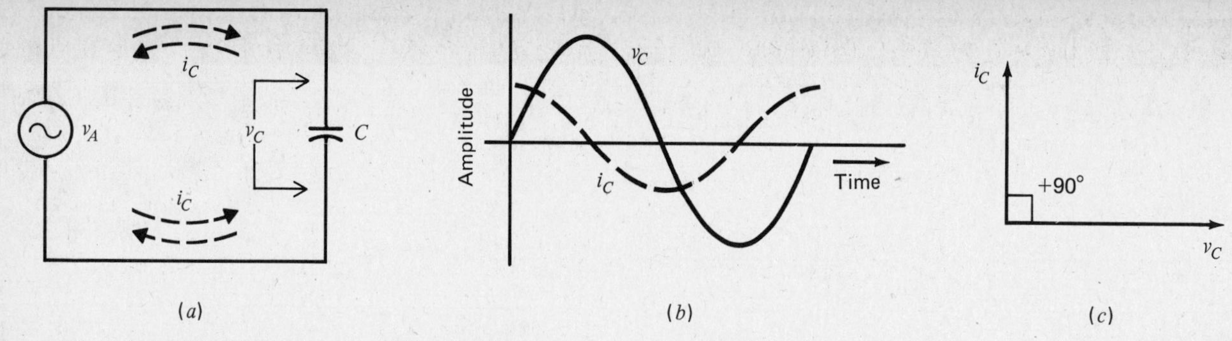

Fig. 27-1. *RC* circuit. (*a*) Purely capacitive circuit. (*b*) Waveshapes of i_C lead v_C by 90°. (*c*) Phasor diagram.

Again, similar to an inductor, the voltages must be added by phasor addition because of the 90° phase shift due to the lagging capacitive voltage. Figure 27-2 illustrates these points. In accordance with the triangles in Fig. 27-2, the total voltage in an *RC* series circuit is

$$V_T = (V_R^2 + V_C^2)^{1/2} \text{ or } V_T = \sqrt{V_R^2 + V_C^2}$$

and the total impedance (opposition to current flow) is

$$Z_T = (R^2 + X_C^2)^{1/2} \text{ or } Z_T = \sqrt{R^2 + X_C^2}$$

Remember that the circuit phase angle is between the series current and the generator ac voltage as follows:

$$\text{Inverse tan } \frac{X_C}{R} = \text{circuit phase angle}$$

This is because the phase angle is negative due to V_C lagging I_C.

The following procedures will allow you to validate the information given in this introduction.

EQUIPMENT

Audio signal generator
Oscilloscope

COMPONENTS

(1) 4.7-kΩ resistor
(1) 47-kΩ resistor
(1) 0.0068-μF capacitor
(1) 0.1-μF capacitor
(1) 0.01-μF capacitor

PROCEDURE

1. Connect the circuit of Fig. 27-3.

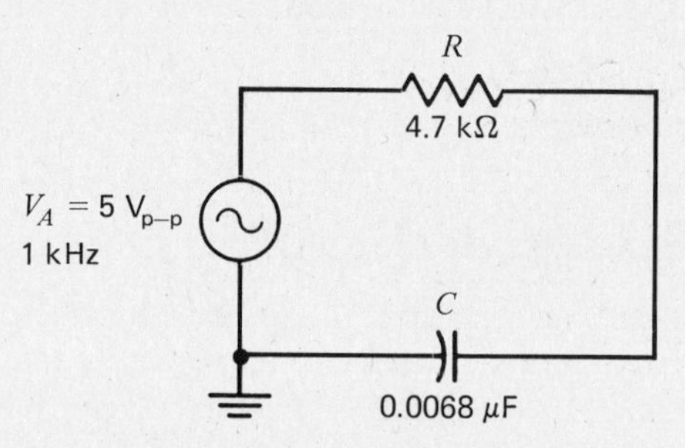

Fig. 27-3. *RC* circuit for measurement.

Note: Remember, the applied voltage (5 $V_{p\text{-}p}$) must be adjusted and maintained only after the entire circuit is connected. The signal generator has its own internal resistance and *IR* voltage drop that can subtract from the applied voltage. Therefore, constantly monitor the input voltage whenever you change frequency in the following steps.

2. Calculate X_C for this circuit in Table 27-1 as

$$X_C = \frac{1}{2\pi fC}$$

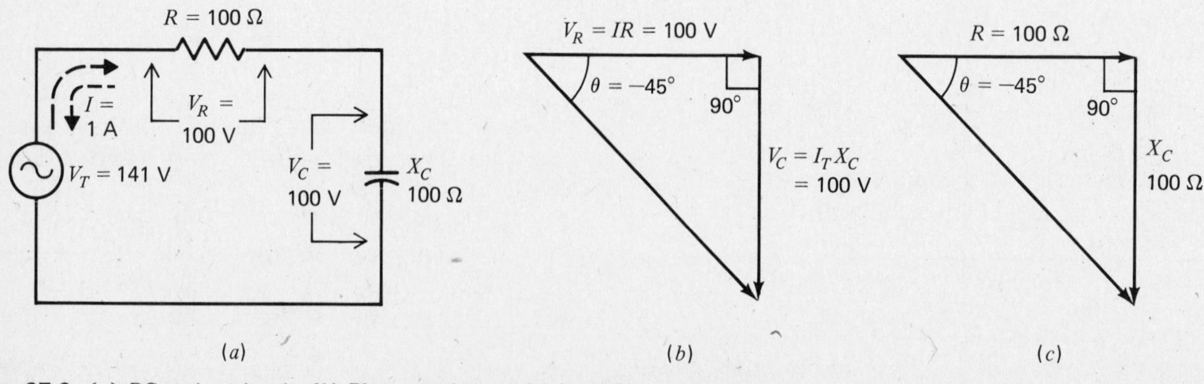

Fig. 27-2. (*a*) *RC* series circuit. (*b*) Phasor voltages. (*c*) Impedance triangle.

3. Measure and record in Table 27-1 the peak-to-peak voltage across the capacitor. Be sure the ground points are kept in the same place.
4. Measure and record in Table 27-1 the peak-to-peak voltage across the resistor. If necessary, exchange places with the capacitor to keep the ground points in a common place.
5. Calculate and record the circuit current I_T in Table 27-1, using the Ohm's law formula. Do this by using the measured voltage for both components.

(a) $I_T = \dfrac{V_R}{R}$

(b) $I_T = \dfrac{V_C}{X_C}$

6. Calculate and record in Table 27-1 the total circuit impedance, using the formula

$$Z_T = (R^2 + X_C^2)^{1/2}$$

7. Refer to the circuit of Fig. 27-3. Change the value of the capacitor to 0.1 μF and repeat steps 1 to 6 above.
8. Refer to the circuit of Fig. 27-3. Change the value of the capacitor to 0.01 μF and change the value of the resistor to 47 kΩ.
9. Readjust the signal generator to 200 Hz and increase the frequency in 200-Hz steps from 200 Hz to 1 kHz. Measure and record the voltage across the resistor V_R at each frequency step.
10. Increase the frequency from 1 to 15 kHz in 1-kHz steps and measure and record V_R at each frequency step.
11. Calculate and record X_C at each frequency in steps 9 and 10 above, as $X_C = 1/(2\pi fC)$.
12. Calculate the circuit current I_T as V_R/R for each frequency in step 11 above. Also, calculate V_C as $I_T X_C$ for each frequency.
13. Plot a graph of V_C and V_R versus frequency for the data in steps 10 to 12. Indicate where the V_C and V_R curves intersect at 45°.

OPTIONAL PROCEDURES

Series Capacitance

Connect two capacitors of equal value in series, and repeat steps 9 to 12. Make your own data table and record all values.

Parallel Capacitance

Connect two capacitors of equal value in parallel, and repeat steps 9 to 12. Make your own data table and record all values.

Note: Use $R = 10$ kΩ and $V_A = 5\ V_{p\text{-}p}$.

NAME DATE

RESULTS FOR EXPERIMENT 27

QUESTIONS

Answer the following questions on a separate sheet of paper. Show all work.

1. Using your graphed data, calculate the circuit phase angle at 8 Hz.

2. Using your graphed data, calculate the circuit phase angle at 600 Hz.

3. For a series *RC* circuit, at what frequency would a 10-μF capacitor have $X_C = 100\ \Omega$?

4. For a series *RC* circuit, what value of capacitor would have 31.8 Ω of X_C at 5 Hz?

5. For the circuit in Fig. 27-4, the voltage across $C = 25\ V_{p\text{-}p}$. The circuit phase angle is 45°. What is the voltage across *R* and V_A?

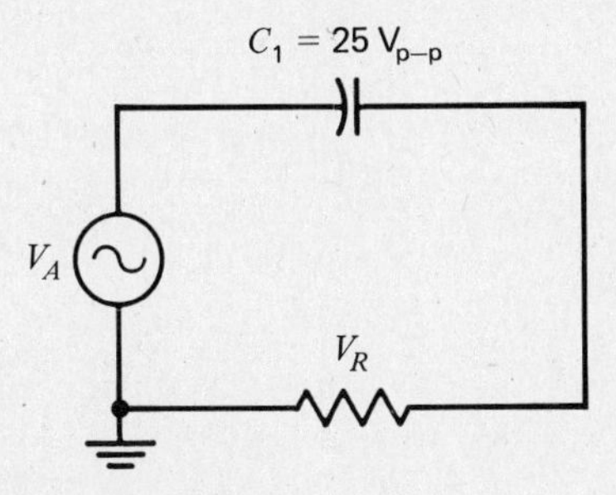

Fig. 27-4. *RC* circuit at 45°.

REPORT

Write a complete report. Discuss the results. Discuss the three most significant aspects of the experiment and write a conclusion.

TABLE 27–1

Procedure Step	Measurement	Value	Calculations
2	X_C		______ Ω
3	V_C	______ $V_{p\text{-}p}$	
4	V_R	______ $V_{p\text{-}p}$	
5a	I_T		______ mA
5b	I_T		______ mA
6	Z_T		______ Ω
7	X_C		______ Ω
	V_C	______ $V_{p\text{-}p}$	
	V_R	______ $V_{p\text{-}p}$	
	I_T(a)		______ mA
	I_T(b)		______ mA
	Z_T		______ Ω

Student is to prepare own tables for $R = 47\ \text{k}\Omega$ (step 8) and steps 9 to 13.

EXPERIMENT 28

FREQUENCY MEASUREMENTS: USING AN OSCILLOSCOPE

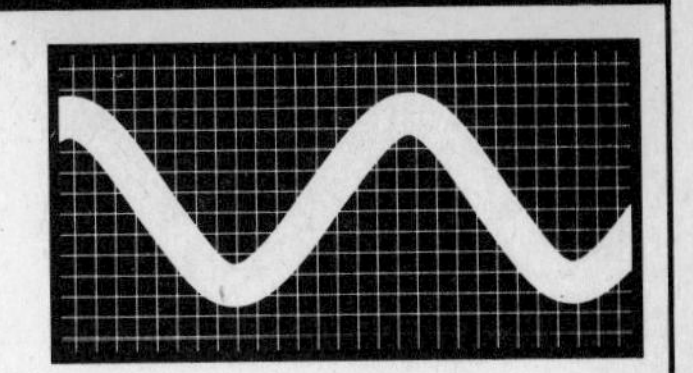

OBJECTIVES

At the completion of this experiment, you will:

- Be familiar with the operation of a frequency counter.
- Be familiar with the XY operation of an oscilloscope.
- Be able to measure frequency by using Lissajous patterns.

SUGGESTED READING

Chapter 16, *Basic Electronics,* B. Grob, Seventh Edition

INTRODUCTION

This experiment demonstrates how a frequency measurement can be made by using different techniques. In previous experiments you learned how to measure the frequency of an ac signal by using the time base of the oscilloscope. That method allowed you to derive the frequency F by observing the time (seconds per division) it took for 1 cycle to occur: $F = 1/\text{time}$. In most cases, that technique is adequate for determining the approximate frequency of an ac signal.

However, the easiest and most accurate method for measuring frequency is to use a digital frequency counter. Digital frequency counters usually have LED (light-emitting diode) displays that are easy to read. In addition, the digital frequency counter is extremely accurate because it uses a crystal oscillator as a reference to count or compare the frequency you are measuring.

Another method for measuring the frequency of an ac signal is to create a Lissajous pattern on an oscilloscope. These patterns are named after Lissajous, a well-known French physicist. The pattern is created by simultaneously applying two ac signals to the horizontal and vertical deflection circuits of an oscilloscope.

To do this, the time base of the oscilloscope must be disabled. Most oscilloscopes have a setting on the time base marked XY. When the time base is set to this feature, it separates the two channels (using a dual-trace oscilloscope) so that one channel goes to the horizontal deflection circuit and the other channel goes to the vertical deflection circuit. During normal operation (not XY), both channel 1 and channel 2 send signals to the horizontal section, and the vertical section is controlled by the time base. However, when the XY mode is set, the oscilloscope will display a pattern that shows the frequency relationship of two ac signals. Figure 28-1 shows an example of four Lissajous patterns where the ratio of two sine-wave signals is determined by the number of horizontal and vertical peaks or lobes.

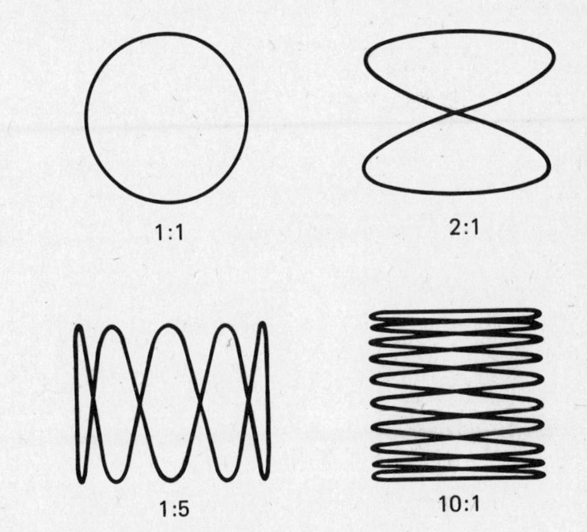

Fig. 28-1. Typical Lissajous patterns.

Although this experiment introduces you to Lissajous patterns and the XY operation of an oscilloscope, you should refer to the oscilloscope's operating manual for detailed information about XY operation. This experiment will simply introduce you to the concepts.

Figure 28-2 shows the three techniques discussed in this experiment to make frequency measurements. Notice that each technique shows a 1-kHz sine-wave signal except for the Lissajous pattern technique, where a 1-kHz signal is compared to another 1-kHz signal to create the 1:1 circle pattern as in Fig. 28-1.

Because this experiment concentrates on the Lissajous pattern method for measuring frequency on the oscilloscope, you must have a good idea of what you expect to see on the CRT. Figure 28-3 shows how to read the Lissajous pattern. In addition, remember that Lissajous patterns will not be stable or readable unless there is an integer (whole-number) relationship

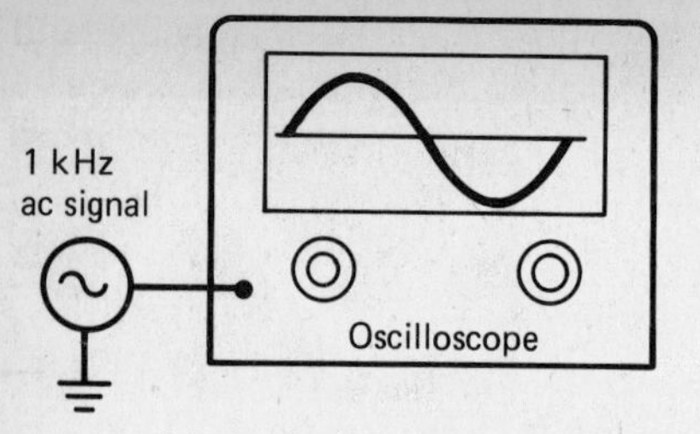

$F = {}^1/\text{Time} = 1$ kHz
When sec/div
is set to 1 ms

(*a*) Oscilloscope time-base method

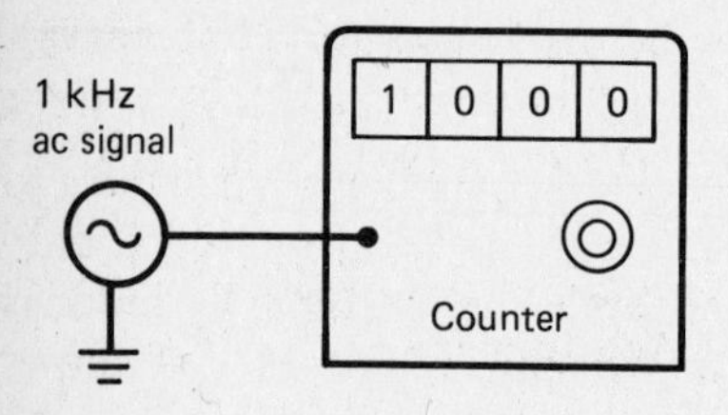

$F = 1$ kHz
Where frequency
counter is
displaying the
measurement

(*b*) Frequency counter method

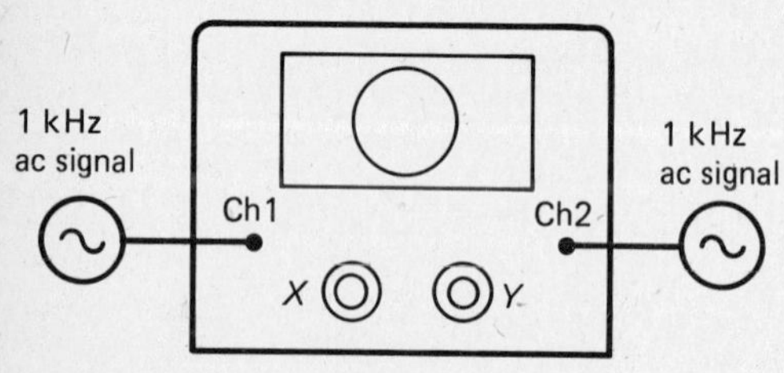

$F = 1$ kHz = circle display
When channel 1 and 2 both have
signals of equal value in magnitude

(*c*) Lissajous pattern (*X–Y*) method

Fig. 28-2. Three techniques for measuring frequency.

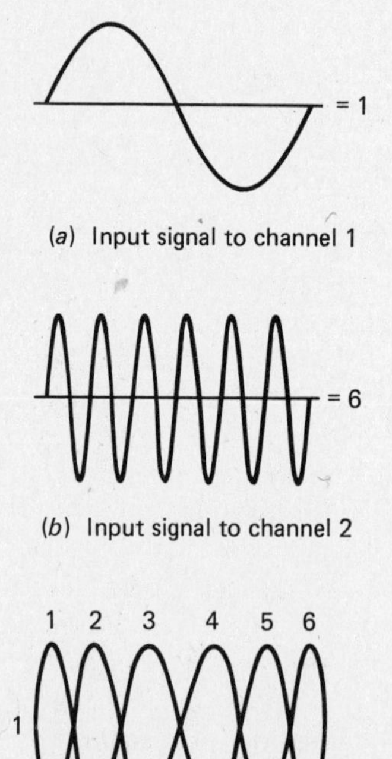

(*a*) Input signal to channel 1

(*b*) Input signal to channel 2

(*c*) Resulting Lissajous pattern

Fig. 28-3. Lissajous pattern 1:6.

between the two frequencies. Also if either channel 1 or channel 2 frequencies are changed even slightly, the Lissajous pattern will rotate and change shape.

Notice that Fig. 28-3 shows the two signals input to channels 1 and 2. If channel 1 is 100 Hz, then channel 2 must be 600 Hz; or channel 1 could be 500 Hz, and channel 2 could be 3 kHz. In every case, the frequency measurement technique, using Lissajous patterns, requires that the frequencies be exact and have low integer ratios. If they do not, the XY relationship will not yield a pattern where the lobes are easy to read or count.

The following procedure does not require that you be an expert on the oscilloscope or frequency counter. It requires only that you have performed the previous ac measurements in this lab manual, using an oscilloscope. However, you can refer to your oscilloscope and frequency counter manuals. These manuals usually contain information about making frequency measurements. More information about XY measurements is given in the phase measurements experiment, Experiment 29.

EQUIPMENT

Oscilloscope with XY capability
Voltmeter
Digital frequency counter
(2) Audio signal generators

COMPONENTS

Leads for connecting equipment as required

PROCEDURE

1. Connect the equipment as shown in Fig. 28-4 and turn on the power.

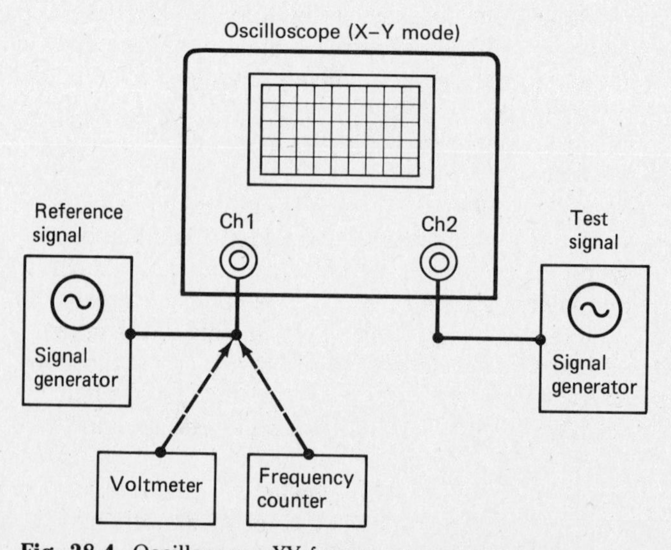

Fig. 28-4. Oscilloscope XY frequency measurement.

2. Set the oscilloscope time base to XY, and set channels 1 and 2 to 0.5 V per division. Make sure that all controls are in the CAL position.

3. Set the two audio generators to approximately 3 V p-p at 500 Hz. Use the voltmeter and frequency counter to verify the settings.

4. Adjust the oscilloscope position controls and the fine tuning (vernier) of the test audio signal generator (channel 2) until the most stable and most circular pattern is displayed on the CRT. Any rotation is due to phase differences between the two signals.

Note: In the following procedures, adjust for the most stable display possible to count and record the peaks of the "bow tie" patterns.

5. Adjust the test signal generator (channel 2) to 1 kHz. Then use the fine tuning to create a stable pattern. Draw the pattern in Table 28-1. Measure and record the two signal generator frequencies, using the frequency counter, in Table 28-1.

6. Adjust the test signal generator (channel 2) to 1.5 kHz. Draw the pattern and record both signal frequencies measured with the counter in Table 28-1.

7. Adjust the reference signal (channel 1) to 200 Hz and the test signal (channel 2) to 1200 Hz. Draw the pattern and record the signal frequencies in Table 28-1.

8. Adjust the test signal (channel 2) to 100 Hz, leaving the reference at 200 Hz. In Table 28-1 draw the pattern, record the measured frequencies, and determine the ratio.

9. Randomly change voltages, frequencies, and oscilloscope settings. Observe any results or consistent occurrences and note them for use in your report.

NAME DATE

RESULTS FOR EXPERIMENT 28

QUESTIONS

1. Which method of measuring frequency is the easiest and most accurate? Why?

2. What happens to the oscilloscope when the XY setting is chosen?

3. Why do you think the Lissajous patterns are often unstable (moving)?

4. Does it matter if the X or Y axis is used as the reference or standard frequency? Why or why not?

5. If the test and reference signals had different amplitudes (voltages), could you still use Lissajous patterns to measure frequency? How?

REPORT

Write a report showing your understanding of the different methods of measuring frequency. Hand in all data.

TABLE 28–1.

Procedure Step		
5	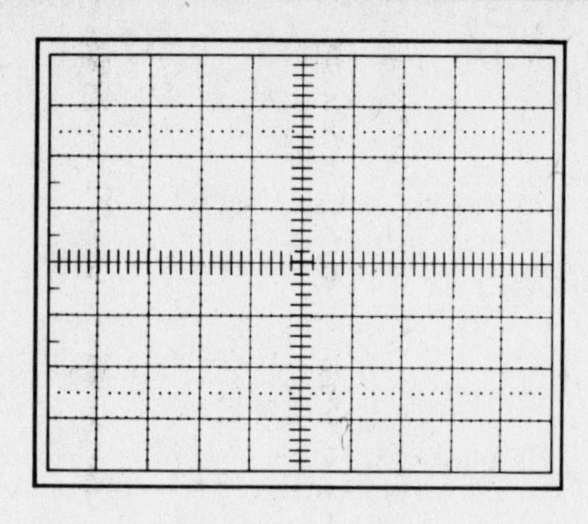	Reference = _______ Hz Test = _______ Hz Ratio = _______ to _______
6	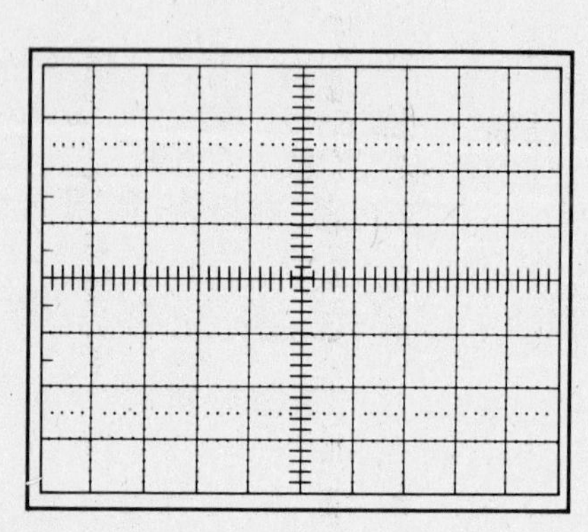	Reference = _______ Hz Test = _______ Hz Ratio = _______ to _______
7	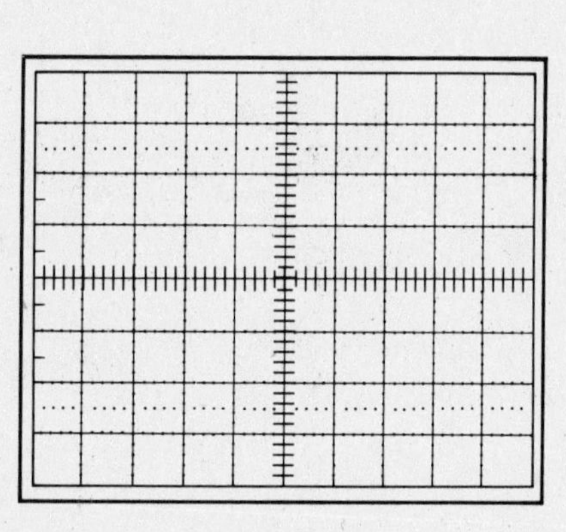	Reference = _______ Hz Test = _______ Hz Ratio = _______ to _______
8	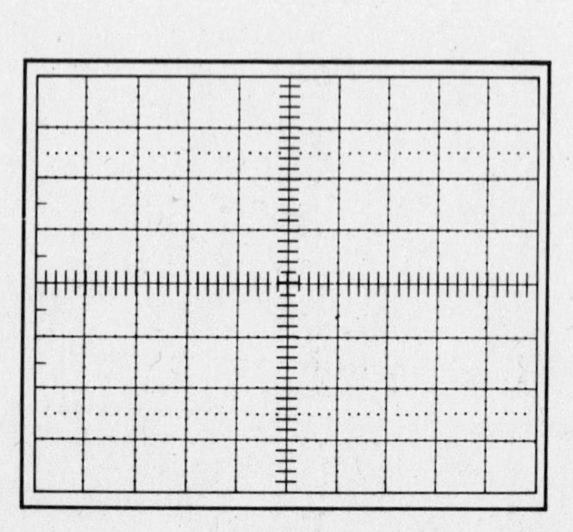	Reference = _______ Hz Test = _______ Hz Ratio = _______ to _______

EXPERIMENT 29

PHASE MEASUREMENTS: USING AN OSCILLOSCOPE

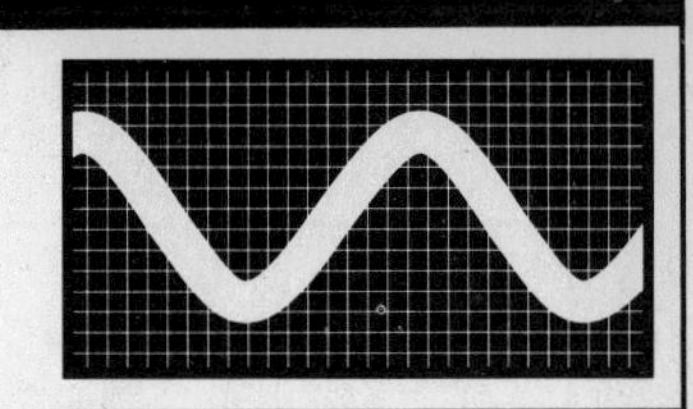

OBJECTIVES

At the completion of this experiment, you will:

- Be able to make phase measurements on the oscilloscope, using the dual-trace method.
- Be able to make phase measurements on the oscilloscope, using Lissajous patterns.
- Be able to determine the phase angle by measuring voltages and calculating the phase shift.

SUGGESTED READING

Chapter 16, *Basic Electronics,* B. Grob, Seventh Edition

INTRODUCTION

This experiment concentrates on the measurement of a phase angle. Unlike frequency measurements, phase measurements compare two sine-wave signals of similar frequency; the result is a phase angle between two waves.

Consider the drawing (Fig. 29-1) where two voltage waveforms are shown 90° out of phase as a dual-trace measurement. Figure 29-1*a* shows the two sine waves as they might be displayed on a dual-trace oscilloscope. Sine wave *A* starts at 0 V and 0°. Therefore, it is considered the reference signal.

Sine wave *B* actually begins 90° before sine wave *A*. In other words, the peak voltage of sine wave *B* occurs when sine wave *A* is at zero. The phasor diagram of Fig. 29-1*b* shows that both signals have about equal magnitude (voltage), indicated by the length of the arrow. The angle is shown with respect to the reference signal *A* on the horizontal axis. Therefore, the phase angle has magnitude and direction; here the direction is +90° compared to 0° of the reference signal. This corresponds to the standard practice of using a counterclockwise rotation as the positive direction of rotation. This also means that signal *B* leads signal *A* by +90°.

Although it is easier to describe phase differences when the signals have the same voltage, consider the drawing in Fig. 29-2. Figure 29-2*a* shows two sine waves with different magnitudes, but having the same frequency. Figure 29-2*b* shows two sine waves 180° out of phase, also with different magnitudes. Notice that both phase measurements can be represented by their corresponding phasor diagrams.

Although using the dual-trace capability of an oscilloscope is the easiest way to measure the phase of like frequencies, Lissajous patterns can also be used. The oscilloscope is used exactly as in Experiment 28, "Frequency Measurements." This means that the XY mode of the time base is selected. As Fig. 29-3 shows, two signals of equal frequency display any phase difference as a Lissajous pattern created by two signals of the exact same amplitude.

Notice that a pattern is created where signal *A* is the maximum vertical number of divisions. Signal *B* is the vertical measurement between the two points where the trace pattern crosses the centerline (vertical) on the graticule. Signal *B* is then divided by signal *A* to obtain the sine of the two signals. And the arcsine is calculated to obtain the phase angle in degrees.

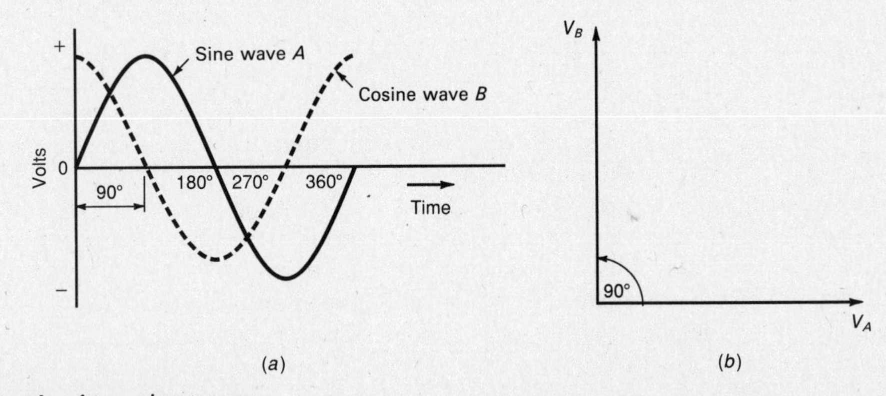

Fig. 29-1. Phase angle of two sine waves.

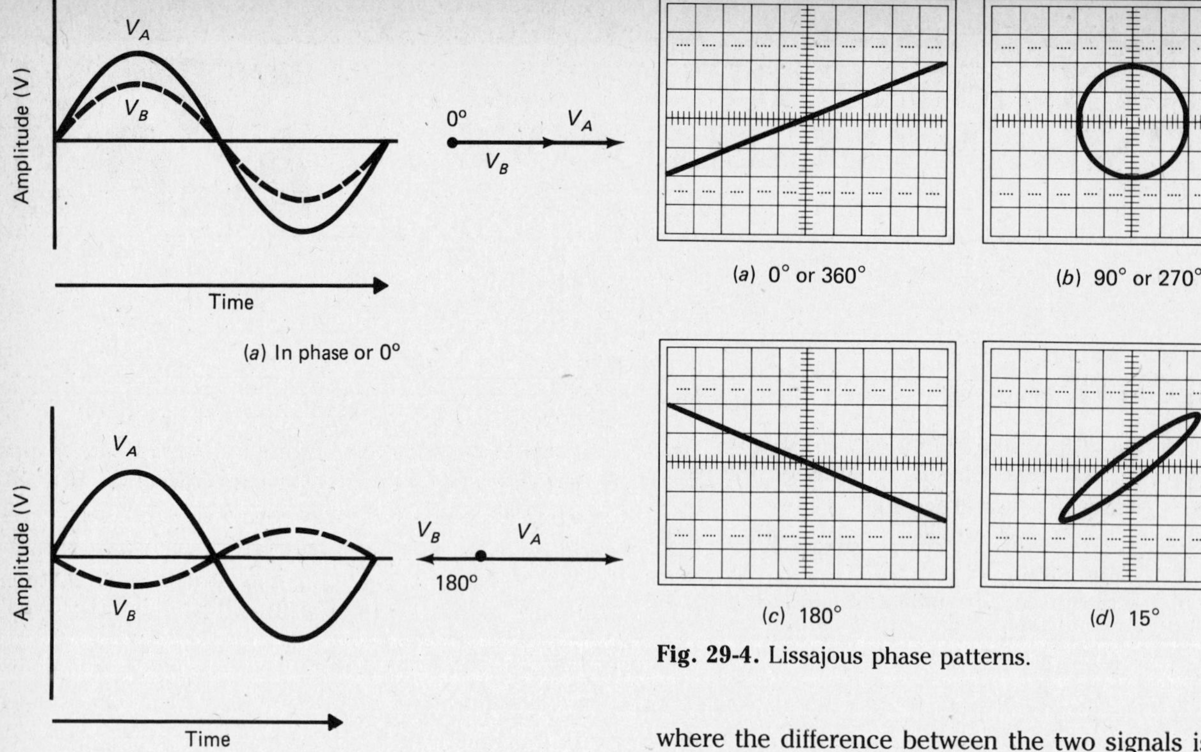

Fig. 29-2. Phase angles with different amplitudes.

Fig. 29-4. Lissajous phase patterns.

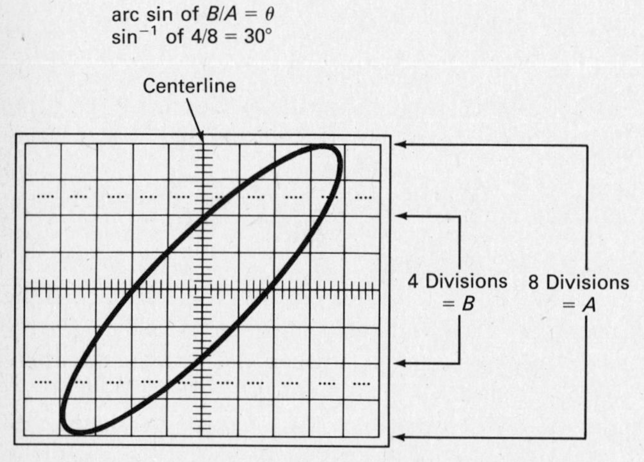

Fig. 29-3. XY phase measurement of two signals with equal frequency and magnitude equals 30°.

Figure 29-4 shows some typical Lissajous patterns for common phase angles. Remember that this type of Lissajous pattern is based on two signals of equal frequency and amplitude measured on an oscilloscope in the XY mode. Because the procedure in this experiment focuses on the use of an oscilloscope to measure phase, you can refer to the oscilloscope's operating manual for more information, if necessary. Also this experiment should be performed after you have completed Experiment 28, since the use of the equipment is similar.

Finally, to measure phase, a phase-shift network will have to be created. This will be a simple *RC* circuit where the difference between the two signals is measured across *R* and the signal generator.

EQUIPMENT

Oscilloscope with XY capability
Voltmeter
Digital frequency counter (optional)
Audio signal generator

COMPONENTS

Leads for connecting equipment as required
(1) 1.2-kΩ ½-W resistor
(1) 0.01-μF capacitor

PROCEDURE

1. Connect the circuit as shown in Fig. 29-5, and turn on the power.
2. Adjust the oscilloscope to XY (time base), and set channels 1 and 2 to 0.5 V per division each.
3. Set the signal generator to 23 kHz at 2 V p-p. Verify this by using the frequency counter and voltmeter (or oscilloscope). Adjust the trace position.
4. The oscilloscope should now display the Lissajous pattern for the phase shift between *R* and the generator. Record the display in Table 29-1 by drawing it as close as you possibly can.
5. Determine the phase angle, using the arcsin (B/A) method described in the Introduction. Record the calculations and results in Table 29-1.
6. Measure the p-p voltage across the resistor, and record the results in Table 29-1. Because the measured phase shift is between *R* and V_G, calculate the

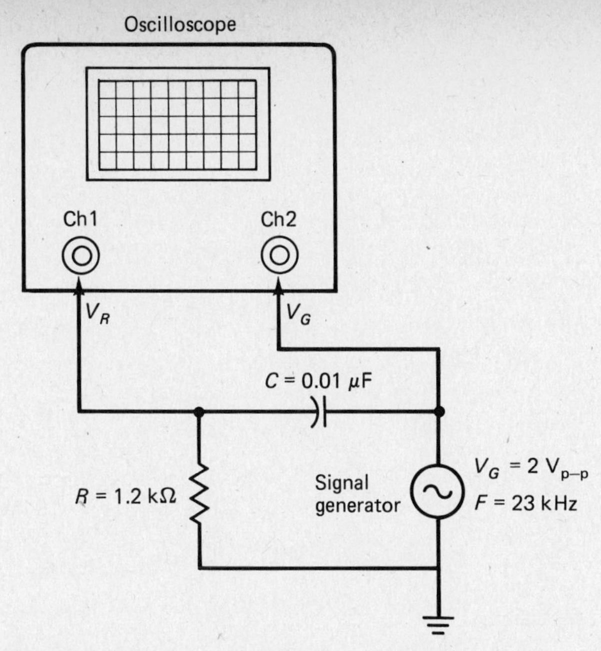

Fig. 29-5. Phase measurement of a phase-shift circuit.

phase shift as the arcsine or inverse cosine of V_R/V_G. Record the result in Table 29-1.

7. Adjust the oscilloscope back to time-base operation, and set it to display one or two complete cycles of each wave. Set both channels to the same ground at the center graticule. Then measure the phase difference between the two signals: channel 2 = V_G and channel 1 = V_R. If necessary, use the time-base CAL adjustment to display one cycle (V_G) over an equal number of divisions. Also you can use the position knobs and, if applicable, choose the CHOP mode instead of the ALTERNATE or ADD mode.

Draw the traces as best you can in Table 29-1. Label the two traces and show the phase difference, in degrees, where the V_G wave crosses the zero or middle graticule lines. Also show the degree per division.

8. Change the frequency of the signal generator to 8 kHz, and repeat steps 1 through 7. Use Table 29-2 to record your results.

NAME DATE

RESULTS FOR EXPERIMENT 29

QUESTIONS

1. Which method of measuring phase is the easiest and fastest? Why?

2. If you had to measure both frequency and phase, could you use the Lissajous pattern method for two signals 500 Hz apart? Why or why not?

3. Why did changing the frequency of V_G cause a different phase angle?

4. If you didn't have any test equipment, could you calculate the phase angle between V_G and V_R? How?

5. Draw the phasor showing the results for a phase angle where $V_G = 12\text{ V}$ and $V_R = 8.4\text{ V}$.

REPORT

No report is required. Hand in all data.

TABLE 29–1.

Procedure Step	V_G = 23 kHz at 2 V p-p
4	
5	A = ______ DIV B = ______ DIV arcsin B/A = ______°
6	V_R = ______ V p-p V_G = ______ V p-p arccos V_R/V_G = ______°
7	
	Degrees per DIV = ______

TABLE 29–2.

Procedure Step	V_G = 8 kHz at 2 V p-p
4	
5	A = ______ DIV B = ______ DIV arcsin B/A = ______°
6	V_R = ______ V p-p V_G = ______ V p-p arccos V_R/V_G = ______°
7	
	Degrees per DIV = ______

EXPERIMENT 30
RC TIME CONSTANT

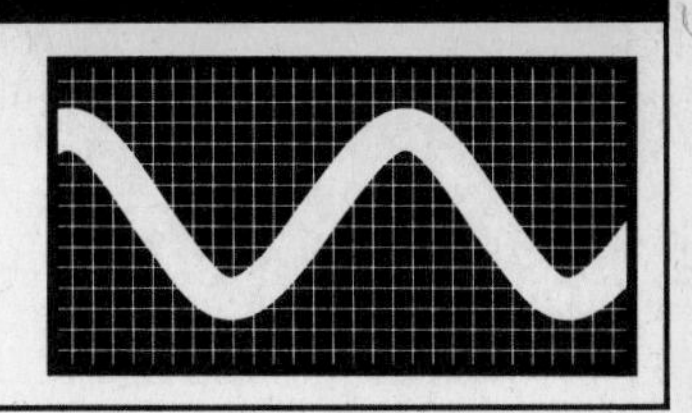

OBJECTIVES

At the completion of this experiment, you will be able to:

- Understand the concept of $T = RC$.
- Validate the decaying charge current in an *RC* series circuit.
- Plot a graph of V_C and I_C versus time for equal values of single, series, and parallel capacitance.

SUGGESTED READING

Chapters 22 and 23, *Basic Electronics,* B. Grob, Seventh Edition

INTRODUCTION

A capacitor in series with a resistance will charge to 63.2 percent of the applied voltage in one time constant. This time constant T (in seconds) is applied to nonsinusoidal waveforms as a transient response, where R is the series resistance in ohms and C is the value of the capacitance in farads, or

$$T = RC$$

For example, the circuit of Fig. 30-1 has the following time constant:

$$\begin{aligned} T = RC &= (1000)(4 \times 10^{-6}) \\ &= 4000 \times 10^{-6} \\ &= 4 \times 10^{-3} \\ &= 4 \text{ ms} \end{aligned}$$

Notice that the applied voltage does not affect the value of T. The time constant T is the time for the voltage across C to change by 63.2 percent. If V_A in Fig. 30-1 were 100 V, then C would change to 63.2 V in 4 ms. In another 4 ms, the capacitor would change to 63.2 percent of the remaining voltage, or 63.2 percent of 36.8 V. This would repeat until the voltage across C equaled the applied voltage. Also, during this time, the circuit current would steadily decay until V_C reached its maximum voltage, or V_A.

Similarly, if the capacitor were discharging, *RC* would specify the time it takes C to discharge 63.2 percent of the way down, to the value equal to 36.8 percent of the initial voltage across C at the start of discharge. A capacitor is usually considered charged after approximately 5 time constants.

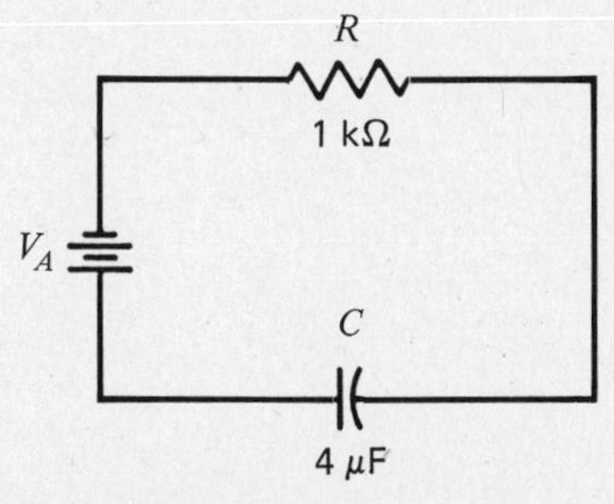

Fig. 30-1. *RC* time constant circuit.

EQUIPMENT

High-voltage power supply
Microammeter (VOM)
Grease pencil (erasable)
Tissue, paper towels, or cloth
Leads
Protoboard

COMPONENTS

Either A: (V_A = 90 V)

(2) 4-μF capacitors and (1) 1.2-MΩ resistor

or B: (V_A = 100 V)

(2) 1-μF capacitors and (1) 3-MΩ resistor
(2) SPST switches

PROCEDURE

1. Connect the circuit of Fig. 30-2. Use either component values (A) or (B). (S_1 can either be a switch, a wire lead, or the power supply on/off switch. S_2 is a wire lead.)

CAUTION: Do not turn power on.

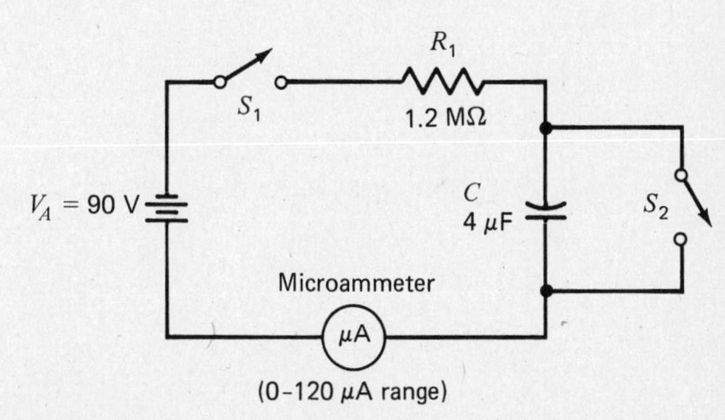

Fig. 30-2. *RC* series circuit for time constant measurement shown with (A) component values. The (B) values are V_A = 100 V, R = 3 MΩ, and C = 1 μF.

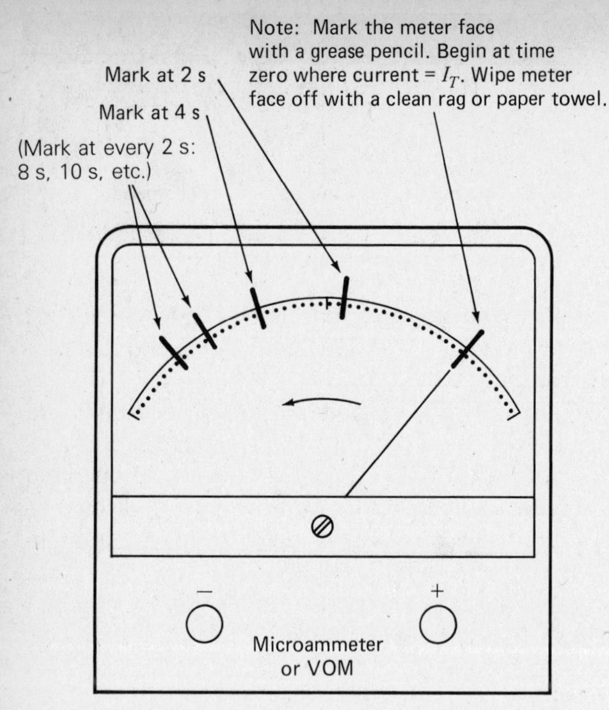

Fig. 30-3. Microammeter in *RC* time constant circuit.

2. Calculate and record in Table 30-1 the amount of circuit current I_T by using Ohm's law. Do not forget to indicate which component values, (A) or (B), you use.
3. Close S_2 in order to discharge C_1 (keep it closed). Close S_1 and measure and record the circuit current I_T in Table 30-1.
4. Refer to Fig. 30-3. Prepare to mark the meter face with a grease pencil by making the first mark at the point of I_T now showing on the meter.

When you start, mark the meter face every 2 s with a fine, straight mark. The 2-s interval can be called out by using the term *mark*. Thus, either your lab partner or a group timekeeper should start by announcing, "ready, go." Every 2 s, when you hear the word *mark*, make your grease pencil mark. Try a practice run first.

Keep time for about 30 s. As the current decays (capacitor charges), the marks will get closer and closer together. Use approximate values when you record the results.

Record the results in a table. Record the value of circuit current at each 2-s interval, beginning at the zero until the 30-s limit. After about 16 s, the grease marks will be difficult to read. Use your best approximation.

5. Repeat the procedure three times. Afterward, add the values for each 2-s interval and divide by 3 to get an average value.
6. Plot a graph of time versus circuit current and V_C (voltage across the capacitor). Refer to Fig. 30-4. The traces should have a similar appearance.
7. Calculate the time constant $T = RC$ and indicate those points on the graph for both V_C and I_T.
8. Repeat steps 1 to 7 with two capacitors in parallel.
9. Repeat steps 1 to 7 with two capacitors in series.

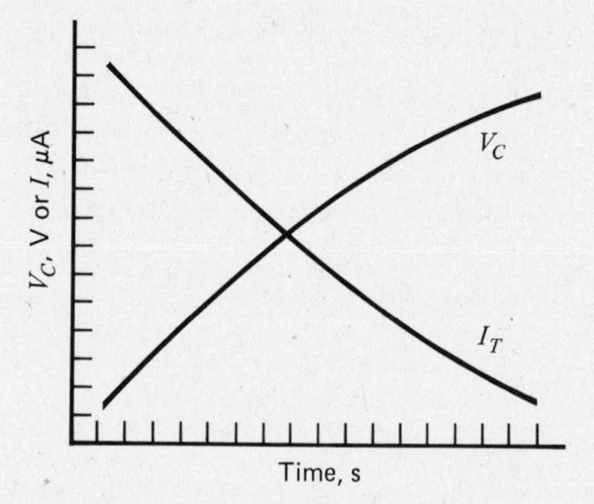

Fig. 30-4. Typical plot of time versus amplitude for *RC* time constant. Label both axes as necessary and indicate *RC* values in title.

NAME ______________________ DATE ______________

RESULTS FOR EXPERIMENT 30

QUESTIONS

Answer TRUE (T) or FALSE (F) to the following:

_____ **1.** The RC time constant is expressed in units of farads.

_____ **2.** The formula for RC time constant is $T = 1/RC$.

_____ **3.** The time constant T is the time for the voltage across R to change by 36.8 percent.

_____ **4.** The change of 63.2 percent is constant for all values of R and C as long as a sine wave is the applied voltage.

_____ **5.** For a 1000-Ω R in series with an 18-μF C, the time constant would be 180 ms.

_____ **6.** The time constant formula is the same for a discharging capacitor.

_____ **7.** After approximately five time constants, a capacitor is considered charged to the applied dc voltage.

_____ **8.** After two time constants in an RC circuit where $R = 2$ kΩ and $C = 15$ μF, the capacitor would be charged to about 86 V with an applied voltage of 100 V.

_____ **9.** For question 8, if C were increased to 30 μF and R were increased to 4 kΩ, the capacitor would charge twice as fast.

_____ **10.** An RC charge or discharge curve usually has the same shape, regardless of the values of R and C.

REPORT

Do not write a formal report. Write a cover sheet that has your name, class, instructor, etc., on it. Attach the graphs behind the cover sheet. Be sure the graphs are neat and well-organized and that all significant data points are labeled clearly on the graph. Also, turn in your data tables.

TABLE 30–1: C = _____, R = _____, V_A = _____

Procedure Step	Measurement	Value	Calculations
2	I_T		_______A
3	I_T	_______A	
7	T		_______s

Note: Remember that you will have three data tables when you are finished:
Table 1. For (1) 4 μF.
Table 2. For (2) 4 μF in series (2 μF total).
Table 3. For (2) 4 μF in parallel (8 μF total).

EXPERIMENT 31

ALTERNATING-CURRENT CIRCUITS: *RLC* SERIES

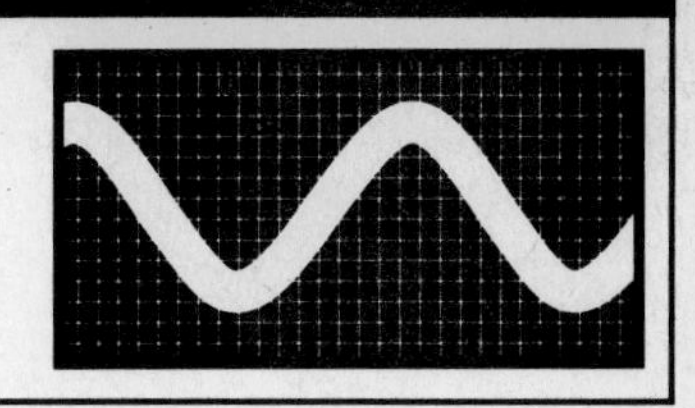

OBJECTIVES

At the completion of this experiment, you will be able to:

- Understand series reactance and resistance.
- Determine the net reactance, phase angle, and impedance of an *RLC* circuit.
- Determine the real power of an ac circuit as opposed to apparent power.

SUGGESTED READING

Chapter 24, *Basic Electronics,* B. Grob, Seventh Edition

INTRODUCTION

Both the previous experiments, *RL* series circuit and *RC* series circuit, were ac circuits. In both cases, the concept of reactance was investigated. The results showed that the reactive component had its own voltage drop that was out of phase with the resistive component, although the circuit current was the same in all parts of the circuit.

In an *RLC* series circuit, the same concepts apply. However, when X_L and X_C are both in the circuit, the opposite phase angles enable one to cancel the effect of the other. For X_L and X_C in series, the net reactance is the difference between the two series reactances, resulting in less reactance than either one alone. In parallel circuits, the branch currents cancel, resulting in less total current.

$I = 6$ A

$X_L = 60\ \Omega$

$IX_L = 360$ V

$V_T = 120$ V

$X_L = 20\ \Omega$

$X_C = 40\ \Omega$

$IX_C = 240$ V

Fig. 31-1. *LC* circuit reactances. Net reactance $= 60\ \Omega - 40\ \Omega = 20\ \Omega$.

Consider the circuit of Fig. 31-1. Notice that the circuit current is found by dividing the applied voltage by the total net reactance of the circuit. Here,

$$\frac{120\text{ V}}{20\ \Omega} = 6\text{ A}$$

The net reactance is the difference between $X_L = 60\ \Omega$ and $X_C = 40\ \Omega$. In the same manner, the difference between the two voltages is equal to the applied voltage, because the IX_L and IX_C voltages are opposite. If the values were reversed, the net reactance would be 20-Ω X_C. The current would still be 6 A, but it would have a lagging (−90°) instead of a leading (+90°) phase angle.

When resistance is added to the circuit, the total effect is determined by phasors. The phasor for the circuit of Fig. 31-1, if *R* was added, would be as shown in Fig. 31-2. Or, if X_L were greater, it would be as shown in Fig. 31-3.

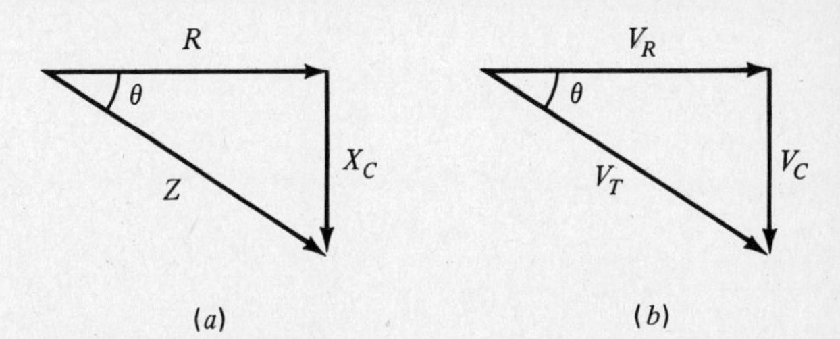

Fig. 31-2. Phasor diagrams for net capacitive reactance. (*a*) Impedance. (*b*) Voltage.

When X_L and X_C are equal, the net reactance is 0 Ω, and the result is called *resonance.* Because resonance has a specific application, it will be studied separately in later experiments.

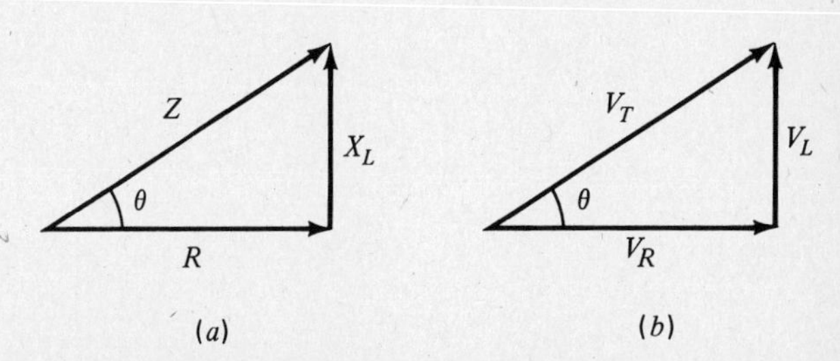

Fig. 31-3. Phasor diagrams for net inductive reactance. (*a*) Impedance. (*b*) Voltage.

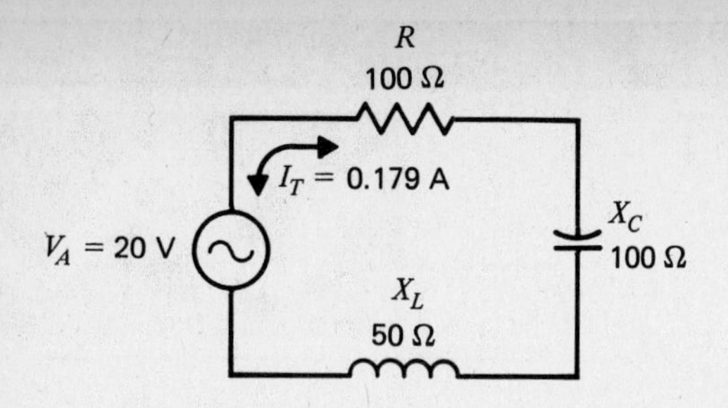

$$\text{Net reactance} = X_C - X_L = 100\ \Omega - 50\ \Omega = 50\ \Omega$$

$$\text{Apparent power} = VI = 20\ \text{V} \times 0.179\ \text{A} = 3.58\ \text{VA}$$

$$\text{Real power} = I^2R = 3.20\ \text{W}$$

Fig. 31-4. *RLC* series circuit.

The procedure that follows will investigate an *RLC* circuit where there will be a net reactance. In that case, there will be a difference between real power and apparent power. Consider the example of circuit Fig. 31-4. With *V* and *I* out of phase because of reactance, the product of the applied voltage and the circuit current is *apparent power,* and the unit is the voltampere (VA). However, the *real power* is the resistive power, dissipated as heat, and can always be found as I^2R. Finally, the ratio of real power over apparent power is called the *power factor* and is equal to the cosine of the phase angle of the circuit. The power factor is used commercially to bring about efficient distribution of energy in power lines.

EQUIPMENT

Signal generator
Oscilloscope
VTVM
Leads
Protoboard

COMPONENTS

(1) 1-kΩ 0.5 W resistor
(1) 0.01-μF capacitor
(1) 33-mH inductor

PROCEDURE

1. Connect the circuit of Fig. 31-5.
2. Use an oscilloscope. Measure and record in Table 31-1 the peak-to-peak voltage across the capacitor (V_C), the inductor (V_L), and the resistor (V_R).

Note: Avoid ground loops when measuring voltage with the oscilloscope by moving the resistor and inductor to keep the same ground point.

3. Calculate and record X_L and X_C in Table 31-1.
4. Calculate and record the circuit current I_T for V_R/R, V_C/X_C, and V_L/X_L.
5. Calculate and record in Table 31-1 the net reactance and the impedance *Z*. On a separate sheet, draw a phasor diagram for these values and determine the phase angle.
6. Determine the apparent power, real power, and power factor and record in Table 31-1.
7. Decrease the frequency to 100 Hz and repeat steps 1 to 6. Record in Table 31-1.

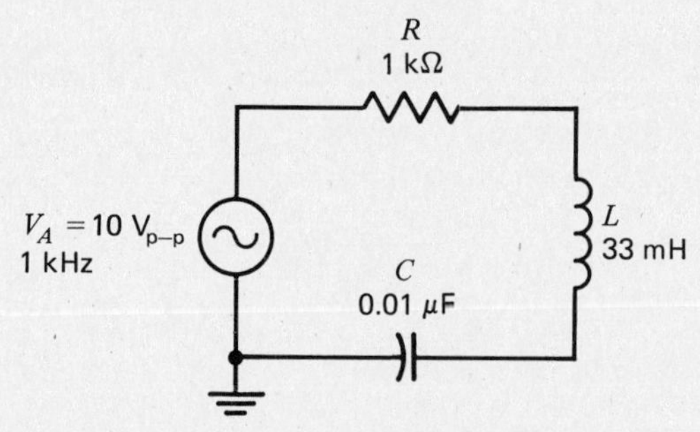

Fig. 31-5. *RLC* series circuit for measuring net reactance.

OPTIONAL PROCEDURE

Use different values of *R*, *L*, and *C*. Record the values and repeat steps 1 to 7. Also, try different frequencies: 10 kHz, 100 kHz, and 1 MHz. Use a separate sheet to record and discuss the results.

NAME DATE

RESULTS FOR EXPERIMENT 31

QUESTIONS

1. In an *RLC* series circuit, why can an inductor have more measured voltage than the applied voltage?

2. If X_L and X_C were zero, what would the phase angle of Fig. 31-5 be?

3. Explain the difference between real power and apparent power.

4. If an *RLC* series circuit had four inductors, five resistors, and seven capacitors, all in series, would the value of current be the same in all parts of the circuit?

5. At what frequency would the circuit of Fig. 31-5 (as shown) have the opposite net reactance?

REPORTS

Write a complete report. Discuss the results. Discuss the three most significant aspects of the experiment and write a conclusion.

TABLE 31–1

Procedure Step	Circuit Component	Steps 2–6: Value at $f = 1$ kHz	Step 7: Value at $f = 100$ Hz
2	V_C Measured	______	______
	V_L Measured	______	______
	V_R Measured	______	______
3	X_L Calculated	______	______
	X_C Calculated	______	______
4	$I_T = V_R/R$	______	______
	$I_T = V_C/X_C$	______	______
	$I_T = V_L/X_L$	______	______
5	X_o Calculated (Net Reactance)	______	______
	Z Calculated	______	______
6	Apparent Power	______	______
	Real Power	______	______
	Power Factor	______	______

EXPERIMENT 32

SUPERPOSING ALTERNATING CURRENT ON DIRECT CURRENT

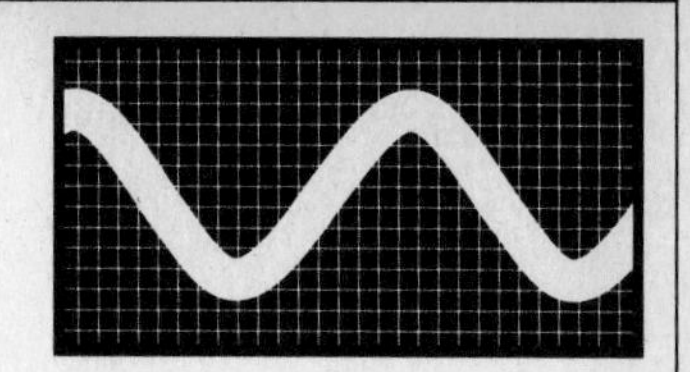

OBJECTIVES

At the completion of this experiment, you will be able to:

- Investigate the effective ac resistance of a power supply.
- Study the function of a bypass capacitor and a coupling capacitor.
- Develop a method for measuring both ac and dc components in the same circuit.

SUGGESTED READING

Chapter 24, *Basic Electronics*, B. Grob, Seventh Edition

INTRODUCTION

Applied voltage will be dropped proportionally across individual resistances, depending upon their ohmic value. The voltage distribution can be solved as shown in Fig. 32-1*a* and *b*.

If a circuit has both an ac and a dc voltage source, as shown in Fig. 32-2, there will be both an ac and a dc voltage distribution. The voltage distribution is solved in the same manner as in Fig. 32-1.

If a bypass capacitor is connected across one of the series resistors, the capacitor can act as a short circuit to the ac component. Only a negligible amount of reactance will be created as a result of the chosen capacitor size and the frequency of the ac signal. The capacitor

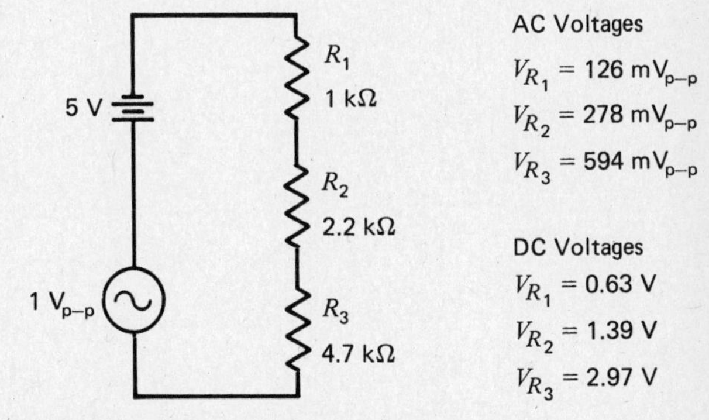

Fig. 32-2. AC and dc resistive circuit.

will become a parallel path for the ac component while blocking the dc component, as shown in Fig. 32-3. Note that the dc component is restricted to R_3 (4.7 kΩ) for a current path. The ac signal will make use of the capacitive path, because the charging and discharging action of the capacitor, in this case, creates only 3.39 Ω of reactance.

Another method of superposing an ac signal onto a dc voltage is through capacitive coupling. In Fig. 32-4, dc voltages are developed across R_1, R_2, and R_3 but are isolated from the signal generator by the coupling capacitor C_1. The dc resistance of the power supply can be considered approximately 0 Ω. The ac signal is applied to the parallel combination of $R_2 + R_3$ and R_1.

Note: When measuring a dc component while an ac component is present, the oscilloscope will display the ac component at the level of the dc component. This

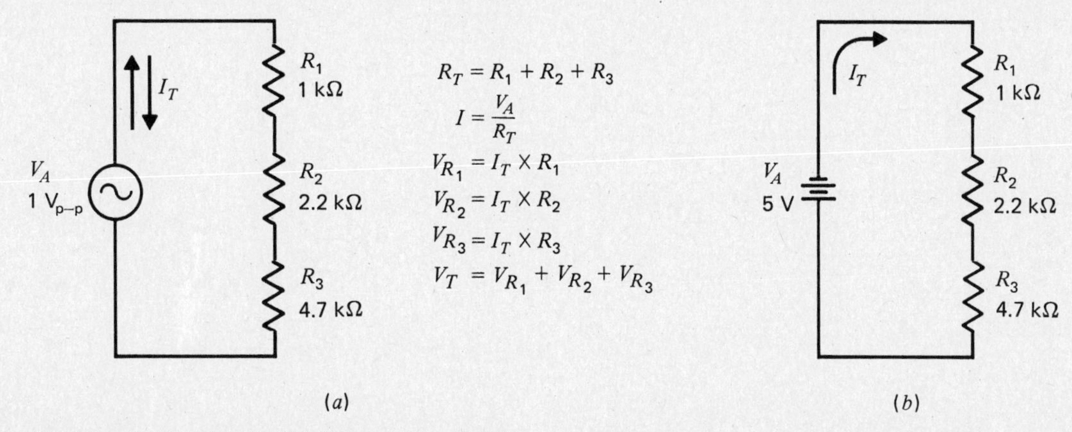

Fig. 32-1. (*a*) AC resistive circuit. (*b*) DC resistive circuit.

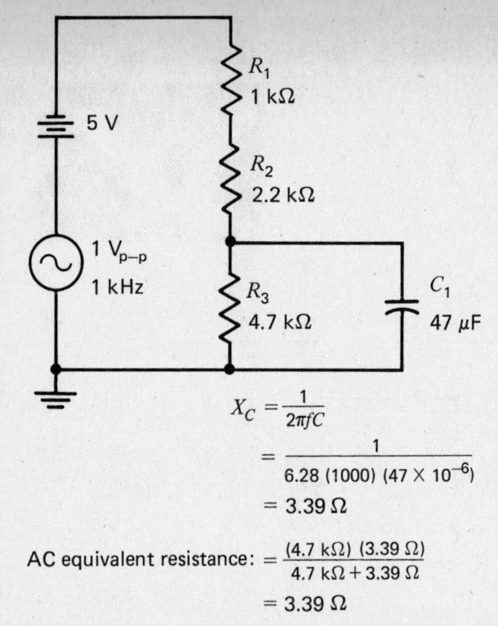

Fig. 32-3. Bypassing for alternating current.

means that a sine wave will appear above or below the zero reference when the input coupling is switched to measure direct current. It will be necessary to develop a technique to read the dc component. Therefore, the sine-wave zero (X axis) value must be determined in relation to the dc zero reference. This will require adjusting the oscilloscope for each reading. Be sure that you develop this technique. Refer to Fig. 32-5.

EQUIPMENT

Oscilloscope
Audio-frequency signal generator
VTVM
Protoboard or springboard
Test leads

COMPONENTS

Resistors (all 0.5 W):

(1) 1 kΩ (1) 4.7 kΩ
(1) 2.2 kΩ

(1) 47-μF electrolytic capacitor

PROCEDURE

1. Connect the circuit of Fig. 32-1*a*. Measure and record the voltages across each resistance as indicated in Table 32-1.
2. Connect the circuit of Fig. 32-1*b*. Measure and record the voltages across each resistance as indicated in Table 32-1.
3. Repeat procedures 1 and 2 for the circuit shown in Fig. 32-2, and complete the information requested in Table 32-2.
4. After recording the measured voltages, calculate the same voltages in accordance with the techniques reviewed in the introduction and record in Tables 32-1 and 32-2. Compare the measured and calculated values.

Note: When measuring a dc component while an ac component is present, the oscilloscope will display the ac component at the level of the dc component. This means that a sine wave will appear above or

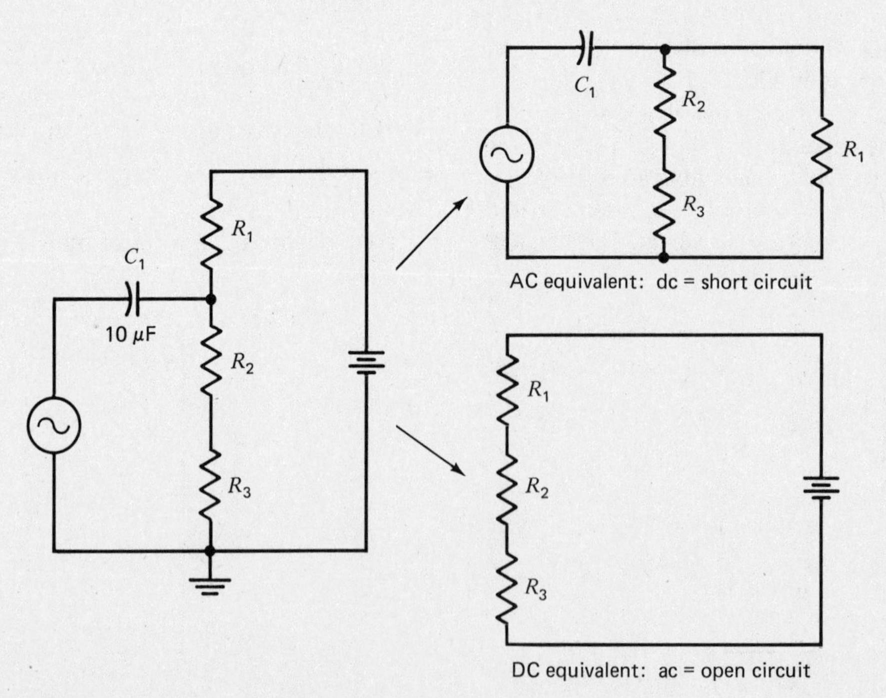

Fig. 32-4. Superposing an ac signal on a dc voltage by capacitive coupling with equivalent circuits shown.

below the zero reference when the input coupling is switched to measure direct current. It will be necessary to develop a technique to read the dc component. Therefore, the sine-wave zero (X axis) value must be determined in relation to the dc zero reference. This will require adjusting the oscilloscope for each reading. Be sure you develop this technique. Again, see Fig. 32-5.

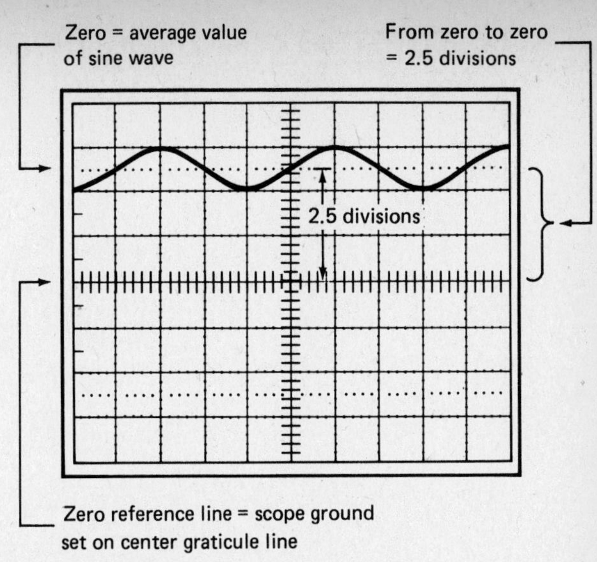

Fig. 32-5. Oscilloscope graticule display.

Optional Procedure Steps

Note: Be sure to avoid ground loops.

5. Connect the circuit of Fig. 32-3. Measure and record all ac and dc voltages. Prepare your own data table, similar to Table 32-2, and label it Table 32-3.
6. Connect the circuit of Fig. 32-4. Measure and record all ac and dc voltages. Prepare your own data table, and label it Table 32-4.

NAME DATE

RESULTS FOR EXPERIMENT 32

QUESTIONS

1. What function does the coupling capacitor perform?

2. What function does the bypassing capacitor perform?

3. Was the effective ac resistance of the power supply equal to 0 Ω?

4. Draw the ac and dc equivalent circuit diagrams for the circuit shown in Fig. 32-2.

5. Is ac resistance the same as dc resistance? Explain.

REPORT

Write a complete report. Discuss the results. Discuss the three most significant aspects of the experiment and write a conclusion. Write your own conclusion by summarizing the concept of superposing alternating current on direct current. Include your calculations in your report.

TABLE 32–1. Measured and Calculated Voltages for Fig. 32-1*a* and *b*

Procedure Step		V_{R_1}	V_{R_2}	V_{R_3}	V_T
1	AC Measured				
2	DC Measured				
4	AC Calculated				
	DC Calculated				

TABLE 32–2. Measured and Calculated Voltages for Fig. 32-2

Procedure Step		V_{R_1}	V_{R_2}	V_{R_3}	V_T
3	AC Measured	________	________	________	________
	DC Measured	________	________	________	________
4	AC Calculated	________	________	________	________
	DC Calculated	________	________	________	________

EXPERIMENT 33

SERIES RESONANCE

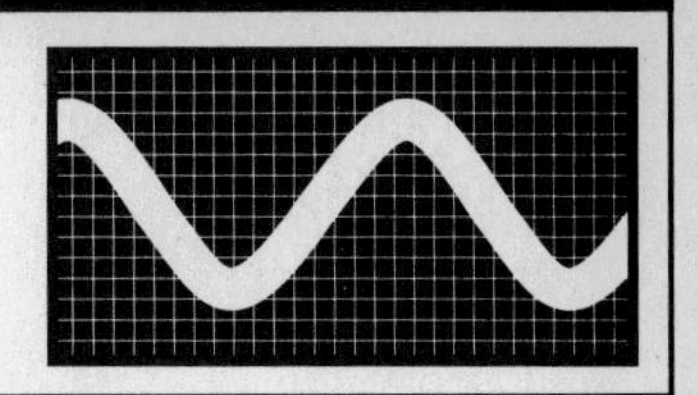

OBJECTIVES

At the completion of this experiment, you will be able to:

- Understand the concepts of series resonance.
- Validate the formula for the resonant frequency $f_r = 1/[2\pi(LC)^{1/2}]$.
- Plot a graph of frequency versus circuit current.

SUGGESTED READING

Chapter 26, *Basic Electronics,* B. Grob, Seventh Edition

INTRODUCTION

The resonant effect takes place when $X_L = X_C$ for a series inductor and capacitor. The frequency at which the opposite reactances are equal is called the *resonant frequency,* and it can be calculated as

$$f_r = \frac{1}{2\pi(LC)^{1/2}} \text{ or } f_r = \frac{1}{2\pi\sqrt{LC}}$$

Because of the canceling effect of X_L and X_c (opposite in phase), the resonant effect can only occur at one specific frequency for a given combination of L and C. In general, large values of L and C provide a relatively low resonant frequency. Smaller values of L and C provide a higher value for f_r. Figure 33-1 shows (*a*) a series resonant circuit and (*b*) its response curve.

Notice that X_L and X_C are equal at 1000 kHz. This can be proven by using the formulas for $X_C = 1/(2\pi fC)$ and $X_L = 2\pi fL$. If the frequency of the signal generator (V_T) were to change, the circuit would no longer be resonant. Although the opposite reactances (180° out of phase) would still give a canceling effect, there would still be some net reactance remaining: X_C for a decrease in frequency and X_L for an increase in frequency. In Fig. 33-1, the only opposition to current flow is the resistance r_S, which is the resistance of the coil.

The main characteristic of a series resonant circuit is that the circuit current is maximum at the resonant frequency, as shown in Fig. 33-1*b*. This is called a resonant rise in current. Also, because of the canceling of X_C and X_L, the circuit impedance is minimum at the resonant frequency. And the circuit current is in phase with the generator voltage, meaning that the circuit phase angle is 0° at resonance.

Finally, because the circuit current is maximum, the voltage across either L or C is maximum at resonance. Thus, if the output of Fig. 33-1 were taken across either L or C, the result would be maximum voltage.

In addition, the quality of a resonant circuit is determined by the sharpness of the resonant rise in voltage across L or C, called Q, where Q is calculated as X_L/r_S. The greater the ratio of the reactance to the series resistance, the higher the Q and the sharper the resonant effect. Note that the X_L reactance, rather than X_C, is used to determine Q, due to the dc resistance of the coil. Also, the greater the L/C ratio, the greater the circuit Q. Thus, increasing L and decreasing C can produce a higher Q for any given resonant frequency. Q can also be measured as $Q = V_{in}/V_{out}$, where V_{out} is equal to the voltage across either L or C and V_{in} equals the generator voltage.

Remember, the main purpose of resonant circuits is

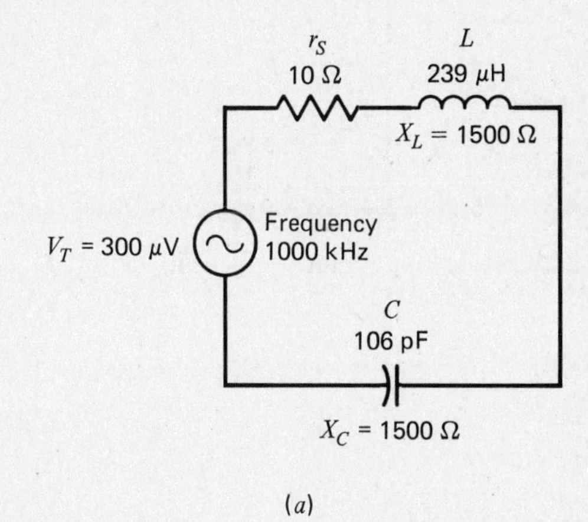

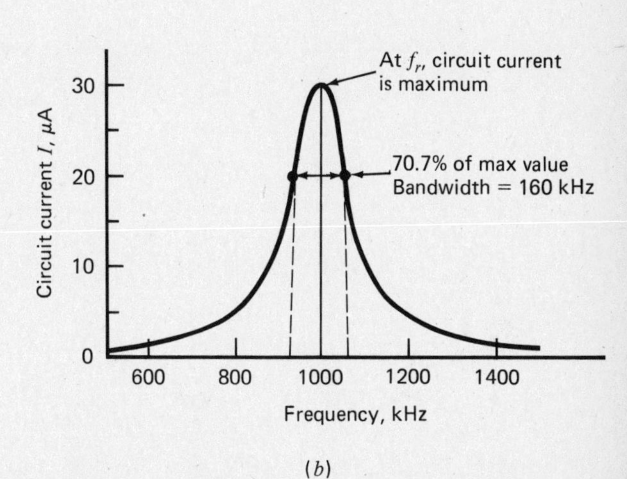

Fig. 33-1. Series resonant circuit. (*a*) Schematic diagram. (*b*) Response curve.

for tuning, that is, to tune to a desired frequency, as in radio, television, and other forms of communication instrumentation.

EQUIPMENT

Audio signal generator
Oscilloscope
Protoboard
Leads

COMPONENTS

(1) 56-Ω 0.5-W resistor
(1) 0.1-μF capacitor
(1) 33-mH inductor

PROCEDURE

1. Connect the circuit of Fig. 33-2.

Note: Avoid ground loops by moving R and L as required.

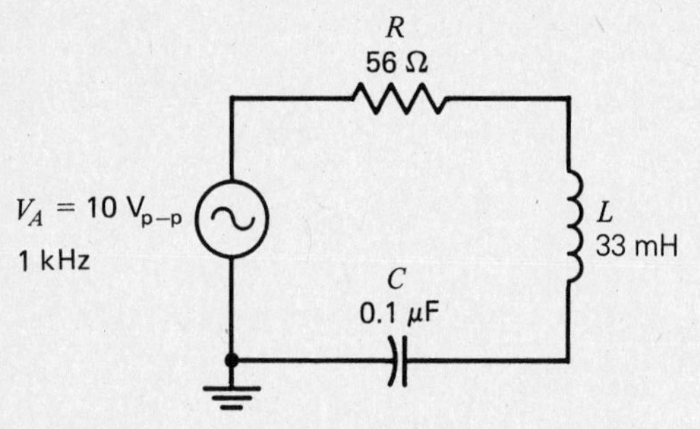

Fig. 33-2. *RLC* series circuit for measurement data.

2. Calculate and record the resonant frequency in Table 33-1.

3. Measure the voltages across R, L, and C, using an oscilloscope, and record the results in Table 33-1.

Note: For steps 4 through 8, create your own data tables and graph the results in a response curve.

4. Increase the frequency in 250-Hz steps from 1 to 5 kHz and repeat step 3 at each frequency. Be sure to keep the applied voltage constant at each step. Thus, measure V_A at each step to be sure there is no change.

Note: Knowing the calculated resonant frequency, measure and record V_R, V_L, and V_C at several (3 or 4) extra frequency points on either side of that frequency.

5. Calculate the Q of the circuit. Now determine Q by measurement, and compare both values.

6. Calculate and record the circuit current for each frequency step as $I = V_R/R$.

7. Calculate and record the values of X_L and X_C at the two edges of the bandwidth. This is the same as the two half-power points on either side of the resonant frequency (f_r), where the voltage decreases to 70.7 percent.

8. Calculate and record the circuit impedance Z and the phase angle at f_r and at both half-power points.

9. Plot a graph of frequency versus the voltages V_R, V_L, and V_C. Indicate the bandwidth (half-power points on either side of f_r), where each half-power point is 70.7 percent of I_{max}.

NAME ______________________ DATE ____________

RESULTS FOR EXPERIMENT 33

QUESTIONS

Answer TRUE (T) or FALSE (F) to the following:

_____ **1.** The resonant effect occurs when X_L is greater than X_C.

_____ **2.** The resonant frequency of a series L_C circuit must always be above 60 Hz.

_____ **3.** The circuit impedance Z of a series circuit is maximum at f_r.

_____ **4.** X_L and X_C are 90° out of phase at resonance.

_____ **5.** The highest value of series circuit current is at resonance.

_____ **6.** The greater the circuit Q, the higher the resonant frequency.

_____ **7.** $f_r = 1/[2\pi(LC)^{1/2}]$ and $Q = X_L/X_C$.

_____ **8.** The resonant frequency of a series RLC circuit with $R = 100\ \Omega$, $L = 8\ \text{H}$, and $C = 4\ \mu\text{F}$ is 28 Hz.

_____ **9.** The Q of the circuit values in question 8 is 14.

_____ **10.** The resonant frequency of the circuit values of question 8 would not change if $L = 4\ \text{H}$ and $C = 8\ \mu\text{F}$.

REPORT

Write a complete report. Discuss the results. Discuss the three most significant aspects of the experiment and write a conclusion.

TABLE 33–1

Procedure Step	Measurement	Measured Value	Calculation
2	f_r	________Hz	________Hz
3	V_R	________V at 1 kHz	
	V_L	________V at 1 kHz	
	V_C	________V at 1 kHz	

EXPERIMENT 34

PARALLEL RESONANCE

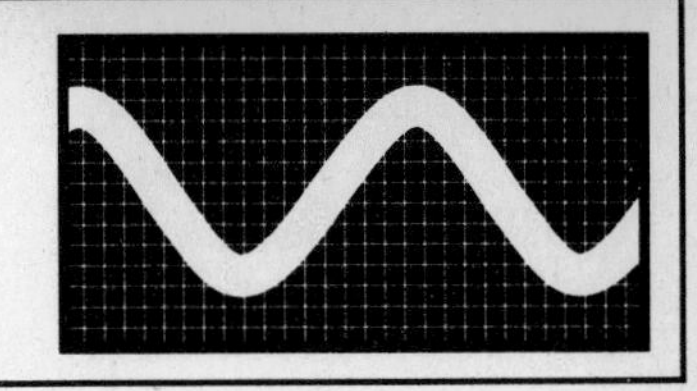

OBJECTIVES

At the completion of this experiment, you will be able to:

- Understand the concept of parallel resonance.
- Note the differences between a series and a parallel *LC* resonant circuit.
- Plot a graph of frequency versus amplitude.

SUGGESTED READING

Chapter 26, *Basic Electronics,* B. Grob, Seventh Edition

INTRODUCTION

Similar to a series resonant circuit, the resonant effect takes place in a parallel *LC* circuit, when $X_L = X_C$. The formulas for f_r and $Q = X_L/R$ are also the same. The cancellation of X_L and X_C, due to the 180° phase difference at the resonant frequency, is similar for both series and parallel circuits. However, the parallel *LC* circuit has differences due to the source being outside the parallel branches of *L* and *C*.

At resonance, the reactive branch currents cancel in the main line and produce minimum current only in the main line. Because the main-line current is minimum, its impedance is maximum at f_r. The parallel branches of *L* and *C* (also called a *tank circuit*), however, have maximum current at resonance, because they are in separate branches. In a circulating manner, the capacitor discharges into the inductor, which in turn has its field current collapse into the other side of the capacitor. This is called the *flywheel effect* and is only possible because of the ability of *L* and *C* to store energy.

Therefore, a parallel resonant circuit, or tank circuit, has both main-line and tank currents. Because the tank current is maximum at resonance, the greatest voltage drop will occur across either *L* or *C* at f_r. In the circuit of Fig. 34-1*a*, note that the resistances r_{S_1} and r_{S_2} are used to measure either line current or tank current for comparison ($V/R = I$).

Remember that in a parallel resonant circuit, low frequencies will take the path of least resistance, or the *L* branch. Likewise, high frequencies will take the path of least reactance, the *C* branch. Therefore, when X_L and X_C are equal, the tank impedance will be maximum. Thus, as a tuning circuit, the maximum voltage can be taken across the tank at the resonant frequency.

A higher circuit *Q* will result in a sharper response curve or narrower bandwidth. Because the bandwidth is determined by the half-power points (70.7 percent of I_{max}) on either side of the resonant frequency, the bandwidth is equal to f_r/Q. Thus, for f_1 and f_2 in Fig. 34-1b, the difference between the two frequencies is the bandwidth.

EQUIPMENT

Audio signal generator
Oscilloscope
Protoboard
Leads

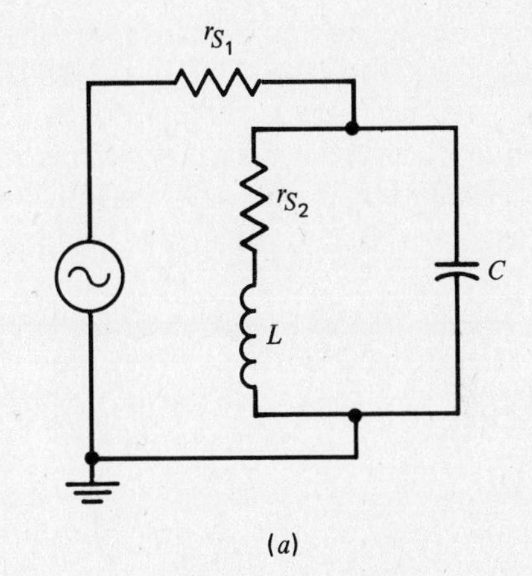

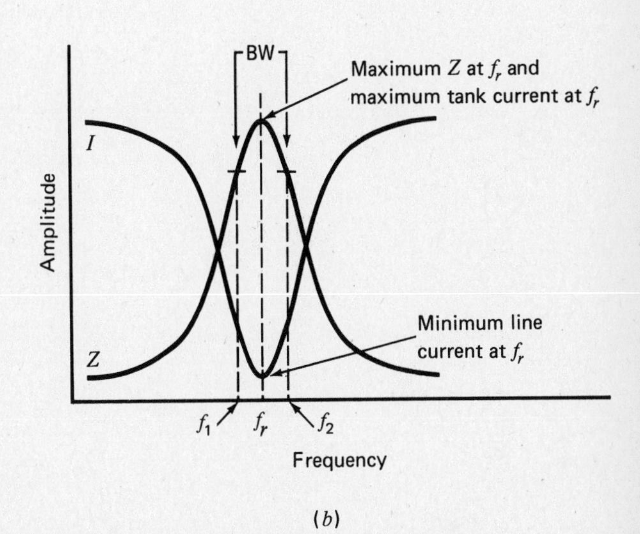

Fig. 34-1. Parallel resonant circuit. (*a*) Schematic. (*b*) Plot.

COMPONENTS

(1) 1-MΩ potentiometer (connected as rheostat)
(1) 100-Ω 0.5-W resistor
(1) 47-kΩ 0.5-W resistor
(1) 0.01-μF capacitor
(1) 100-mH inductor

PROCEDURE

1. Calculate and record in Table 34-1 the resonant frequency of the circuit of Fig. 34-2. Show your work.
2. Calculate and record in Table 34-1 the Q. Show your work.
3. Calculate and record in Table 34-1 the bandwidth (equals f_r/Q).
4. Connect the circuit of Fig. 34-2.

Note: Keep the input voltage constant for all frequencies, and avoid ground loops by moving the components as necessary.

5. Measure and record in Table 34-1 the voltage across r_{S_1}, r_{S_2}, L, and C. Calculate and record main-line and tank currents ($V/R = I$).
6. Increase the frequency in 250-Hz steps from 3500 to 6500 Hz, and repeat step 5 for each frequency step. Create your own data table to record the values.
7. Determine Z_T as follows and record the results in Table 34-1. Replace r_{S_1} (main-line resistor) with a 1-MΩ rheostat. Adjust the circuit to the resonant frequency and adjust r_{S_1} so that its voltage equals the voltage across the tank. Record the value. ***Note:*** When the voltage across the rheostat equals the voltage across the tank, its ohmic value is equal to the tank impedance. This is due to the laws of series circuits, where equal resistances or impedances divide the applied voltage equally.
8. Plot a graph of frequency versus amplitude for both tank current and main-line current. Indicate the bandwidth and any other significant points on the graph.

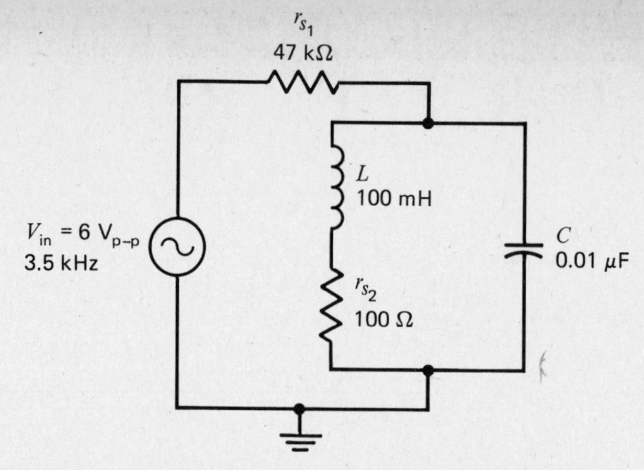

Fig. 34-2. Parallel resonant circuit for measurement.

NAME ______________________ DATE ______________

RESULTS FOR EXPERIMENT 34

QUESTIONS

Answer TRUE (T) or FALSE (F) to the following:

_____ **1.** The formula for resonance is $1/(2\pi fC)$.

_____ **2.** The higher the Q, the sharper the bandwidth response.

_____ **3.** One difference between a series resonant circuit and a parallel resonant circuit is that the parallel resonant circuit has minimum impedance at f_r, while the series has maximum impedance at f_r.

_____ **4.** Greater values of L and C will result in increased bandwidth.

_____ **5.** If a tank circuit had a resonant frequency of 8 kHz with a 2-kHz bandwidth, the bandwidth could not be decreased without changing the resonant frequency.

_____ **6.** At the resonant frequency, a tank circuit requires minimum input power from the source.

_____ **7.** In a tank circuit, line current is maximum and tank current is minimum at resonance.

_____ **8.** If a voltage were applied to a tank circuit, like Fig. 34-1, no current would flow.

_____ **9.** For a tank circuit with $L = 2$ H and $C = 10\ \mu$F, the resonant frequency is approximately 36 Hz.

_____ **10.** For question 9, a resonant frequency of 3600 Hz could be obtained by increasing L to equal 200 H.

REPORTS

Write a complete report. Compare the measured and calculated values, and discuss the relationship between the tank current and the main-line current. Discuss the three most significant aspects of the experiment and write a conclusion.

TABLE 34–1

Procedure Step	Measurement	Measured Value	Calculation
1	f_r		________Hz
2	Q		________(X_L = ____ Ω)
3	BW		________Hz
5	$V_{r_{S1}}$	________$V_{p\text{-}p}$ at 3.5 kHz	
	$V_{r_{S2}}$	________$V_{p\text{-}p}$ at 3.5 kHz	
	V_L	________$V_{p\text{-}p}$ at 3.5 kHz	
	V_C	________$V_{p\text{-}p}$ at 3.5 kHz	
	I_{tank}		________A
	I_{line}		________A
7	Z_T	________Ω	

EXPERIMENT 35
FILTERS

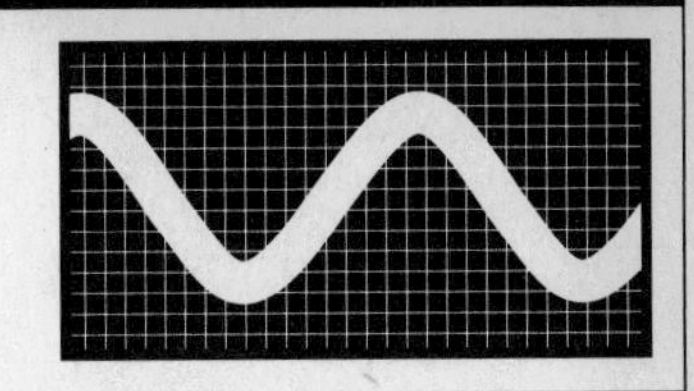

OBJECTIVES

At the completion of this experiment, you will be able to:

- Understand how filters separate different frequency components.
- Learn the difference between high-pass, low-pass, bandstop, and bandpass filters.
- Plot a graph of frequency versus amplitude for different filter types.

SUGGESTED READING

Chapter 27, *Basic Electronics,* B. Grob, Seventh Edition

INTRODUCTION

Filters are used to separate wanted from unwanted signals. For example, a radio can receive many different station broadcasts. But somehow, it must isolate the desired broadcast signal and filter out the frequencies that are not part of the broadcast. Thus, filters are used to allow only those desired frequencies to pass through certain parts of a circuit.

Inductors and capacitors are used, in various configurations, to build filters. Generally, capacitors allow high frequencies to pass, while inductors allow low frequencies to pass through them with very little reactance or opposition to current flow. Also, the manner in which a filter's components are placed determines the type of filter it is.

Low-pass filters allow low frequencies to pass through them, while higher frequencies are sent to ground. High-pass filters allow only high frequencies to pass, while blocking low frequencies by sending them to ground.

The following procedures examine four simple filter types.

EQUIPMENT

Audio signal generator
Oscilloscope
Protoboard
Leads

COMPONENTS

(1) 1-kΩ resistor
(1) 56-Ω resistor
(1) 0.01-μF capacitor
(1) 0.1-μF capacitor
(1) 10-μF capacitor
(2) 33-mH inductors

PROCEDURE

1. Connect the circuit of Fig. 35-1.

Note: Remember to keep the input voltage constant at each frequency for all steps in this procedure.

2. Increase the frequency from 1 kHz to 10 kHz in 1-kHz steps. Measure and record in Table 35-1 the peak-to-peak output voltage at each step.

3. At 10 kHz, place another 33-mH inductor across points C and D. Measure and record the output voltage.

4. At 10 kHz, remove one of the inductors and replace the 0.01-μF capacitor with a 0.1-μF capacitor. Measure and record the output voltage.

5. Connect the circuit of Fig. 35-2.

6. Decrease the frequency from 10 kHz to 100 Hz in the steps shown in Table 35-1. Measure and record the output voltage at each step.

7. Place a second capacitor of 10 μF across points A and B. Measure and record the output voltage for each step shown in Table 35-1.

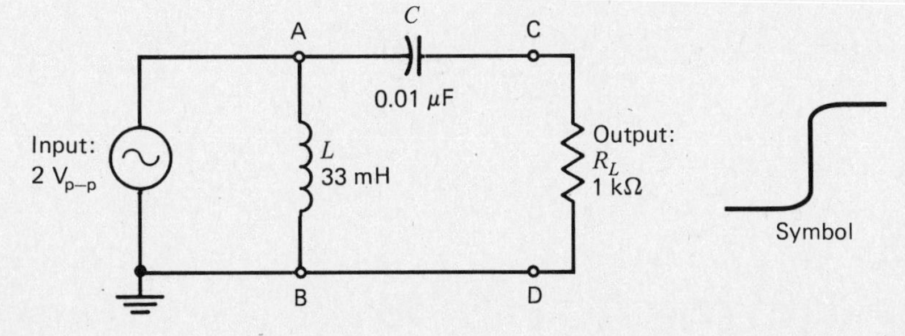

Fig. 35-1. High-pass filter.

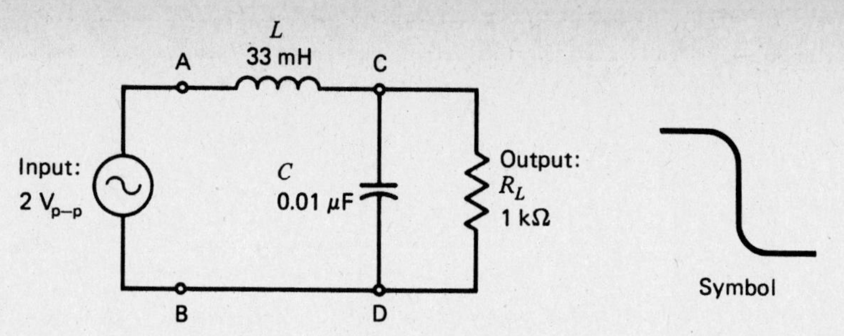

Symbol

Fig. 35-2. Low-pass filter.

8. Connect the circuit of Fig. 35-3 below.
9. Increase the frequency from 2 to 18 kHz in the steps shown in Table 35-1. Measure and record the output voltage at each step.
10. Connect the circuit of Fig. 35-4.
11. Increase the frequency from 2 to 18 kHz in the steps shown in Table 35-1. Measure and record the output voltage at each frequency step.
12. Plot a graph of frequency versus load voltage or amplitude for each of the four filter circuits above.

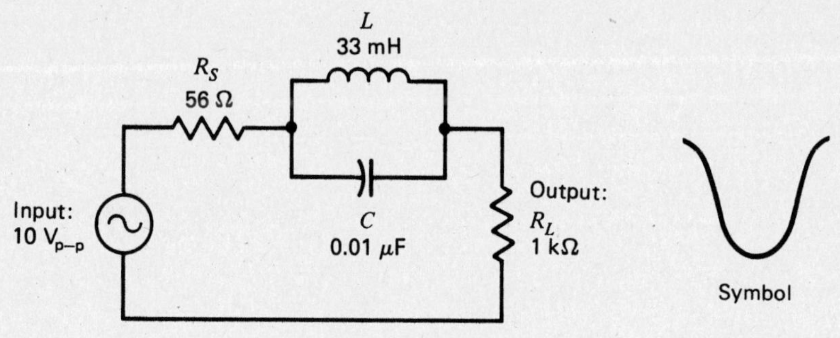

Symbol

Fig. 35-3. Bandstop filter.

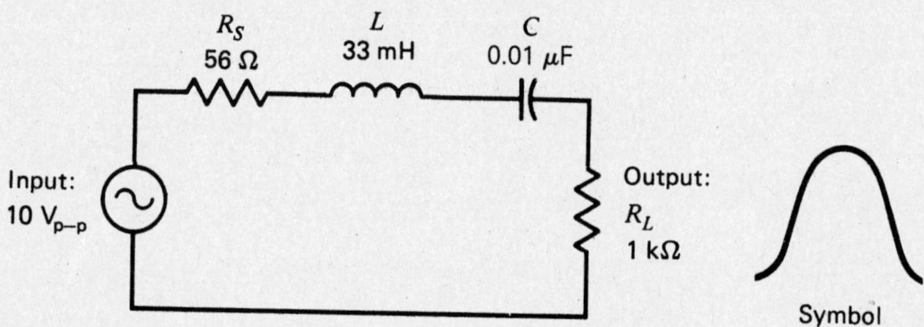

Symbol

Fig. 35-4. Bandpass filter.

NAME DATE

RESULTS FOR EXPERIMENT 35

QUESTIONS

1. Explain what happens to low frequencies in the circuit of Fig. 35-1. Why don't they reach the load R_L?

2. Explain what happens to high frequencies in the circuit of Fig. 35-2. Why don't they reach the load R_L?

3. Explain how the resonant effect works as a filter in the circuit of Fig. 35-3.

4. Explain how the resonant effect works as a filter in the circuit of Fig. 35-4.

5. Design a filter circuit that (*a*) passes frequencies from approximately 8 kHz to 12 kHz and (*b*) stops or rejects frequencies from 8 kHz to 12 kHz. Show all values and calculations, where $V_A = 1\ V_{p\text{-}p}$ and $R_L = 8\ \Omega$ (similar to a radio speaker).

REPORT

Write a complete report. Discuss the results. Discuss the three most significant aspects of the experiment and write a conclusion.

TABLE 35–1

Procedure Step	Measurement Frequency	Load Voltage, $V_{p\text{-}p}$
2	1 kHz	
	2 kHz	
	3 kHz	
	4 kHz	
	5 kHz	
	6 kHz	
	7 kHz	
	8 kHz	
	9 kHz	
	10 kHz	
3	10 kHz	
4	10 kHz	
6	10 kHz	
	9 kHz	
	8 kHz	
	7 kHz	
	6 kHz	
	5 kHz	
	4 kHz	
	3 kHz	
	2 kHz	
	1 kHz	
	500 Hz	
	200 Hz	
	100 Hz	
	50 Hz	
7	10 kHz	
	5 kHz	
	2 kHz	
	1 kHz	
	500 Hz	
	100 Hz	
	50 Hz	

TABLE 35–1 (continued)

Procedure Step	Measurement Frequency	Load Voltage, $V_{p\text{-}p}$
9	2.0 kHz	
(C = 0.01 μF)	3.0 kHz	
$F_c \approx$ 9 kH	4.0 kHz	
	5.0 kHz	
	6.0 kHz	
	6.5 kHz	
	7.0 kHz	
	7.5 kHz	
	8.0 kHz	
	8.5 kHz	
	9.0 kHz	
	9.5 kHz	
	10.0 kHz	
	12.0 kHz	
	14.0 kHz	
	16.0 kHz	
	18.0 kHz	
11	2.0 kHz	
(C = 0.01 μF	3.0 kHz	
$F_o \approx$ 9 kHz)	4.0 kHz	
	5.0 kHz	
	6.0 kHz	
	7.0 kHz	
	8.0 kHz	
	8.5 kHz	
	9.0 kHz	
	9.5 kHz	
	10.0 kHz	
	11.0 kHz	
	12.0 kHz	
	13.0 kHz	
	14.0 kHz	
	15.0 kHz	
	18.0 kHz	

EXPERIMENT 36
PN JUNCTION

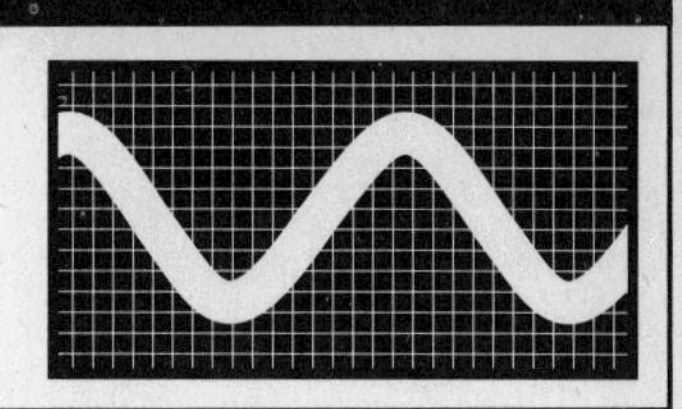

OBJECTIVES

At the completion of this experiment, you will be able to:

- Determine the forward-bias conditions of a diode.
- Determine the reverse-bias conditions of a diode.
- Develop the characteristic curve for a diode.

SUGGESTED READING

Chapter 28, *Basic Electronics,* B. Grob, Seventh Edition

INTRODUCTION

This experiment will introduce the characteristics of a silicon diode. The diode will be ohmmeter-tested and then connected to a dc power supply and its operating characteristics graphed.

Basically, a 1N4004 diode will conduct in the forward direction (forward bias) with a low value of internal forward resistance. In the reverse-bias condition, the diode will not conduct and has an almost infinite resistance. However, under extreme conditions of high voltage or current, in either direction, the diode will break down and conduct or else be destroyed.

A manufacturer's specification sheet for this diode has been included in Appendix A. Note the ratings and characteristics for these diodes. These terms should be consistent with the explanations found in most theory textbooks.

EQUIPMENT

Oscilloscope
VTVM
DC power supply, 0–10 V
DC power supply, 0–150 V
Protoboard or springboard
Test leads

COMPONENTS

(1) 1N4004
Resistors (all 0.5 W):

(1) 330 Ω	(1) 12 kΩ
(1) 680 Ω	(1) 1.0 MΩ

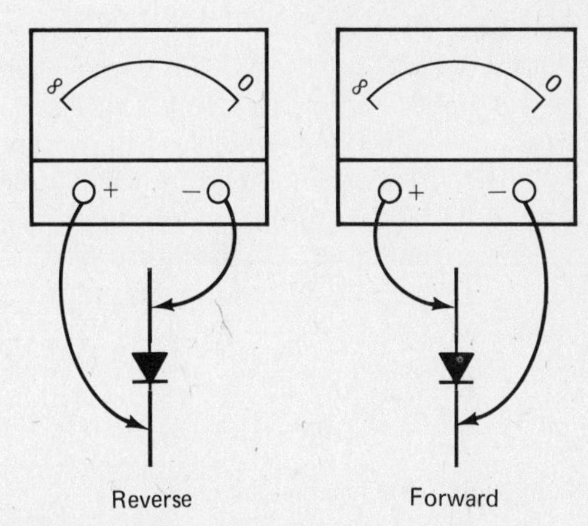

Fig. 36-1. Circuit connection for the measurement of forward and reverse resistance.

PROCEDURE

1. With the VTVM on the ohmmeter function, measure and record in Table 36-1 the forward and reverse resistance of the diode. See Fig. 36-1.
2. Connect the circuit of Fig. 36-2, with R_{L_1} equal to 12 kΩ. By use of the oscilloscope, measure and record in Table 36-1 the voltage of the power supply (V_T), voltage across the diode (V_D), and the voltage across the load resistor (V_{R_1}). Also, with the oscilloscope across the diode, slowly increase the power supply from 0 to 4 V and note that the voltage across the diode should increase to approximately 0.65 V and remain constant regardless of any further increase in supply voltage.
3. Repeat step 2 for a load resistance of 680 Ω. Be sure to record this information in Table 36-1.
4. Reverse the power supply leads so that the polarity is reversed for the circuit of Fig. 36-2. The diode is now reversed-biased. For both values of load resistance (12 kΩ and 680 Ω), measure and record in Table 36-1 the values of V_T, V_{R_1}, and V_D.

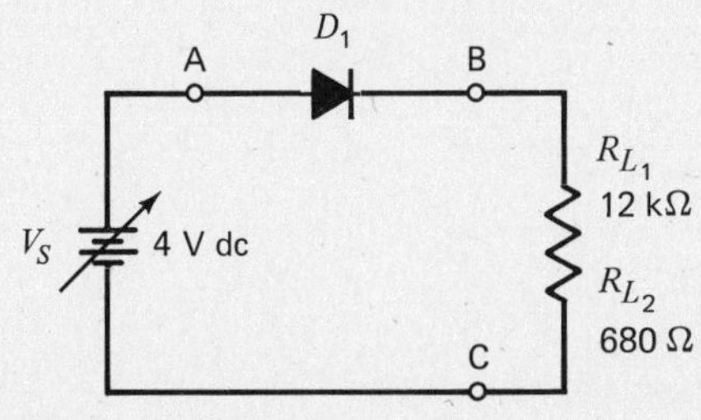

Fig. 36-2. Forward-biasing the diode.

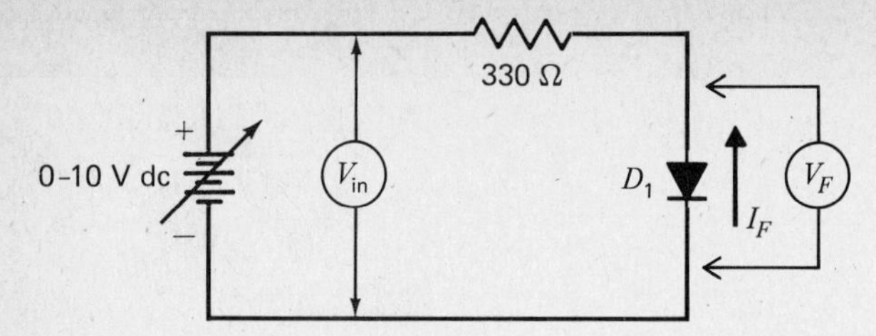

Fig. 36-3. Forward-bias condition.

5. Connect the forward-biasing circuit of Fig. 36-3.
6. Using the VTVM, measure and record in Table 36-1 the resistance value of the 330-Ω resistor.
7. Turn on the power supply and adjust the applied voltage to 0.2 V. The forward-biasing voltage measure across the diode is given the symbol V_F. Measure and record in Table 36-2 the value of V_F. Calculate and record the forward-biasing current I_F by the relationship of

$$I_F = \frac{V_F}{R}$$

8. Repeat step 7 for the V_T values of:

0.25 V	0.65 V
0.30 V	0.70 V
0.35 V	0.75 V
0.40 V	0.80 V
0.45 V	0.85 V
0.50 V	0.90 V
0.55 V	0.95 V
0.60 V	1.00 V

9. Using the data obtained in steps 7 and 8, plot the forward characteristics of this diode in Fig. 36-4 (quadrant 1).

Note: If desired, draw your own plot similar to Fig. 36-4 to hand in with your report.

10. Connect the circuit shown in Fig. 36-5.
11. Increase the power supply voltage from 0 to 150 V in 50-V steps. This voltage will reverse-bias the diode. The reverse-bias voltage measured across

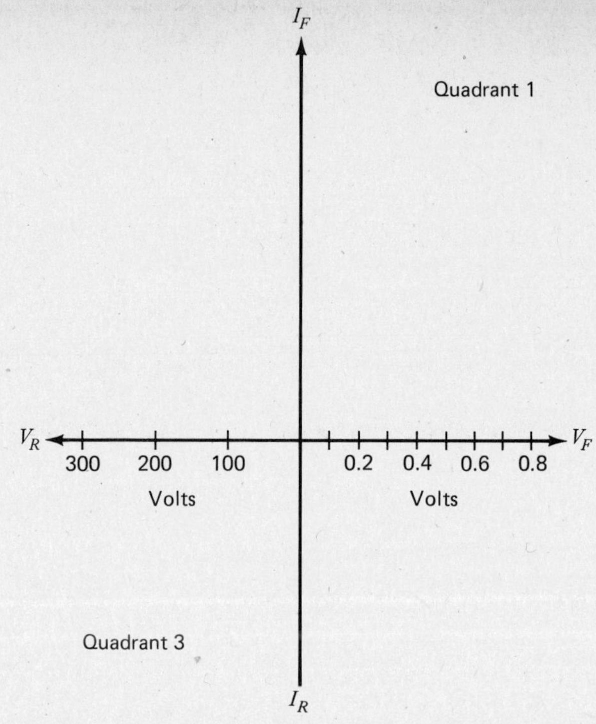

Fig. 36-4. Silicon diodes' forward and reverse characteristic curves.

the diode is given the symbol V_R. Measure and record in Table 36-3 the corresponding voltage dropped across R_1.
12. Calculate and record the reverse current I_R for each voltage level applied in step 11.
13. Plot I_R versus V_R on Fig. 36-4 (quadrant 3).

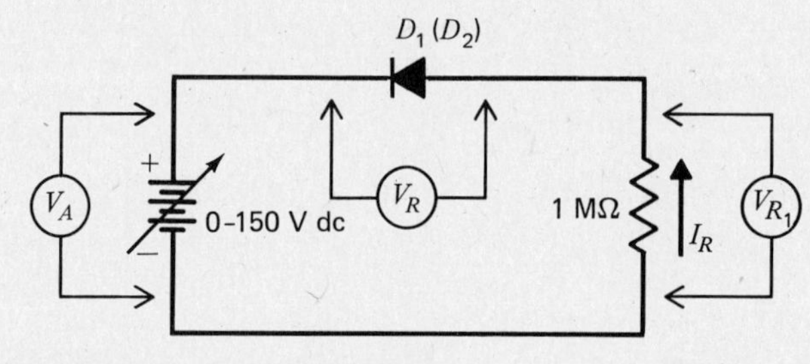

Fig. 36-5. Reverse-bias condition.

NAME ____________________ DATE ____________

RESULTS FOR EXPERIMENT 36

QUESTIONS

1. What general statement could be made for the forward and reverse resistances of a good diode?

2. Under what conditions can the diode be forward-biased?

3. Under what conditions can the diode be reversed-biased?

4. Describe the significance of the curve found in quadrant 1 of Fig. 36-4.

5. Describe the significance of the curve found in quadrant 3 of Fig. 36-4.

REPORT

Write a complete report. Discuss the results. Discuss the three most significant aspects of the experiment and write a conclusion.

TABLE 36–1

Procedure Step		R_L	V_T	V_D	V_R
1	Fwd. bias R = ______Ω				
	Rev. bias R = ______Ω				
2	Fwd. bias	12 kΩ	______	______	______
3	Fwd. bias	680 Ω	______	______	______
4	Rev. bias	12 kΩ	______	______	______
		680 Ω	______	______	______
6	330 Ω R (nominal)				
	= ______Ω (measured)				

TABLE 36–2

Procedure Step	V_T	V_F	Calculated I_F
7	0.20 V	________	________
8	0.25 V	________	________
	0.30 V	________	________
	0.35 V	________	________
	0.40 V	________	________
	0.45 V	________	________
	0.50 V	________	________
	0.55 V	________	________
	0.60 V	________	________
	0.65 V	________	________
	0.70 V	________	________
	0.75 V	________	________
	0.80 V	________	________
	0.85 V	________	________
	0.90 V	________	________
	0.95 V	________	________
	1.00 V	________	________

TABLE 36–3

Procedure Step	V_T	V_D	Calculated I_R
11 and	50 V	________	________
12	100 V	________	________
	150 V	________	________

EXPERIMENT 37

DIODE CHARACTERISTICS—RECTIFICATION

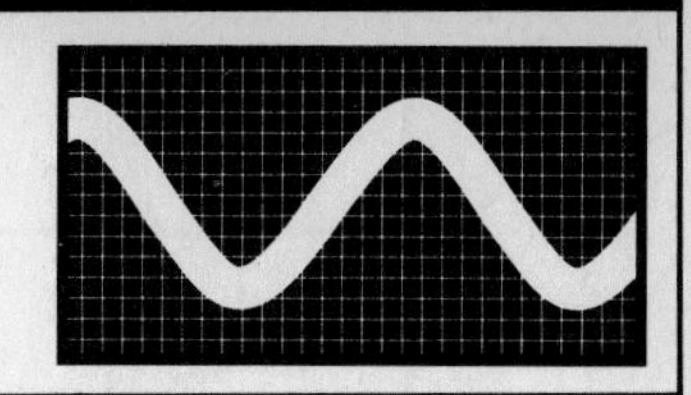

OBJECTIVES

At the completion of this experiment, you will be able to:

- Construct a half-wave rectifier circuit.
- Construct a full-wave rectifier circuit.
- Compare a bridge circuit configuration with a center-tap circuit configuration.

SUGGESTED READING

Chapters 28 and 29, *Basic Electronics,* B. Grob, Seventh Edition

INTRODUCTION

Only a battery can supply direct current to a circuit. However, ac voltage can be rectified (turned into dc). For practical purposes, rectified ac voltage will always contain some amount of the ac component, regardless of circuit design.

At this point, the diode has been forward-biased and reversed-biased under dc bias conditions. Therefore, the characteristics under ideal circumstances have been established. The following procedure will use ac voltages in rectifier circuits. The diode characteristics are similiar, while the results are different.

EQUIPMENT

Oscilloscope
1:1 probe
VTVM
120:12.6-V center-tapped @ 6.3-V filament transformer
Protoboard or springboard
Test leads

COMPONENTS

(4) 1N4004 silicon diodes
Resistors (all 0.5 W):

(2) 560 Ω (1) 12 kΩ

PROCEDURE

1. Connect the circuit of Fig. 37-1.

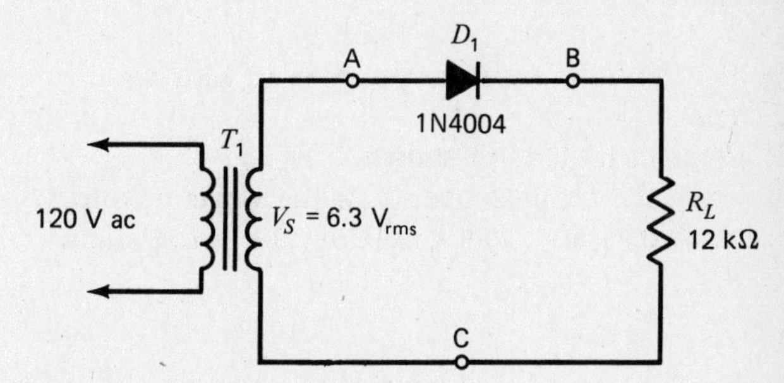

Fig. 37-1. Half-wave rectifier circuit.

2. Using the oscilloscope, draw the waveforms from points A to C and from B to C on the two graticule displays in Fig. 37-5 (at the end of this experiment). On these graticule displays be sure to include the peak-to-peak voltages read on the oscilloscope. Also include the load resistor voltage read with the VTVM.

3. Connect the circuit shown in Fig. 37-2.

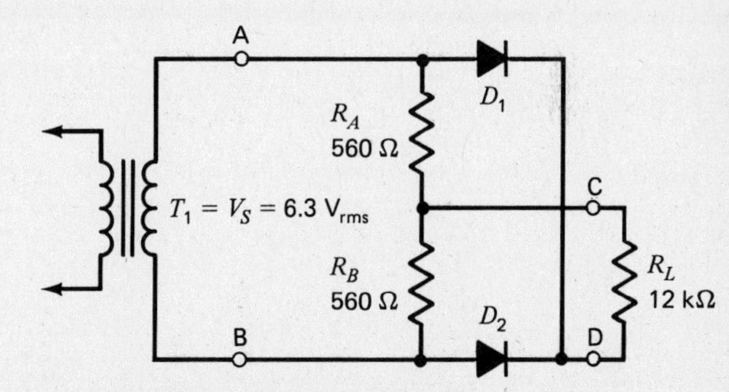

Fig. 37-2. Full-wave rectifier circuit.

4. Using the oscilloscope draw the waveforms from points A to B and from C to D on the two graticule displays in Fig. 37-6 (at the end of this experiment). On these graticule displays be sure to include the peak-to-peak voltages read on the oscilloscope. Also include the load resistor voltage read with the VTVM.

5. Connect the circuit shown in Fig. 37-3.

6. Using the oscilloscope, draw the waveforms from points A to B and from C to D on the two graticule displays in Fig. 37-7 (at the end of this experiment). On these graticule displays be sure to include the peak-to-peak voltages read on the oscilloscope. Also

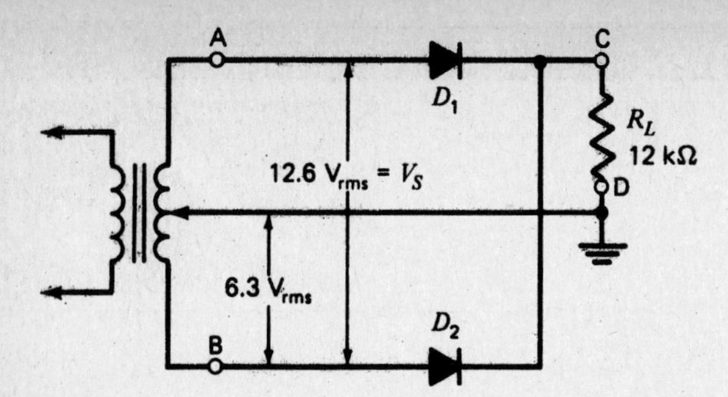

Fig. 37-3. Full-wave rectifier circuit using a center-tapped transformer.

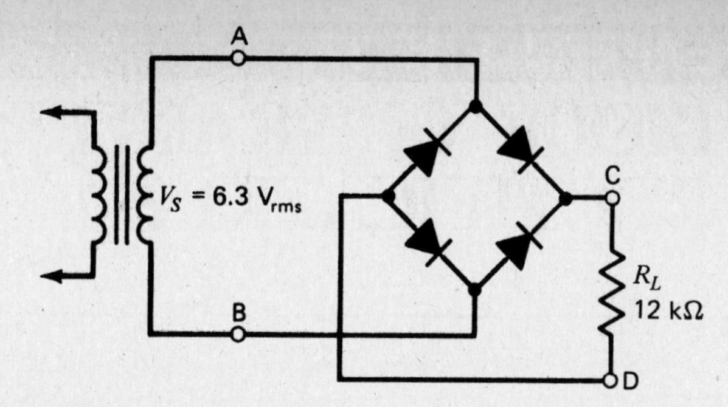

Fig. 37-4. Full-wave diode connected in a four-diode bridge arrangement.

include the load resistor voltage read with the VTVM.

7. Connect the circuit shown in Fig. 37-4.

8. Using the oscilloscope, draw the waveforms from points A to B and from C to D on the two graticule displays in Fig. 37-8 (at the end of this experiment). On these graticule displays be sure to include the peak-to-peak voltages read on the oscilloscope. Also include the load resistor voltage read with the VTVM.

NAME DATE

RESULTS FOR EXPERIMENT 37

QUESTIONS

1. How does a full-wave bridge rectifier circuit compare with a center-tap circuit?

2. Concerning each of the circuits studied, how do the graticule displays compare?

3. How does the oscilloscope input coupling affect the displayed waveform?

4. Make a general statement concerning the resistance of a PN junction diode.

5. Which of the studied circuits would be best as a dc power supply?

REPORT

Write a complete report. Discuss the results. Discuss the three most significant aspects of the experiment and write a conclusion.

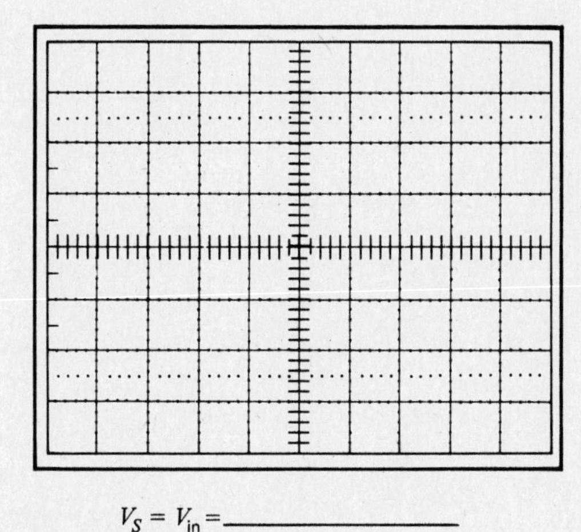

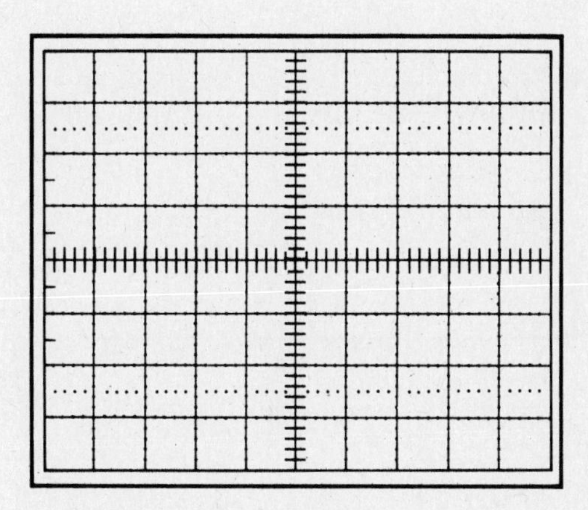

$V_S = V_{in} =$ ______________ $V_{R_L} = V_{out} =$ ______________

$V_{R_L} =$ __________(VTVM) dc value

Fig. 37-5. Graticule for step 2.

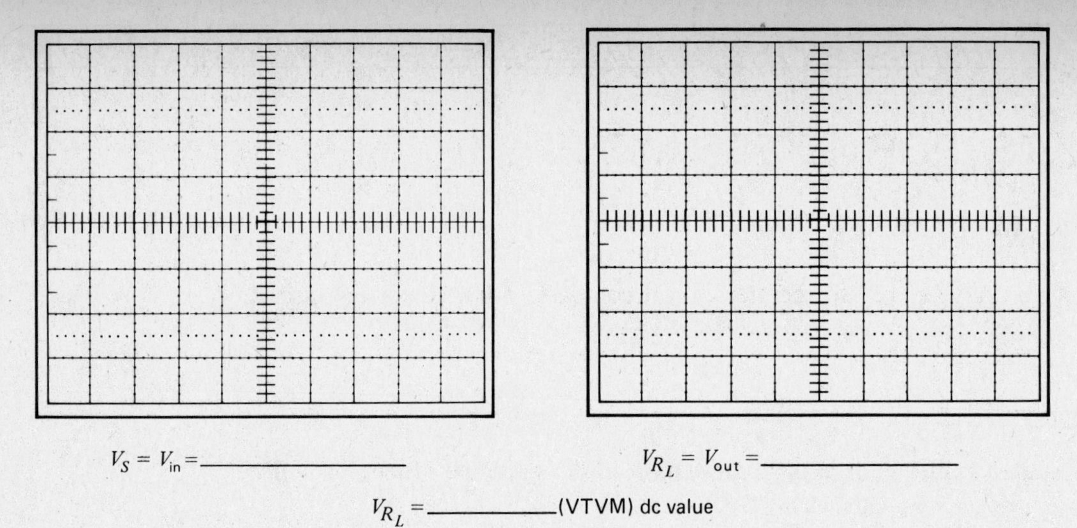

$V_S = V_{in} =$ ____________ $V_{R_L} = V_{out} =$ ____________

$V_{R_L} =$ ____________ (VTVM) dc value

Fig. 37-6. Graticule for step 4.

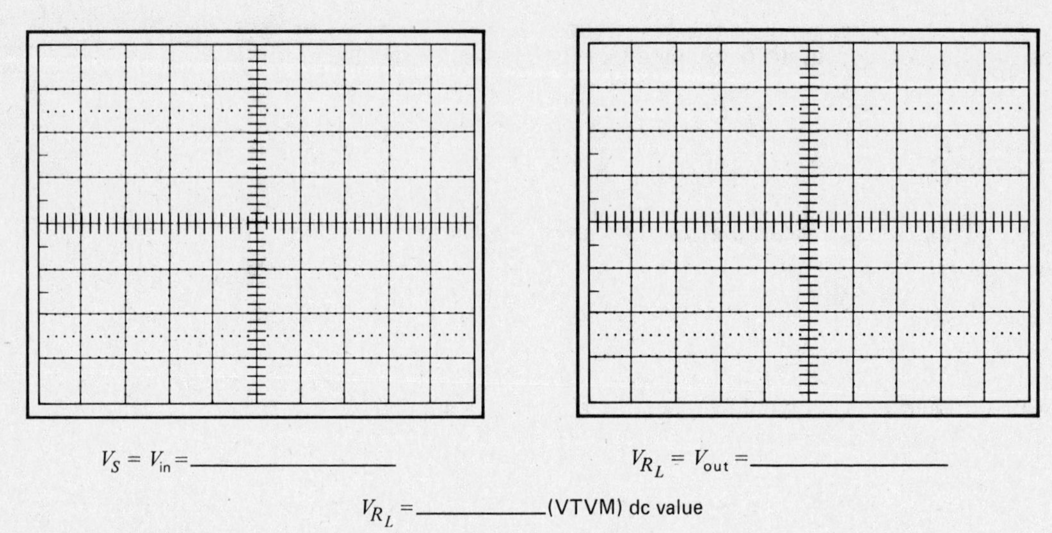

$V_S = V_{in} =$ ____________ $V_{R_L} = V_{out} =$ ____________

$V_{R_L} =$ ____________ (VTVM) dc value

Fig. 37-7. Graticule for step 6.

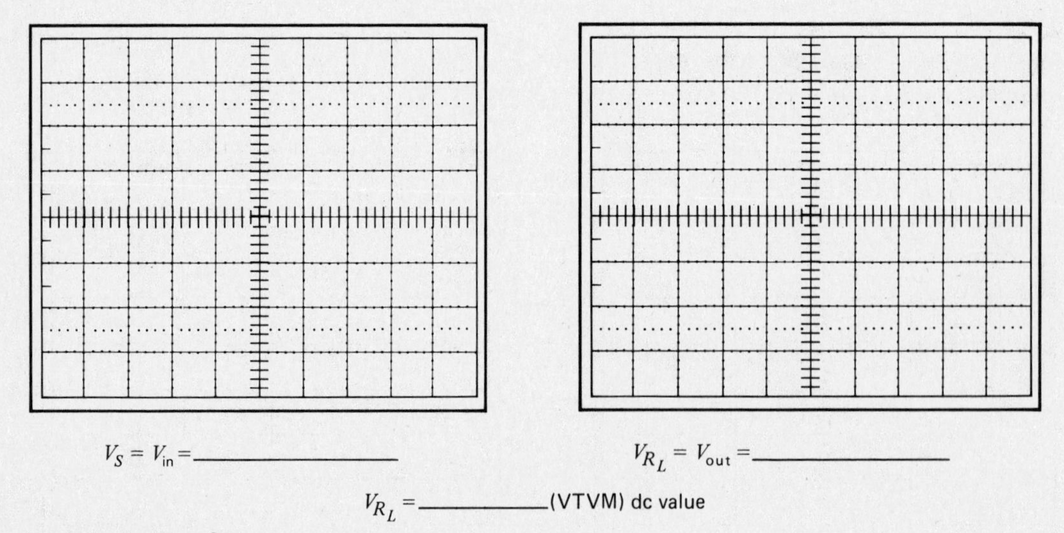

$V_S = V_{in} =$ ____________ $V_{R_L} = V_{out} =$ ____________

$V_{R_L} =$ ____________ (VTVM) dc value

Fig. 37-8. Graticule for step 8.

EXPERIMENT 38

ZENER DIODES FOR REGULATION AND PROTECTION

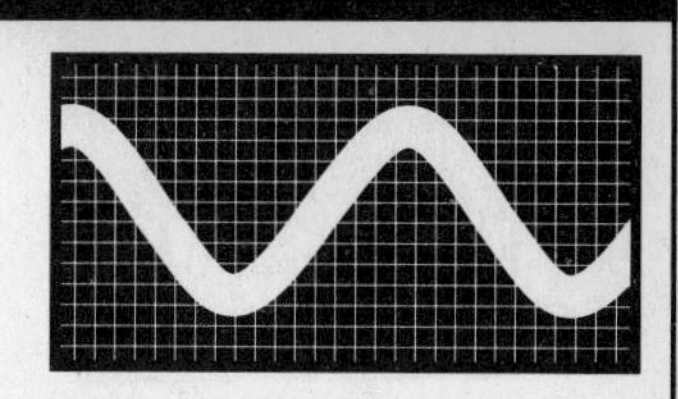

OBJECTIVES

At the completion of this experiment, you will:

- Understand how zener diodes operate.
- Be able to use a zener diode to protect circuits from excessive current or voltage.
- Be able to use a zener diode as a simple voltage regulator.

SUGGESTED READING

Chapter 28, *Basic Electronics,* B. Grob, Seventh Edition

INTRODUCTION

Zener diodes are similar to, but also different from, ordinary silicon PN junction diodes. Zener diodes are similar because both these types of diodes are made from P and N material. However, they differ because in the zener diode, the P and N material is "doped" so that it can be used in the reverse direction. Recall that a typical PN junction diode (silicon) allows current to flow in the forward direction when about 0.7 V is across it, causing forward bias.

Generally, a PN junction diode is found in a rectifier circuit where it passes only one direction of current flow from an ac signal. Current is blocked when the PN junction diode is reverse-biased, unless the reverse voltage reaches the breakdown level, which is usually around 5 V. However, the zener diode is designed to perform a different task. It is used in its reverse direction so that current will flow through it when a particular voltage reverse-biases the zener diode. Like the PN junction diode, the zener diode allows current to flow when about 0.7 V is across it in the forward direction. The characteristics of an 8-V zener diode are shown in Fig. 38-1.

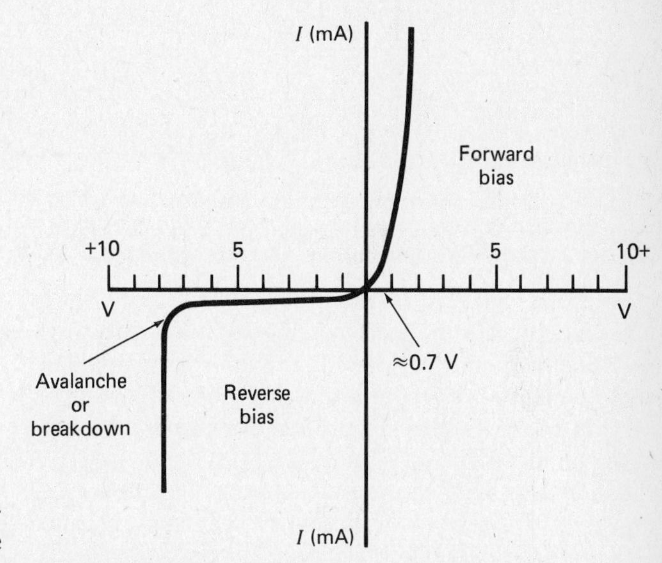

Fig. 38-1. The 8-V zener diode characteristics.

Notice that when 8 V is across the zener diode, in the reverse direction, an "avalanche" or breakdown of the PN junction occurs and the zener is then like a closed switch. Before 8 V is across the zener, only a small leakage current flows. Figure 38-2 shows the symbol for a zener diode where current will flow when 8 V is across the zener.

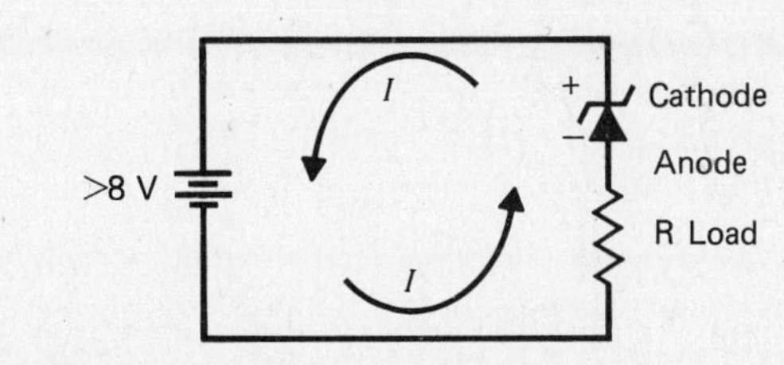

Fig. 38-2. Zener diode used as a switch.

The zener diode cathode (−) is on the positive side of the battery, and its anode (+) is on the negative side. When 8 V or more is across the diode, current will flow. Of course, the load has some resistance, which is why more than 8 V is required here.

However, zener diodes are not usually used in this configuration. They are usually put in parallel with other components requiring voltage regulation or protection. Figure 38-3 shows a zener diode used to protect a circuit which would be destroyed if too much voltage were across it.

In Fig. 38-3, the zener was selected to protect the load from too much current. The load circuit will be destroyed if the voltage across it is 15.5 V or more, resulting in too much current: $I_L = V_Z/R_L$. While V applied is less than 15 V, the zener is like an open switch and the current I_L is all flowing through the load. (See Fig. 38-3*a*.) However when V applied reaches 15 V, the zener acts as a closed switch (0 Ω) so that very little current

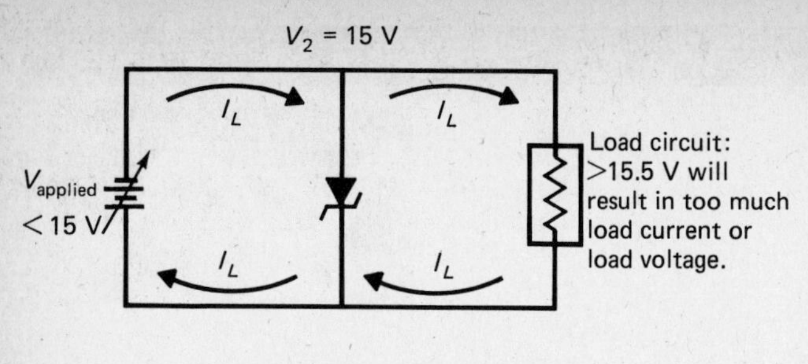

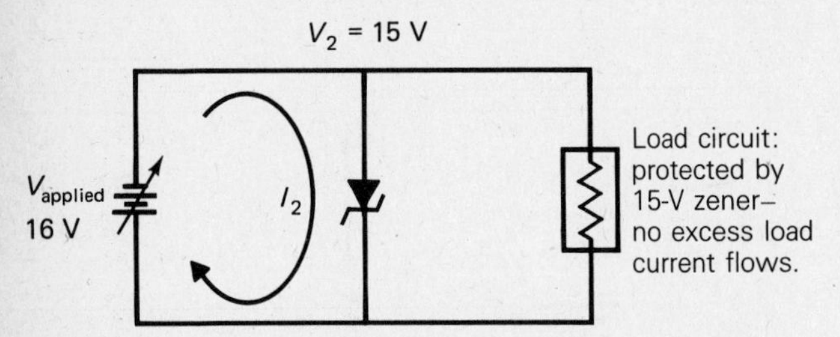

Fig. 38-3. Zener protection circuit (*a*) Zener is like an open switch—current flows through load. (*b*) Zener is like a closed switch—minimal current flows through load.

can flow through the load (see Fig. 38-3*b*). In this way, the zener protects the load circuit. In other words, the zener acts as a protection switch to keep the load protected in case there is a short that draws more current than the load can withstand.

Finally, Fig. 38-4 shows a zener diode used to keep a constant voltage across a load. This is a simple regulator where R_S limits the total circuit current and, regardless of how much current is drawn by the load, the zener diode keeps the load voltage constant. However, this circuit assumes that V applied remains high enough that the voltage across the zener allows current I_Z to flow.

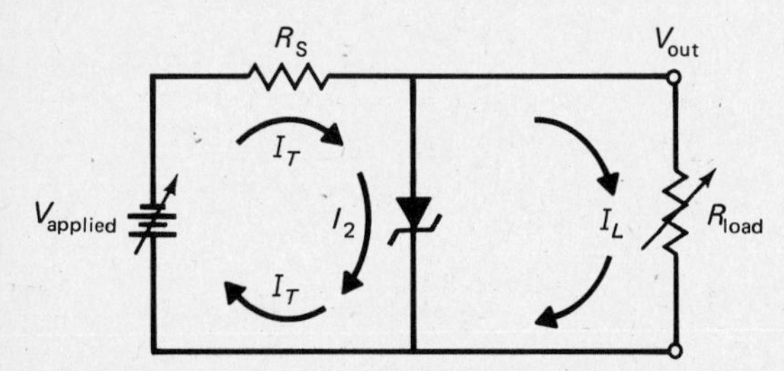

Fig. 38-4. Simple zener diode voltage regulator.

Although the zener diode protection circuit and the zener diode regulator circuit may appear similar, remember that the difference is simple. In Fig. 38-3, the protection zener is *not* biased on during normal operation—it conducts only when too much voltage is seen across it or the load. In Fig. 38-4, the zener is purposely biased on during normal operation so that it keeps a constant voltage across the load.

The following procedure will validate the basic operation of a zener diode. However, keep in mind that not all aspects of a zener diode are covered here. A zener diode, like any semiconductor device, has numerous properties and characteristics which require extensive testing to verify. This experiment will introduce you to the most widely used application of a zener diode.

EQUIPMENT

Variable dc power supply (0–30 V)
VOM, VTVM, or DVM
Ammeter

COMPONENTS

(1) 1-kΩ resistor (½-W)
(1) 330-Ω resistor (1-W)
(1) Zener diode, 5-V (1-W)
(1) 4.7 kΩ resistor (½-W)
(1) 100 Ω resistor (½-W)

PROCEDURE

1. Connect the circuit shown in Fig. 38-5.

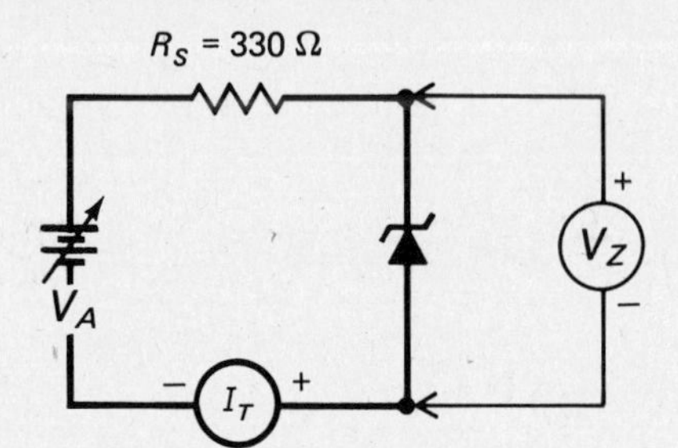

Fig. 38-5. Zener diode circuit.

2. Turn on the power supply, and slowly increase the applied voltage until 5 V appears across the zener diode. Measure and record the applied voltage at this point in Table 38-1.

3. Measure and record the current I_T in Table 38-1.

4. Measure the voltage across R_S and record the results in Table 38-1. Calculate the current through R_S and record the results next to I_T (measured) in Table 38-1.

5. Increase the applied voltage by 2 V, and repeat steps 3 and 4. Use the voltmeter to verify that V applied is 2 V greater. Be sure to measure and record V_Z and V_A.

6. Increase the applied voltage by 2 V more (4 V greater than step 2), and repeat steps 3 and 4. Be sure to measure and record V_Z.

7. Connect a 4.7-kΩ resistor across the zener diode, keeping the rest of the circuit the same as shown in Fig. 38-5.

8. Repeat steps 2 through 4 for the circuit with the 4.7-kΩ load connected. Turn off the power when you have finished.

9. Repeat steps 2 through 4 with a 100-Ω load resistor. Turn off the power when you have finished.

10. Remove any load resistors and connect the zener in the opposite polarity. Vary the power supply voltage between 0 and 8 V and measure the voltage across the zener at 1-V intervals. Also measure I_T and record the results in Table 38-2.

NAME ____________ DATE ____________

RESULTS FOR EXPERIMENT 38

QUESTIONS

1. How does a zener diode act when it is connected in the forward-bias manner?

2. How does a zener diode operate when it is connected as a voltage regulator?

3. How does a zener diode operate when it is connected as a protection device?

4. Explain the difference between a zener diode and a PN junction diode.

5. What was the purpose of R_S in the circuit in Fig. 38-5?

REPORT

Write a complete report. Describe how the zener diode acts both above and below its zener voltage.

TABLE 38–1

Step(s)	V_Z Measured, V	V Applied (V_a) Measured, V	I_T Measured, mA	I_T Calculated, mA	V_{RS} Measured, V
2–4	5	______	______	______	______
5 $V_a + 2$ V	______	______	______	______	______
6 $V_a + 4$ V	______	______	______	______	______
8 $R_L = 4.7$ kΩ	5	______	______	______	______
9 $R_L = 100$ Ω	5	______	______	______	______

TABLE 38–2. Step 10

V Applied (V_a), V	V_Z, V	I_T, mA
0		
1		
2		
3		
4		
5		
6		
7		
8		

EXPERIMENT 39
LIGHT-SENSITIVE DIODES

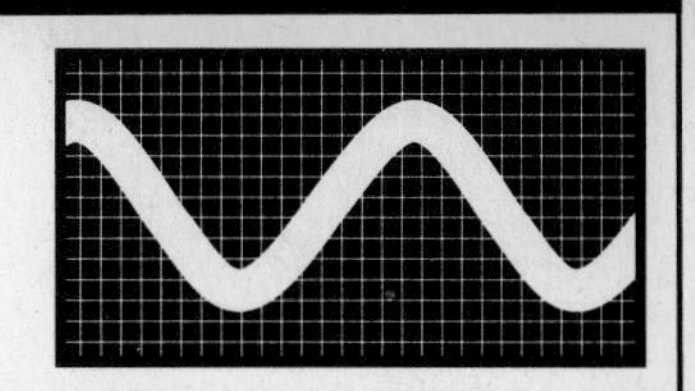

OBJECTIVES

At the completion of this experiment, you will:

- Be able to understand how light energy can be converted to electric current.
- Be familiar with the use of a photodiode.

SUGGESTED READING

Chapter 28, *Basic Electronics,* B. Grob, Seventh Edition

INTRODUCTION

Although many devices are sensitive to light, the most common types are light-sensitive diodes. These diodes, or photo-sensitive cells, have been used for years to convert or transduce energy from one form to another. For example, a photodiode converts light to electric current, and a light-emitting diode (LED) converts electric current to light.

Changes in the communications industry have created a demand for light-sensitive diodes to be used with fiber-optic cable. This cable, made of a special type of glass, is replacing copper wire in many applications throughout the world.

Instead of varying (modulating) the frequency and amplitude of an ac signal, the intensity of light is modulated and sent through this fiber-optic cable. Thus, diodes are used to convert the light to electric current so that audio, visual, and data transmission can be used in electronic systems.

Although the light in fiber-optic cable is provided by a laser, light-emitting diodes are similar because they convert, or use, electric energy to create light. Light-sensitive diodes, or photodiodes, are almost always found in the receiving end of fiber cable.

Figure 39-1 shows the typical construction and use of a light-sensitive diode. This diode, also called a solar cell, converts light to electric current. Notice that the symbol is like a battery or cell, except that the Greek letter λ ("lambda") is shown next to the cell to indicate light or light wave. Also notice that the P-type material is very thin so that the light energy can penetrate and excite the PN junction, causing electrons to flow, as the following procedure demonstrates.

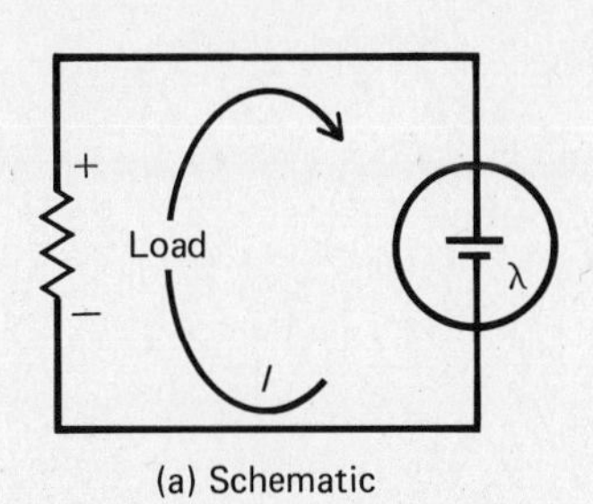

(a) Schematic

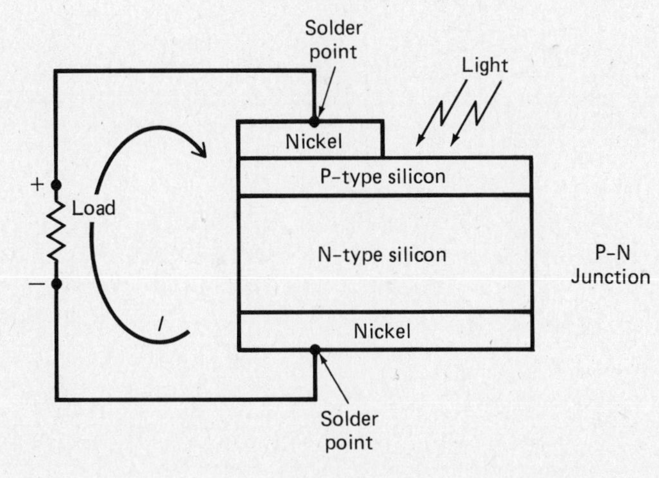

(b) Cross section

Fig. 39-1. Solar cell or light sensitive diode schematic and cross section.

EQUIPMENT

Ammeter
Voltmeter
Lamp (light source)
Leads or solder and soldering iron

COMPONENTS

(2) Solar cells (IR#S1M) or any similar PN silicon photodiodes
(1) 330-Ω $\frac{1}{2}$-W resistor
(4) Bulbs, one each, 25, 60, 100, and 150 W

PROCEDURE

1. Connect the circuit shown in Fig. 39-2, using the 150-W bulb.

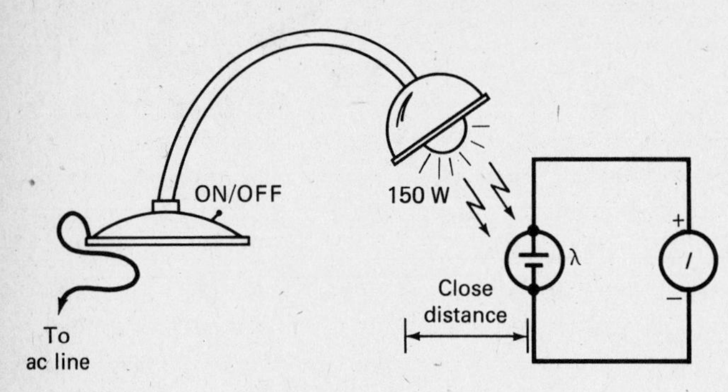

Fig. 39-2. Light sensitive circuit.

Note: Any variable light source can be used instead of four bulbs, as long as it supplies enough light (variable) to produce adequate results.

2. Position the lamp (ON) as close as possible to the diode or cell so that the maximum current flows without overheating the circuit. Record the current in Table 39-1.

3. Remove the ammeter and put a 330-Ω resistor in place of the ammeter. Measure and record in Table 39-1 the voltage across the resistor. Keep the lamp in the same position.

4. Turn off the lamp, and replace the 150-W bulb with a 100-W bulb; repeat steps 2 and 3.

5. Turn off the lamp, and replace the 100-W bulb with a 60-W bulb. Repeat steps 2 and 3.

6. Turn off the lamp, and replace the 60-W bulb with a 25-W bulb. Repeat steps 2 and 3.

7. Connect a second diode or cell in series (as close as possible) with the first cell. Repeat steps 2 to 6.

8. Connect a second diode or cell in parallel (as close as possible) to the first cell. Repeat steps 2 to 6.

NAME ______________________ DATE ______________

RESULTS FOR EXPERIMENT 39

QUESTION

Describe how these types of photocells could be used in three different applications. Describe the difference between a zener diode and a regular diode.

REPORT

No report is required. Submit all data (Table 39-1).

TABLE 39–1

Step(s)	Light Bulb or Wattage	Current *I*	Volts across 330/Ω
2 and 3	______	______	______
4	______	______	______
5	______	______	______
6	______	______	______
Two Devices in Series			
	______	______	______
7	______	______	______
	______	______	______
	______	______	______
Two Devices in Parallel			
	______	______	______
8	______	______	______
	______	______	______
	______	______	______

EXPERIMENT 40

RECTIFICATION AND FILTERS

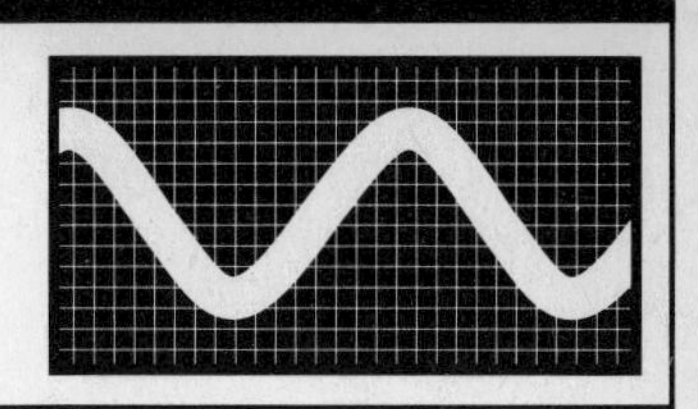

OBJECTIVES

At the completion of this experiment, you will be able to:

- Compare the average value, also called the dc value, of the half-wave and the full-wave voltages across the load resistor.
- Observe the effects of filtering on a rectified ac voltage.

SUGGESTED READING

Chapters 27, 28, and 29, *Basic Electronics,* B. Grob, Seventh Edition

INTRODUCTION

Rectified ac voltage can be filtered to resemble the unidirectional quality of battery dc voltage, although it will always contain some small amount of the ac component, regardless of filtering. This ac voltage component is called *ripple,* and it is a measure of the quality of rectification and filtering.

This experiment investigates the use of reactive components as filters for smoothing the rectified ac voltage. Capacitive reactance, inductive reactance, *RC* time constant, and circuit configuration all affect the smoothing of the ripple.

EQUIPMENT

Oscilloscope
VTVM
AC filament power supply (12.6 V, center tap at 6.3 V)
Test leads

COMPONENTS

(4) 1N4004 diodes or equivalent
(1) 1-H inductor (choke) or larger up to 8 H
(1) 47-μF capacitor
(1) 10-μF capacitor
(1) 680-Ω 0.5-W resistor
(1) 2.2-kΩ 0.5-W resistor

PROCEDURE

1. Connect the circuit shown in Fig. 40-1 without the capacitor.
2. Using an oscilloscope, measure and record in Table 40-1 the ac output voltage across R_L.

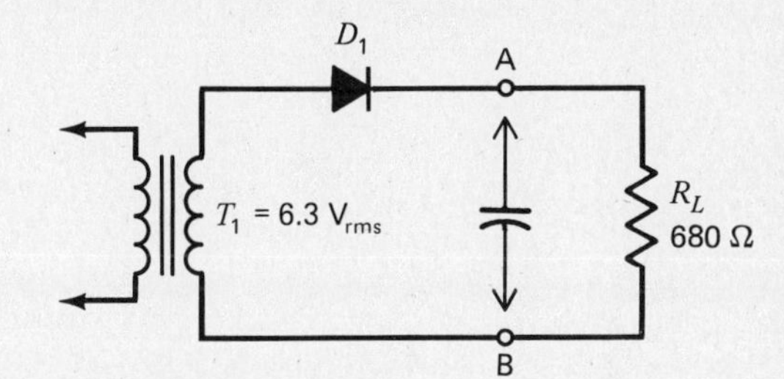

Fig. 40-1. Half-wave rectifier.

3. Using a VTVM, measure and record the dc value of the voltage across R_L.
4. Connect a 47-μF filtering capacitor across points A and B in the Fig. 40-1 circuit. Repeat steps 2 and 3 for measuring the output voltage.
5. Determine and record in Table 40-1 the percent of ripple for the filtered circuits only, using the formula

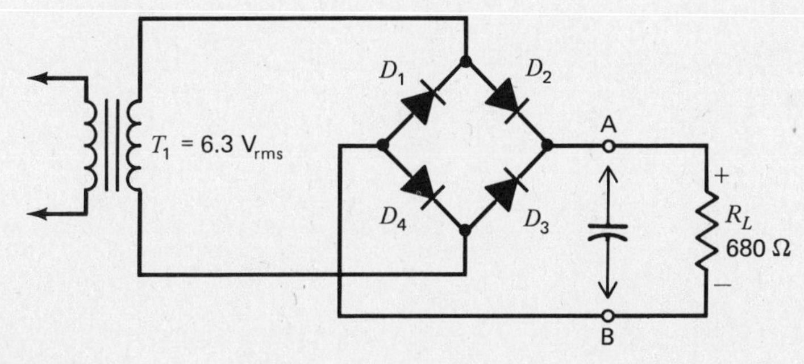

Fig. 40-2. Full-wave bridge circuit.

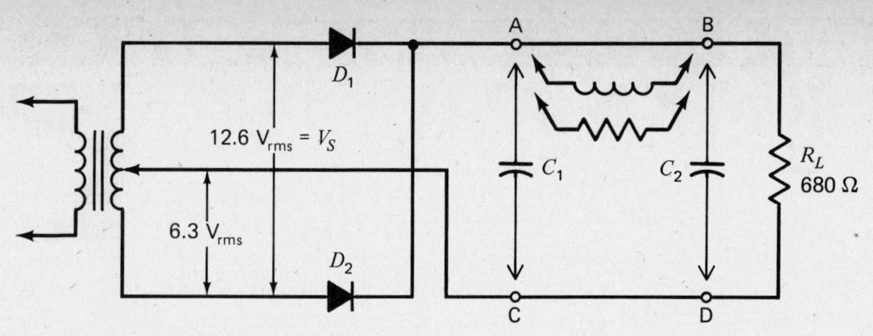

Fig. 40-3. Full-wave center-tap circuit.

$$\% \text{ ripple} = \frac{\text{rms value of ripple voltage across } R_L}{\text{dc component across } R_L} \times 100$$

6. Connect the full-wave rectification bridge circuit of Fig. 40-2.
7. Repeat steps 2 to 5 for the full-wave rectifier bridge of Fig. 40-2.
8. Connect the full-wave center-tap circuit shown in Fig. 40-3 with the load resistor only (omit the "arrowed" components).

Note: Procedures 9 to 14 refer to circuit Fig. 40-3 as altered for each specific filter configuration. Do not remove a filter component unless you are instructed to do so. The inductor is also called a "choke."

9. Using the oscilloscope, measure and record the ac output voltage across R_L.
10. Using the oscilloscope, measure the dc value of the voltage across R_L and, only if filtered, compute the percent of ripple.
11. Connect a 47-μF capacitor (C_1) across points A and C and repeat steps 9 and 10.
12. Remove the jumper between points A and B, and in its place (in series with the load), connect the choke. Repeat steps 9 and 10.
13. Connect a 10-μF capacitor (C_2) across R_L (points B and D) and repeat steps 9 and 10.
14. Replace the choke with a 2.2-kΩ resistor, leaving the capacitors in place, and repeat steps 9 and 10.

NAME ______ DATE ______

RESULTS FOR EXPERIMENT 40

QUESTIONS

1. How does a full-wave bridge rectifier compare with a center-tap circuit in this experiment?

2. Concerning each of the circuits studied in this experiment, how do each of the displayed oscilloscope waveforms compare?

3. How does the oscilloscope input coupling affect the displayed waveform?

4. Does the resistance of the PN junction diodes affect the operation of the rectified circuits studied?

5. With respect to this experiment, what would be a definition of filtration?

REPORT

Write a complete report. Discuss the results. Discuss the three most significant aspects of the experiment and write a conclusion.

TABLE 40–1

Procedure Step	Measurement	Value	Ripple Calculation
2	V_{R_L} (oscill.)	______ $V_{p\text{-}p}$	
3	V_{R_L} (VTVM)	______ V	
4	V_{R_L} (oscill.)	______ $V_{p\text{-}p}$	
	V_{R_L} (VTVM)	______ V	
5	% ripple		______ %
7	V_{R_L} (oscill.)	______ $V_{p\text{-}p}$	
	V_{R_L} (VTVM)	______ V	
	V_{R_L} (oscill.)	______ $V_{p\text{-}p}$	
	V_{R_L} (VTVM)	______ V	
	% ripple		______ %
9	V_{R_L} (ac)	______ $V_{p\text{-}p}$	
10	V_{R_L} (dc)	______ V	
11	V_{R_L} (ac)	______ $V_{p\text{-}p}$	
	V_{R_L} (dc)	______ V	______ %
12	V_{R_L} (ac)	______ $V_{p\text{-}p}$	
	V_{R_L} (dc)	______ V	______ %
13	V_{R_L} (ac)	______ $V_{p\text{-}p}$	
	V_{R_L} (dc)	______ V	______ %
14	V_{R_L} (ac)	______ $V_{p\text{-}p}$	
	V_{R_L} (dc)	______ V	______ %

EXPERIMENT 41
CAPACITIVE COUPLING

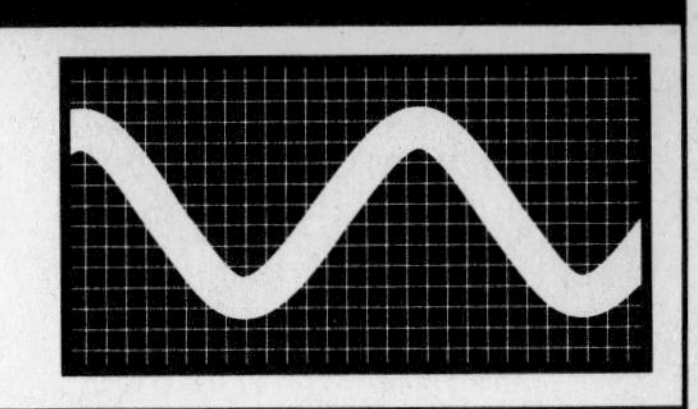

OBJECTIVES

At the completion of this experiment, you will be able to:

- Describe capacitor coupling action.
- Measure the demonstrated effects of capacitive coupling.

SUGGESTED READING

Chapter 22, *Basic Electronics,* B. Grob, Seventh Edition

INTRODUCTION

The most common type of coupling in amplifier circuits (Fig. 41-1) is *capacitive.* In this type of coupling, the output of one stage is connected to the input of another stage while the signals are sent through a capacitor. As in any coupling circuit which has sensitivity to frequency changes, the main goal of a good capacitive coupling circuit is to pass the signal with little attenuation of desired signals while attenuating the undesired frequencies. An advantage of the capacitive coupling circuit is its ability to block the dc voltage which may be present in the output signal of the first stage. This is important in that, as in transistor amplifiers, the dc operational characteristics of the second stage would be affected if its input signal were not free of possible dc components.

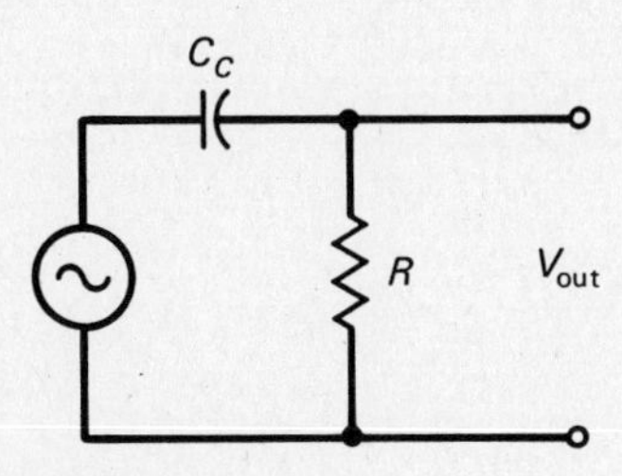

Fig. 41-1. AF coupling circuit.

EQUIPMENT

AF signal generator
Oscilloscope
Protoboards
Test leads

COMPONENTS

(1) 0.01-μF capacitor
(1) 100-kΩ $\frac{1}{2}$-W resistor

PROCEDURE

1. Connect the circuit of Fig. 41-2. The capacitor in this circuit is used in a coupling application. The relative low reactance of this capacitor allows practically all the generated ac voltage to be dropped across the resistor. Ideally very little of the generated ac voltage is dropped across the coupling capacitor.

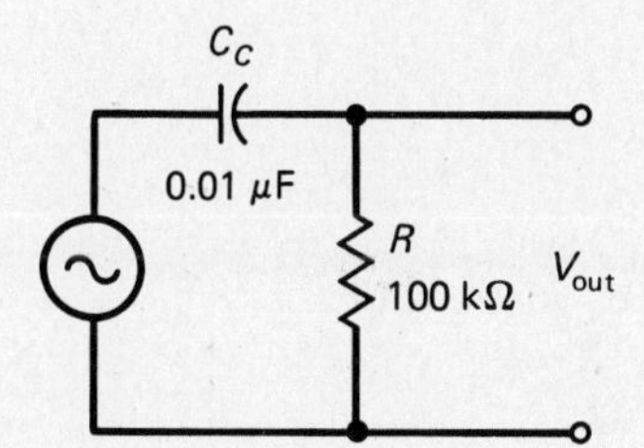

Fig. 41-2. AF coupling circuit.

2. Turn on the AF signal generator. Adjust the generator to 5 V peak to peak at 100 Hz. Measure and reord in Table 41-1 the following ac voltages: source generator, coupling capacitor, and resistor.

3. Complete the measurements, and record the data in Table 41-1.

4. The dividing line for C_c to be a coupling capacitor at a specific frequency can be determined when X_c equals one-tenth (or less) of R. At this point the series RC circuit becomes primarily resistive, and all the voltage drop of the AF signal generator is across the series resistance, with very little voltage dropped across the coupling capacitor. Using the one-tenth rule described above and the information contained in Table 41-1, determine the frequency at which the capacitor becomes a *coupling* capacitor. Adjust the signal generator to this frequency, and complete the measurements required in Table 41-2.

NAME

DATE

RESULTS FOR EXPERIMENT 41

QUESTIONS

1. After reading the capacitor coupling section in your book, describe what a coupling capacitor is used for.

2. What importance does the one-tenth rule have, and where is it used?

REPORT

Write a complete report. Discuss the measured and calculated results. Discuss the three most significant aspects of the experiment, and write a conclusion.

TABLE 41–1

Frequency	V_s p-p	V_c p-p	V_r p-p
100 Hz	______	______	______
200 Hz	______	______	______
300 Hz	______	______	______
400 Hz	______	______	______
500 Hz	______	______	______
600 Hz	______	______	______
700 Hz	______	______	______
800 Hz	______	______	______
900 Hz	______	______	______
1.0 kHz	______	______	______
1.1 kHz	______	______	______
1.2 kHz	______	______	______
1.3 kHz	______	______	______
1.4 kHz	______	______	______
1.5 kHz	______	______	______
1.6 kHz	______	______	______
1.7 kHz	______	______	______
1.8 kHz	______	______	______
1.9 kHz	______	______	______
2.0 kHz	______	______	______
3.0 kHz	______	______	______
4.0 kHz	______	______	______
5.0 kHz	______	______	______
6.0 kHz	______	______	______
7.0 kHz	______	______	______
8.0 kHz	______	______	______
9.0 kHz	______	______	______
10 kHz	______	______	______
15 kHz	______	______	______
20 kHz	______	______	______

TABLE 41–2

Frequency Where C_c Becomes a Coupling Capacitor	V_s p-p	V_c p-p	V_r p-p

EXPERIMENT 42

FET AMPLIFIER

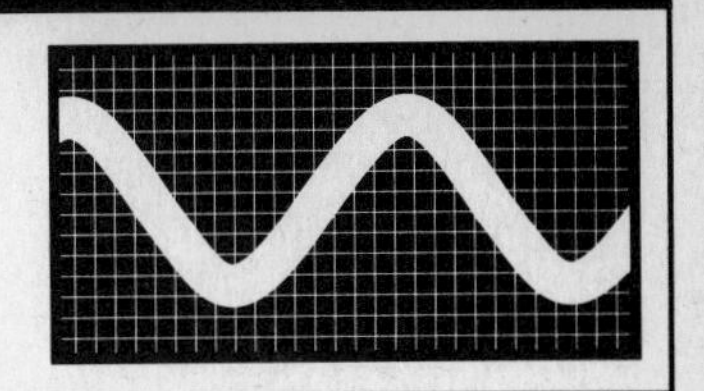

OBJECTIVES

At the completion of this experiment, you will be able to:

- Identify the polarity of a JFET.
- Study the circuit components and power supply polarity necessary to permit a JFET to function as an amplifier.
- Evaluate the performance of a JFET amplifier.

SUGGESTED READING

Chapters 28 and 29, *Basic Electronics,* B. Grob, Seventh Edition

INTRODUCTION

A junction field-effect transistor (JFET) is a semiconductor device that is controlled like a vacuum tube. In its simplest form, the JFET is a layer or channel of N-type material that acts like a resistor between its two end terminals called the *source* and the *drain*. On both sides of the channel, P-type material forms a gate through which electrons flow (drain current) from source to drain. With current flowing through the N channel, the JFET is considered to be a *normally on* device, which is also referred to as in an *enhancement mode*. However, when a control voltage is applied to the gate which is a reverse-bias gate-to-source condition, the channel is depleted and the current through the channel decreases. Once the channel has been depleted enough, the JFET is said to be *pinched off*; that is, no further drain current will flow. It is important to note, however, that current can still flow while the channel is pinched off. The point is that no further amount of current will flow. See the family of curves in Fig. 42-1. It is also important to note that a JFET amplifier (common source) will have a Q point of operation well into the pinch-off region, allowing for a symmetrical (positive to negative) output voltage swing.

EQUIPMENT

Audio signal generator
Oscilloscope
DC power supply, 0–15 V
VTVM
Test leads

COMPONENTS

(1) 2N3823 FET, or equivalent
Resistors (all 0.5 W):

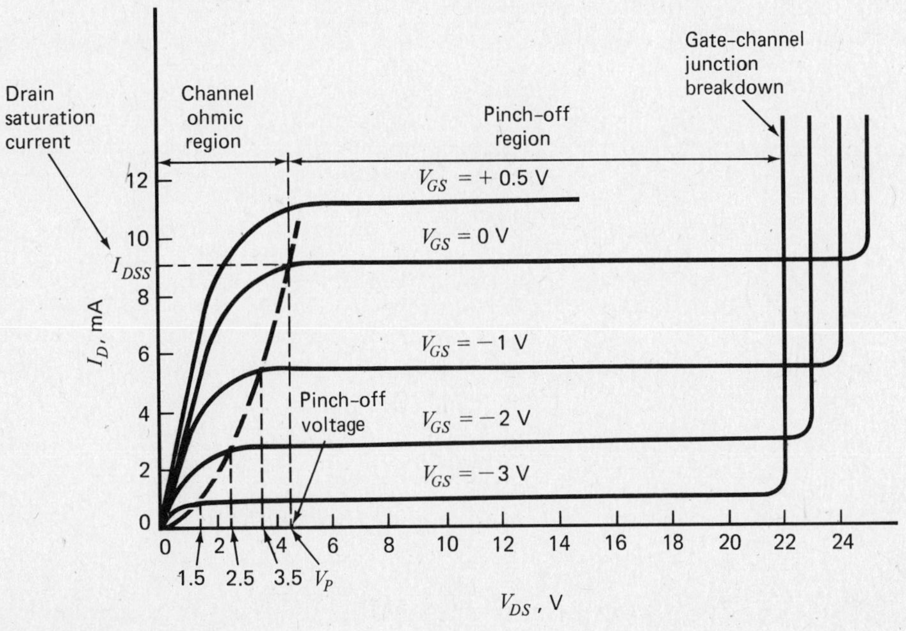

Fig. 42-1. N-channel JFET drain curves.

(1) 1 kΩ (1) 5.6 kΩ
(1) 2.2 kΩ (1) 8.2 kΩ
(1) 4.7 kΩ (1) 470 kΩ

(2) 0.01-μF capacitors
(1) 100-μF capacitor

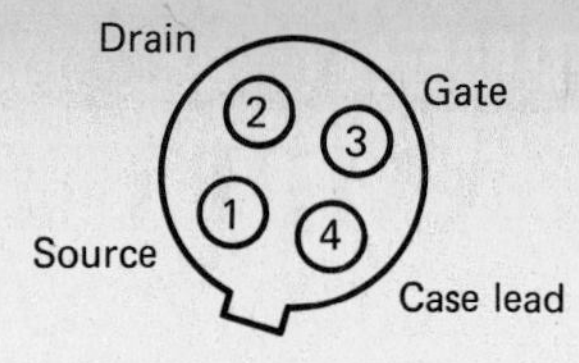

2N3823 bottom view

PROCEDURE

1. Use an ohmmeter on the R × 1000 range and measure $S - G$, $G - D$, and $S - D$ in both directions of lead polarity. Record results in Table 42-1. A diode junction should be indicated when measuring $S - G$ and $G - D$. The polarity of the JFET should be indicated by the forward-bias indication of the ohmmeter. A P channel will require a positive-ground, negative-drain potential. An N-channel JFET will require a negative-ground, positive-drain potential. See Fig. 42-2 for the correct lead orientation for N- and P-channel FETs.

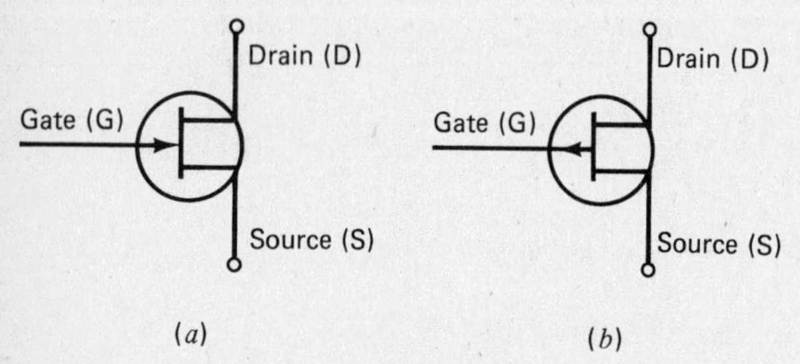

Fig. 42-2. (*a*) N-channel FET. (*b*) P-channel FET.

2. Connect the circuit shown in Fig. 42-3, being sure to observe proper V_{DD} polarity and also electrolytic capacitor polarity. Apply 1 kHz at 100 mV p-p. Use a 2.2-kΩ resistor as the load (drain) resistance. Measure and record in Table 42-2 all dc and peak-to-peak voltages. Determine the voltage gain (A_V).

3. Replace the load resistor with 4.7-kΩ, 5.6-kΩ, and 8.2-kΩ resistances and record all dc and ac peak-to-peak values around the circuit. Determine A_V. Record in Table 42-2.

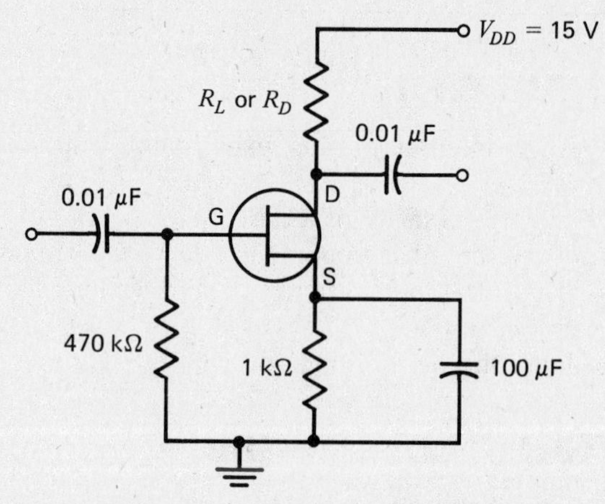

Fig. 42-3. N-channel FET amplifier using a 2N3823 FET.

4. Using the 8.2-kΩ resistor as a drain load, increase the power supply voltage to 30 V dc, and note any effects (in Table 42-3) upon the amplifier.

5. Remove the source bypass capacitor and note the effect (in Table 42-3) upon gain and dc parameters.

Note: Use V_{DD} = 15 V and R_L = 5.6 kΩ.

6. While the source bypass capacitor is removed, make a frequency check of this amplifier, determine the low-frequency cutoff value, and calculate the input resistance based on that value and the value of the coupling capacitor.

Note: Use semi-log graph paper in Appendix E.

NAME

DATE

RESULTS FOR EXPERIMENT 42

QUESTIONS

1. What was the polarity of the JFET used in this experiment?

2. What is the function of the bypass capacitor? What happens when it is removed from the amplifier circuit?

3. What factors influence the input resistance of the amplifier?

4. What is the overall effect that is created by increasing the value of load resistance?

5. What effects to the amplifier are created by increasing the value of the power supply voltage?

REPORT

Write a complete report. Discuss the results. Discuss the three most significant aspects of the experiment and write a conclusion.

TABLE 42–1

Procedure Step	Measurement	R, Ω
1	S to G	
	G to D	
	S to D	
Leads reversed	S to G	
	G to D	
	S to D	

TABLE 42–2

Procedure Step	R_D, kΩ	V_G(dc)	V_D(dc)	V_S(dc)	V_G(ac)	V_D(ac)	V_S(ac)	A_V
2	2.2							
3	4.7							
	5.6							
	8.2							

TABLE 42–3

Procedure Step	V_{DD}	Noted Effects
4	30 V dc	
5	15 V dc	Capacitor removed:

EXPERIMENT 43

TRANSISTOR AMPLIFIER

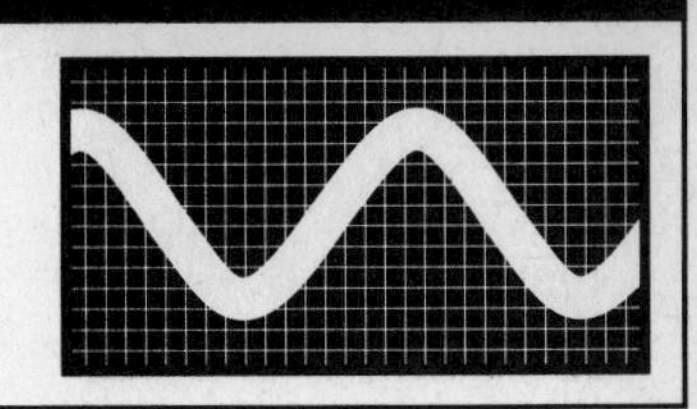

OBJECTIVES

At the completion of this experiment, you will be able to:

- Study the phase relationship of the input signal versus the output signal of a common-emitter amplifier.
- Investigate the purpose of biasing in a common-emitter amplifier.
- Determine the polarity of a transistor emitter-base junction.
- Study the concept of biasing a transistor.
- Plot a load line on the characteristic curves of a transistor.
- Compare the actual operation of a transistor amplifier to the operation indicated by the load-line graph.

SUGGESTED READING

Chapter 30, *Basic Electronics*, B. Grob, Seventh Edition

INTRODUCTION

This experiment uses a bipolar junction transistor (BJT) to amplify a small value of audio-range signal. The transistor polarity will be verified (PNP or NPN), and the bias supply will be adjusted to allow a maximum value of undistorted output voltage for the sine-wave input. The concept of gain (A_V) will be examined by varying the value of the collector resistor. The fundamental aspects of the common-emitter configuration can be determined from an analysis of the data obtained in the experiment.

Bias is defined as electrical force applied to a semiconductor transistor for the purpose of establishing a reference level for the operation of the device. In this experiment, a common-emitter PNP transistor amplifier is biased for operation under different levels of reference. The reference levels are established with respect to the load line and the Q point of operation. Consistent with the principles discussed in theory textbooks, amplification of a sine wave will have varying output results depending upon the biasing of the transistor. The results obtained in this experiment should provide sufficient data for analysis of the concept of transistor bias.

EQUIPMENT

Audio signal generator
Oscilloscope
Low-voltage power supply
Microammeter
Test leads
VTVM

COMPONENTS

(1) 100-kΩ potentiometer
(2) 10-μF at 25-V capacitors
(1) 2N3638 transistor or equivalent
Resistors (all 0.5 W):

(1) 470 Ω	(1) 10 kΩ
(1) 1 kΩ	(1) 4.7 kΩ

PROCEDURE

1. Using a VTVM (range = R × 1000), test the transistor to determine the emitter-to-base and base-to-collector junctions. This is a test of the diode action of the transistor. Forward bias is positive potential to the P material and negative potential to the N material, indicated by a low-resistance reading with the ohmmeter. Emitter-to-collector junctions will have a high resistance reading.

2. Connect the circuit shown in Fig. 43-1.

3. With the power supply (V_{CC}) adjusted at -10 V, obtain -5 V at the collector by adjusting the potenti-

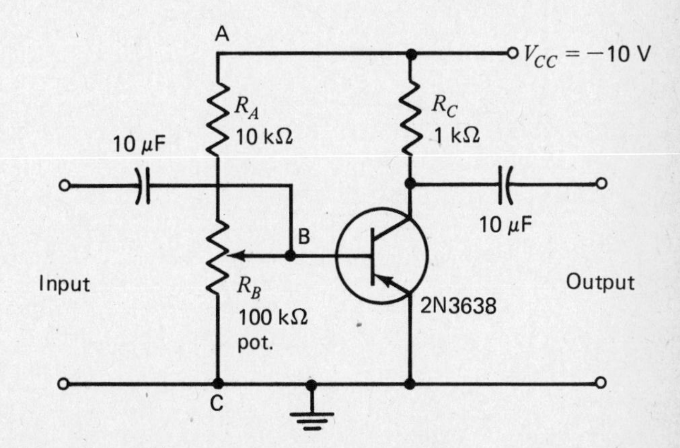

Fig. 43-1. PNP common-emitter amplifier.

ometer (R_B). This means that one-half the voltage (V_{CC}) of the supply is developed both across the collector to ground and across the collector resistor itself.

Note: The wiper of the potentiometer divides the potentiometer so that part of its resistance is added to the value of R_A. The remaining resistance is equal to R_B. This is called *voltage divider bias.*

4. Apply a 30-mV p-p input signal to the amplifier.

Note: It may be necessary to attenuate the output of the signal generator. This can be done by creating a voltage divider network across the terminals of the signal generator. For example, a 10-kΩ resistor in series with a 1-kΩ resistor will divide the voltage into 11 equal parts, 1 part across 1 kΩ and 10 parts across 10 kΩ. The frequency of the input signal should be set at 1 kHz.

5. Measure and record in Table 43-1 the output voltage from the amplifier.
6. Measure and record in Table 43-1 the dc voltage from the base to the emitter.
7. Calculate the gain of the amplifier (A_V):

$$A_V = \frac{V_{\text{out}}}{V_{\text{in}}}$$

Record the results in Table 43-1.

8. Disconnect the power supply from the circuit (V_{CC} and ground). Carefully remove the potentiometer wiper lead (point B) from the circuit and measure the resistance of R_A (points A and B) and R_B (points B and C). Record the results in Table 43-1.
9. Change the value of R_C to 470 Ω and repeat steps 2 to 8. Record the results in Table 43-2.

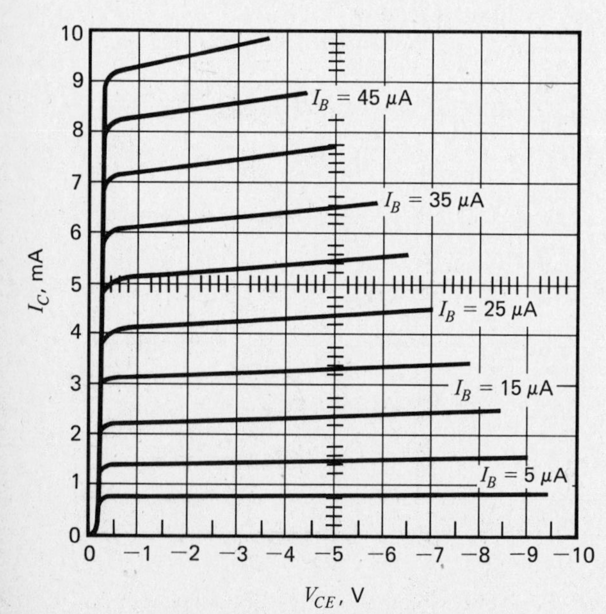

Fig. 43-2. 2N3638 family of curves (approximate).

10. Change the value of R_C to 4.7 kΩ and repeat steps 2 to 8. Record the results in Table 43-3.
11. Determine the phase relationship of the input and output signals by obtaining a second oscilloscope lead and using the dual-trace capabilities of the oscilloscope. Note the results in your report.
12. Draw a dc load line on the family of characteristic curves for a 2N3638 shown in Fig. 43-2, where R_C = 1 kΩ and V_{CC} = 10 V.

Note: The circuit for which this load line is drawn appears in Fig. 43-3.

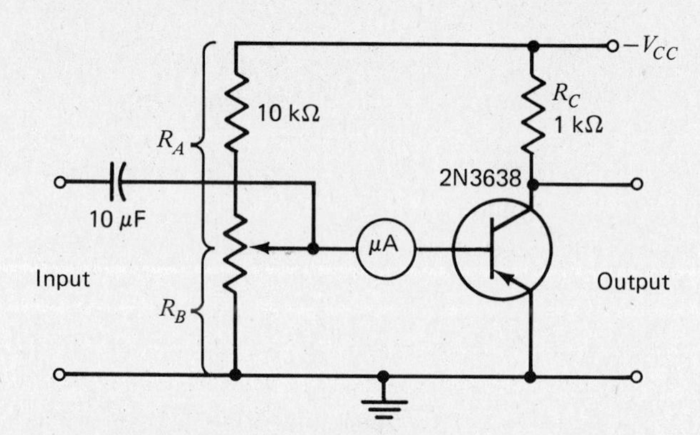

Fig. 43-3. Common-emitter amplifier.

13. Connect the circuit of Fig. 43-3.

Note: Voltages are negative for the PNP transistor, and for convenience, the following abbreviations will be used:

Q = quiescent, the point or conditions under which the currents and voltages in a transistor exist when no signal is applied = dc bias conditions
I_B = base current
I_C = collector current
V_{CC} = power supply voltage
V_{CE} = collector-to-emitter voltage
V_C = collector-to-ground voltage
V_{BE} = base-to-emitter voltage

DC Characteristics

14. Check all connections and polarity. Apply −10 V, which is equal to V_{CC}.
15. Adjust the bias potentiometer to obtain −5.0 V, which is equal to V_{CEQ} = ½ V_{CC}.
16. Measure and record I_{BQ} and I_{CQ} on the characteristic curve sheet shown in Fig. 43-4.

Note: I_{CQ} cannot be measured directly. Use Ohm's law to determine the current through R_C.

17. Measure and record V_{BEQ} in Table 43-4.
18. Disconnect V_{CC} and carefully disconnect the microammeter from the circuit in order to measure the values of R_A and R_B. Measure and record these values in Table 43-4.
19. Reconnect the circuit and repeat steps 16 to 18,

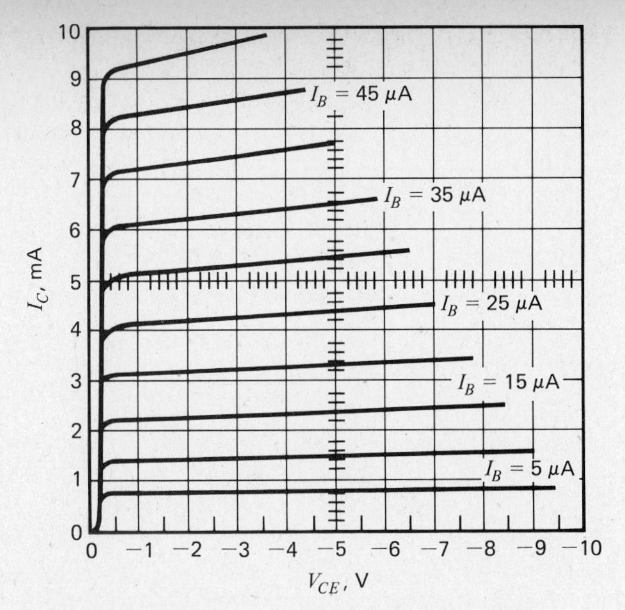

Fig. 43-4. 2N3638 family of curves (approximate).

with the bias potentiometer set at $V_{CEQ} = -8.5$ V and -1.5 V. (Record results in Tables 43-5 and 43-6, respectively.)

AC Characteristics

20. Disconnect or short-circuit the microammeter. Reconnect the power supply and adjust the bias potentiometer to obtain $V_{CEQ} = 5.0$ V.

21. Apply an input signal at 1 kHz and adjust the signal amplitude until the output voltage is equal to 6.0 V p-p. Measure and record the input signal (V_{in}) and use this same value for the following procedures (Table 43-7).

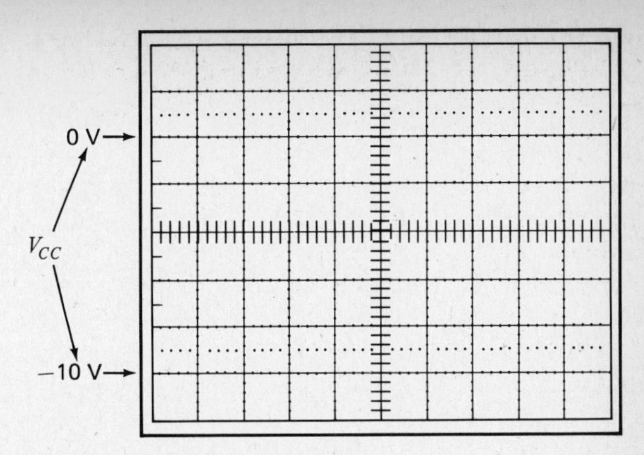

Fig. 43-5. Oscilloscope graticule.

22. Record the output waveform of the amplifier with the oscilloscope set at dc coupling. When drawing the output waveforms, indicate the parameters of the transistor bias as shown on the graticule shown in Fig. 43-5.

23. Repeat steps 21 and 22 with the bias potentiometer adjusted to obtain $V_{CEQ} =$ 8.5 V and 1.5 V. Be sure to use the previous value of input signal. The output signal will change. Use Table 43-7 and Fig. 43-6 for these data.

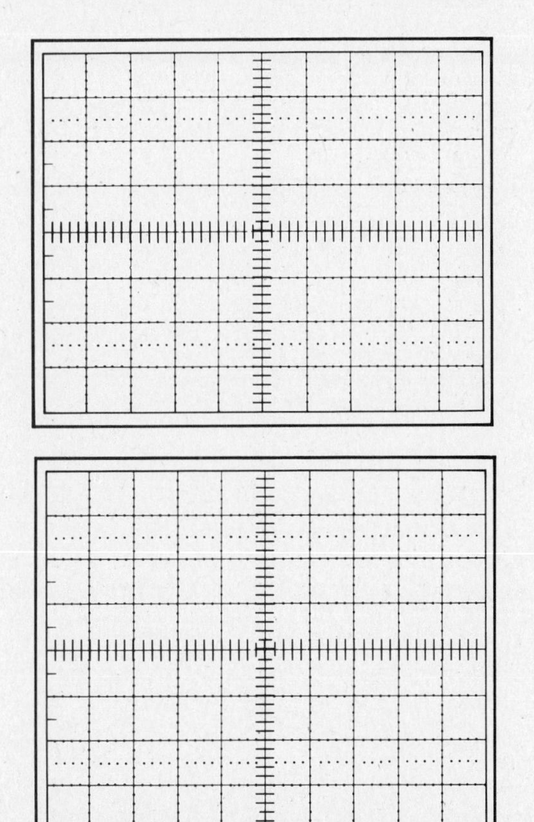

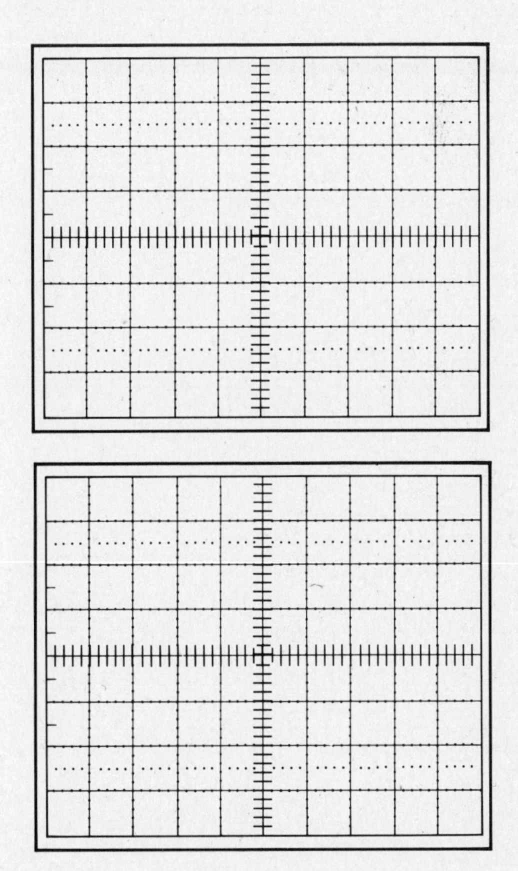

Fig. 43-6. Additional oscilloscope graticules.

NAME DATE

RESULTS FOR EXPERIMENT 43

QUESTIONS

1. What is the purpose of biasing the transistor?

2. What were the values of biasing resistors used?

3. What is the difference between using NPN versus PNP transistors with respect to bias supply voltage?

4. What is the phase relationship of the input voltage versus the output voltage?

5. Explain the differences between dc and ac characteristics.

REPORT

Write a complete report. Discuss the results. Discuss the three most significant aspects of the experiment and write a conclusion.

TABLE 43–1. V_{in} = 30 mV_{p-p}

Procedure Step	Measurement	Value
5	V_{out}	______V
6	V_{BE}	______V
7	A_V	______
8	R_A	______Ω
	R_B	______Ω

TABLE 43–2. R_C = 470 Ω

Procedure Step	Measurement	Value
9	V_{out}	______V
	V_{BE}	______V
	A_V	______
	R_A	______Ω
	R_B	______Ω

TABLE 43–3. R_C = 4.7 kΩ

Procedure Step	Measurement	Value
10	V_{out}	______V
	V_{BE}	______V
	A_V	______
	R_A	______Ω
	R_B	______Ω

TABLE 43–4. V_{CEQ} = −5 V

Procedure Step	Measurement	Value
17	V_{BEQ}	______V
18	R_A	______Ω
	R_B	______Ω

TABLE 43–5. V_{CEQ} = −8.5 V

Procedure Step	Measurement	Value
19	V_{BEQ}	______V
	R_A	______Ω
	R_B	______Ω

TABLE 43–6. V_{CEQ} = −1.5 V

Procedure Step	Measurement	Value
19	V_{BEQ}	______V
	R_A	______Ω
	R_B	______Ω

TABLE 43–7.

Procedure Step	Measurement	Value
21	V_{in}	______V
23	V_{CEQ} = 8.5 V	V_{out} = ______V
	V_{CEQ} = 1.5 V	V_{out} = ______V

EXPERIMENT 44

TRANSISTOR AS A SWITCH

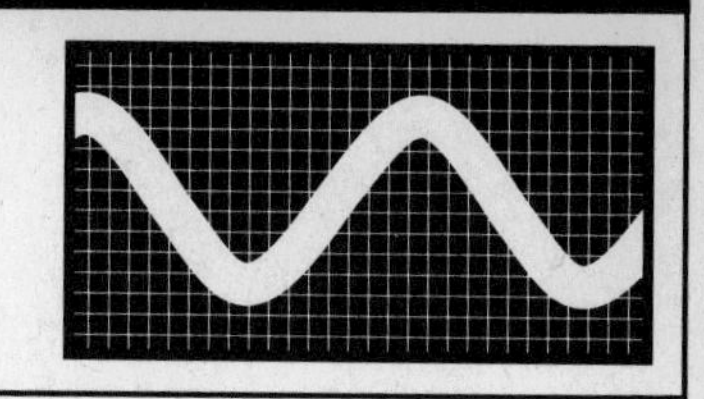

OBJECTIVES

At the completion of this experiment, you will be able to:

- Understand the importance of cutoff and saturation to the operation of a transistor switch.
- Define the purpose of a transistor inverter.
- Identify the function of a transistor switch.

SUGGESTED READING

Chapters 29 and 30, *Basic Electronics*, B. Grob, Seventh Edition

INTRODUCTION

The computers of today do not process numbers in the base 10 (that is, 0, 1, 2, 3, . . . , 9). Computers instead use binary logic of base 2 (0 and 1) to perform their functions. One fundamental circuit is the transistor switch, also known as an *inverter*. Here, a transistor connected in a common-emitter fashion inverts a signal. That is, if a high input signal is applied, a low output signal is created. If a low input signal is applied, then a high output signal is created. The circuit of Fig. 44-1 is an example of a transistor inverter design.

The circuit of Fig. 44-1 is also a transistor switch. In a transistor switch circuit, a voltage level applied to the base terminal will control the potential at the collector. In this fashion the transistor can be used to turn on or off circuitry connected to the collector. This common-emitter circuit is being switched from cutoff and saturation, as shown in the load line of Fig. 44-2.

In this experiment, a transistor will be connected to demonstrate this switching ability.

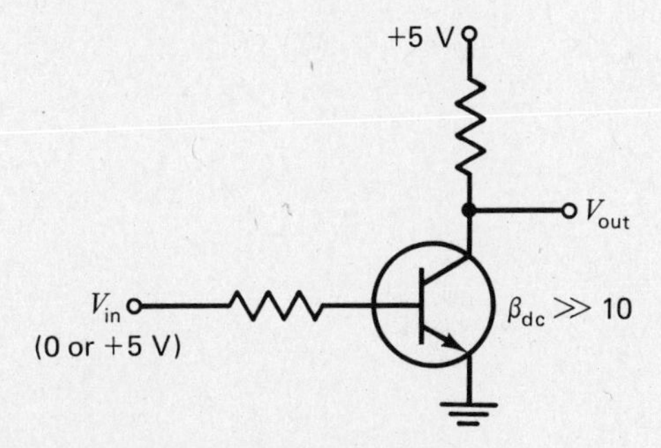

Fig. 44-1. Transistor inverter design.

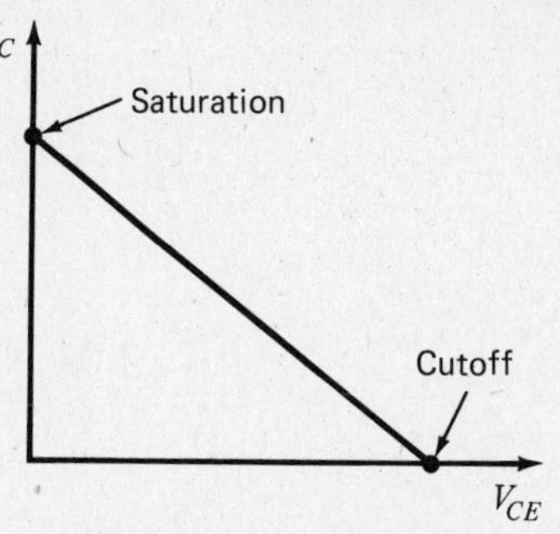

Fig. 44-2. Cutoff and saturation plotted on a load line.

EQUIPMENT

DC power supply, 0–10 V
VTVM
Protoboard or springboard
Test leads

COMPONENTS

(1) NPN transistor
(1) PNP transistor
Resistors (all 0.5 W):

(1) 1 kΩ (1) 10 kΩ

PROCEDURE

1. Connect the circuit in Fig. 44-3. Apply the correct polarity of voltage to V_{CC}.

2. Connect point A to ground. Measure and record in Table 44-1 the voltage from point B to ground.

3. Connect point A to +5 V. Measure and record in Table 44-1 the voltage from point B to ground.

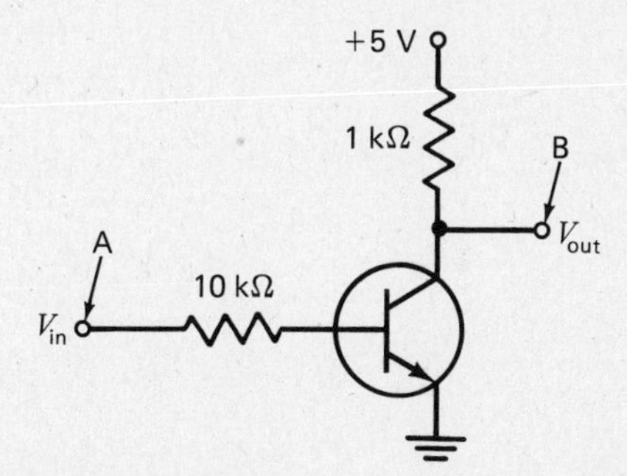

Fig. 44-3. Transistor switch.

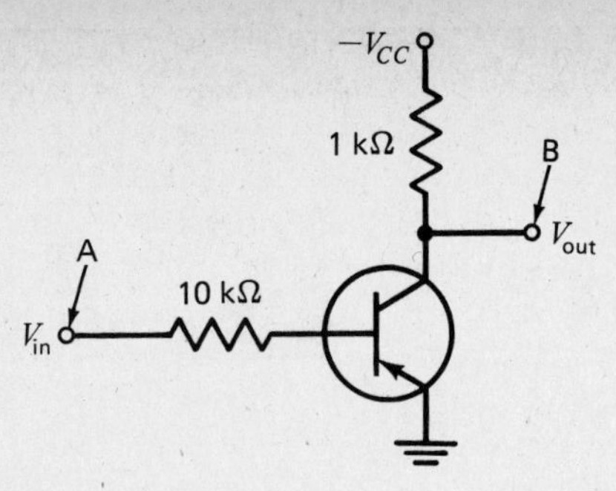

Fig. 44-4. Transistor switch.

4. Connect the circuit of Fig. 44-4. Apply the correct polarity of voltage to V_{CC}.
5. Connect point A to ground. Measure and record in Table 44-1 the voltage from point B to ground.
6. Connect point A to -5 V. Measure and record in Table 44-1 the voltage from point B to ground.
7. Construct a table of your results that will contrast the two circuits.

NAME ______________________ DATE ____________

RESULTS FOR EXPERIMENT 44

QUESTIONS

1. In the above circuits, what voltage level would a binary 1 represent? A binary 0? Are the answers the same for both the circuits of Figs. 44-3 and 44-4?

2. What is saturation? How is it demonstrated in this experiment?

3. What is cutoff? How is it demonstrated in this experiment?

4. Are the saturation and cutoff points the same for both the circuits shown in Figs. 44-3 and 44-4?

5. What are the fundamental differences between the two circuits shown in Figs. 44-3 and 44-4? Do the differences significantly affect overall outcomes? Explain.

REPORT

Write a complete report. Discuss the results. Discuss the three most significant aspects of the experiment and write a conclusion.

TABLE 44–1

Procedure Step	Function and Measurement	Value
2	Point B to ground	_______V
3	Point B to ground	_______V
5	Point B to ground	_______V
6	Point B to ground	_______V

EXPERIMENT 45

TWO-STAGE TRANSISTOR AMPLIFIER

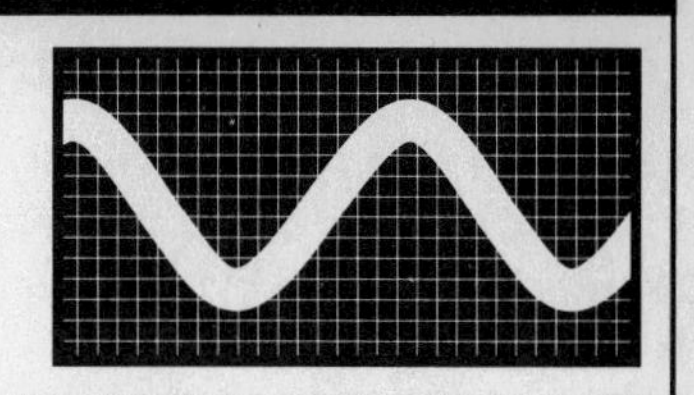

OBJECTIVES

At the completion of this experiment, you will be able to:

- Build and test a two-stage amplifier.
- Design a two-stage amplifier for voltage gain.

SUGGESTED READING

Chapters 29 and 30, *Basic Electronics,* B. Grob, Seventh Edition

INTRODUCTION

Transistors are usually connected in various configurations or stages to perform different tasks. By connecting two transistor amplifiers in the common-emitter configuration, more gain can be obtained from one supply voltage. To accomplish this, the output signal from the first stage is input or coupled into the next stage. This type of amplifier is called a *cascaded amplifier* because it uses more than one transistor.

All the concepts that applied to the single-stage transistor amplifier are applicable in this experiment. The transistor must be properly biased in each stage, and both transistors will have the same polarity: NPN or PNP.

In the circuit shown in Fig. 45-1, note the bias resistors on Q_1, the first-stage transistor. Here R_a and R_b provide voltage-divider bias. And R_c is the collector resistor, and R_e is the emitter resistor.

The bias resistors on Q_2 act exactly the same as those used for Q_1. In fact, they can be the exact same values where R_a and R_b provide voltage divider bias, R_c is the collector resistor, and R_e is the emitter resistor.

Notice also that capacitors C_1 and C_3 block the dc current from the input and output. Capacitor C_2 is a coupling capacitor that keeps the dc currents isolated between stages and passes or couples the ac output signal from stage 1 into stage 2.

Designing a Two-Stage Amplifier

In a previous experiment, you may recall that a single-stage amplifier's voltage gain was equal to V_{out}/V_{in}. In this experiment, the voltage gain is controlled by the relationship of R_c/R_e. For example, if $R_c = 10\ \text{k}\Omega$ and $R_e = 1\ \text{k}\Omega$, then the voltage gain is equal to 10, assuming the transistor is operating in the middle of the load line and the common-emitter configuration is used.

Although an amplifier can be designed (through mathematical calculations) to produce a specific result, for example, a specific voltage gain, the bias resistor values are often adjusted during testing of the prototype to achieve the desired gain. In this experiment, the test amplifier will produce a voltage gain of about $A_V = 70$ after the bias resistors are adjusted. However, the

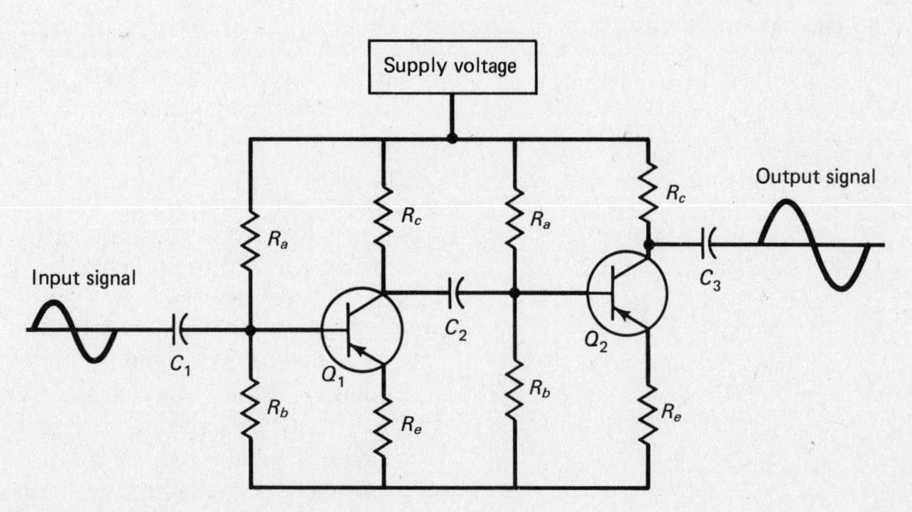

Fig. 45-1. Two-stage or cascaded transistor amplifier.

original design was calculated for each stage to have a voltage gain of 10. Thus, stage 1 gain × stage 2 gain equals 100, as described in the five steps below:

1. A load line was drawn for a single stage where $V_{cc} = 12\text{ V}$, $V_{ce} = 6\text{ V}$, $I_c\text{ (max)} \approx 6\text{ mA}$ where R_c is assumed to be 1.5 kΩ. This load line was drawn on a typical family of curves for a 2N3904, made on a curve tracer.

2. The Q point (middle of the load line) shows $I_c = 3.0\text{ mA}$ and $I_B = 12.5\ \mu\text{A}$ in Fig. 45-2.

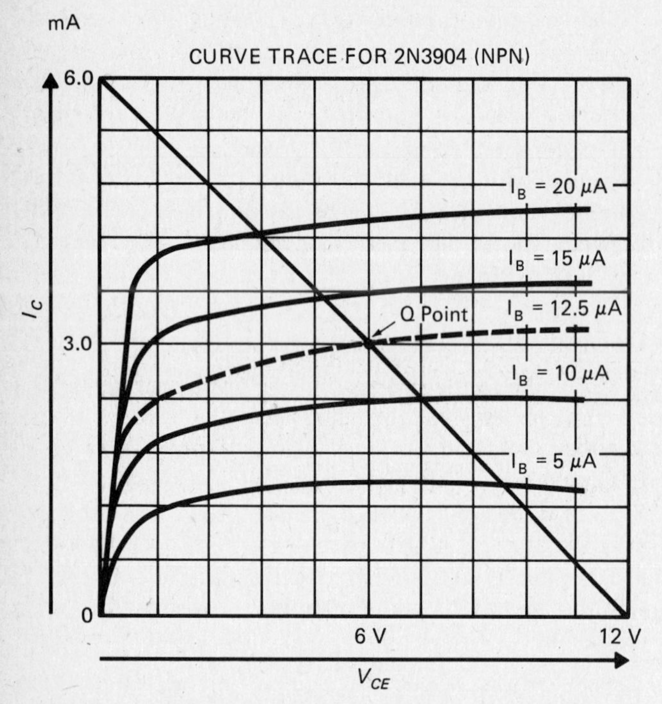

Fig. 45-2. Load line for 2N3904 made on a curve tracer.

3. To get a total gain of $A_V = 100$, each stage has an R_c/R_e ratio of 10/1. Thus, the first-stage A_V times the second-stage $A_V = 100$.
Therefore, if $R_c = 1.5\text{ k}\Omega$, then R_e is chosen to be 150 Ω: 1.5 kΩ/150 Ω ≈ 7/1.

4. To determine the voltage divider bias resistors R_a and R_b, the following calculations were made:

$$R_b = \frac{V_{R_b}}{I_{R_b}}$$

$$V_{R_b} = V_{\text{base}} = V_{BE} + V_{R_e},$$

where $$V_{R_e} = R_e I_{cQ}$$

$$V_{R_e} = 150\ \Omega \times 3.0\text{ mA} = 0.450\text{ V}$$

$$V_{R_b} = 0.7 + 0.450 = \underline{1.15\text{ V}}$$

$$I_{R_b} = I_B \times 10, \text{ where 10 is an approximation}$$

$$12.5\ \mu\text{A} \times 10 = \underline{0.125\text{ mA}}$$

$$R_b = \frac{1.15\text{ V}}{0.125\text{ mA}} = \underline{\underline{9.2\text{ k}\Omega}}$$

$$R_a = \frac{V_{R_a}}{I_{R_a}}$$

$$V_{R_a} = V_{cc} - V_{R_b} = 12 - 1.15 = \underline{10.43\text{ V}}$$

$$I_{R_a} = I_{R_b} + I_b = 0.125\text{ mA} + 12.5\ \mu\text{A} = \underline{0.1375\text{ mA}}$$

$$R_a = \frac{10.43\text{ V}}{0.1375\text{ mA}} = \underline{\underline{76\text{ k}\Omega}}$$

5. Finally, the capacitors are chosen so that their reactance will not create any appreciable voltage drop at the amplifier's lowest operating frequency. A typical approximation is made by using the formula $C = 10/(2\pi F_1 Z_{\text{in}})$, where F_1 is the lowest frequency and Z_{in} is the estimated input impedance of the amplifier. For example,

$$C = \frac{10}{6.28(100\text{ Hz})(1.5\text{ k}\Omega)} = \underline{\underline{10\ \mu\text{F}}}$$

EQUIPMENT

Oscilloscope (dual-trace)
Audio signal generator
Variable dc power supply
Voltmeter or DVM

COMPONENTS

(2) Transistors—2N3904 NPN silicon
Resistors, all ½ W:
(2) 82-kΩ (2) 1.5-kΩ (1) 1-kΩ (2) 150-Ω
(3) 10-kΩ (1) 100 Ω
(3) 10-μF capacitors

PROCEDURE

1. Connect the circuit of Fig. 45-3, but don't connect the ac signal source or the load resistance yet.

2. Measure and record all the dc voltages around the circuit, as indicated in Table 45-1. Also calculate the currents $I = V/R$ through each resistor.

3. Apply an input signal of 100 mV p-p at 5 kHz. Be sure to measure the input signal when it is connected to the amplifier input at C_1. Measure the two-stage output voltage and calculate the gain as $V_{\text{out}}/V_{\text{in}}$ in Table 45-1.

4. Disconnect the ac signal. Disconnect the stages at points A and B; also disconnect C_2 at point D. Measure and record the dc voltages across each bias resistor. Record the results in Table 45-1.

5. Apply an input signal of 100 mV p-p to each stage, and measure the output. The output of stage 1 is at point D. The input to stage 2 is at point C. Record the output voltages and calculate the gain for each stage in Table 45-1.

6. Reconnect both stages, and perform a frequency response test. To do this, apply a 5-kHz signal at

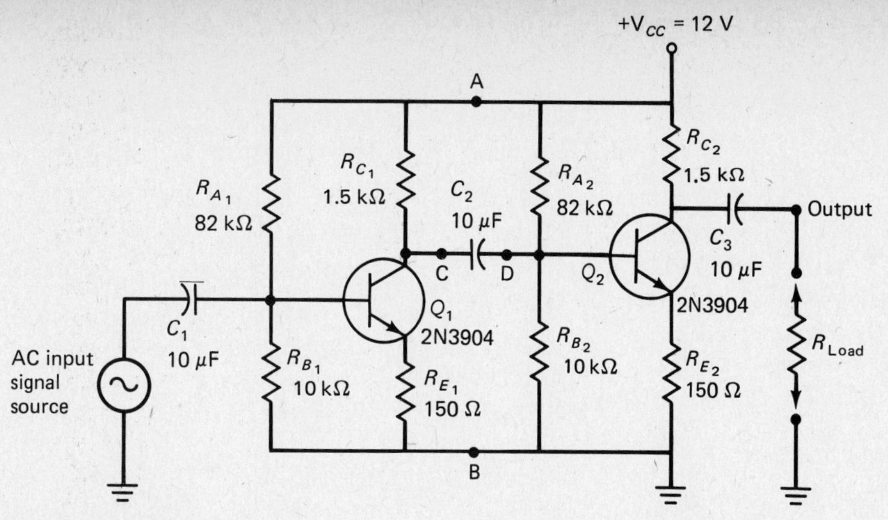

Fig. 45-3. Two-stage voltage amplifier.

100 mV and decrease the frequency until the 3-dB (half-power) point is reached. The output of stage 2 will show about 70 percent of its previous output voltage. Then increase the frequency until the signal output is 70 percent of its value at 5 kHz. For example, if the 5-kHz output is 10 V, it will roll off to 7 V at the low (F_1) and high (F_2) ends of the amplifier's bandwidth. Calculate the bandwidth (BW) as $F_2 - F_1$. Record all values according to Table 45-1.

7. Connect the load resistor $R_L = 1\ \text{k}\Omega$, and measure the output voltage at 10 kHz with $V_{in} = 100$ mV. Record the results in Table 45-1.

8. Replace R_L with a 100-Ω resistor and repeat step 7.

9. Replace R_L with a 10-kΩ resistor and repeat step 7.

OPTIONAL

10. Adjust the bias resistors to obtain a gain of 100 as calculated in the introductory material. This step will require you to use a *trial-and-error* process. Try replacing R_{C_2} as a start. On a separate sheet of paper, describe what you did to obtain and verify a gain of 100.

11. Perform a frequency response (bandwidth) test on the single-stage amplifier. Compare the results with the two-stage bandwidth response in your report.

NAME ______________________ DATE ____________

RESULTS FOR EXPERIMENT 45

QUESTIONS

No questions are required.

REPORT

Write a complete report. Discuss bias, gain, and frequency response. Submit all data and, if required, graph the bandwidth.

TABLE 45–1 Step 4

	Resistor	Nominal Value	Measured Voltage	Calculated Current	Voltages with Stages Disconnected
Step 2	R_{a_1}	82 kΩ	______	______	______
	R_{b_1}	10 kΩ	______	______	______
	R_{c_1}	1.5 kΩ	______	______	______
	R_{e_1}	150 Ω	______	______	______
	R_{a_2}	82 kΩ	______	______	______
	R_{b_2}	10 kΩ	______	______	______
	R_{c_2}	1.5 kΩ	______	______	______
	R_{e_2}	150 Ω	______	______	______

Two Stages

Step 3 V_{in} = 100 mV, V_{out} = ______, A_V = ______

Note: V_{out} = stage 2 output

Single Stages

Step 5 Stage 1: V_{in} = 100 mV, V_{out} = ______ V, A_V = ______

Stage 2: V_{in} = 100 mV, V_{out} = ______ V, A_V = ______

Bandwidth

Step 6 At V_{in} = 100 mV, F_1 = ______ Hz at ______ V_{out}

BW = ______. F_2 = ______ Hz at ______ V_{out}

Step 7 R_L = 1 kΩ, V_{out} = ______ V, A_V = ______

Step 8 R_L = 100 Ω, V_{out} = ______ V, A_V = ______

Step 9 R_L = 10 kΩ, V_{out} = ______ V, A_V = ______

EXPERIMENT 46

LOGIC CIRCUITS

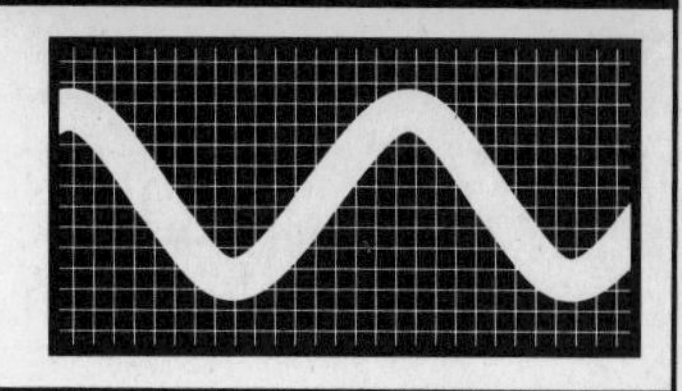

OBJECTIVES

At the completion of this experiment, you will be able to:

- Build a diode OR gate.
- Build a diode AND gate.
- Construct a truth table for a gate circuit.

SUGGESTED READING

Chapter 31, *Basic Electronics*, B. Grob, Seventh Edition

INTRODUCTION

The computers of today do not process numbers in the base 10 (that is, 0, 1, 2, 3, . . . , 9). Computers instead use binary logic of base 2 (0 and 1) to perform their functions. Two fundamental circuits are the OR and AND gates.

The OR gate has two or more input signals but creates only one output signal. If any input signal is high (binary 1), the output signal is high. OR gates can be discrete, that is, made from several components such as diodes and resistors. OR gates can also come packaged as integrated circuits or ICs. The OR gate circuit constructed in this experiment is discrete. Figure 46-1 shows a discrete OR gate and its truth table.

Truth tables are data tables that quickly show the overall circuit action of the gate under investigation. For the circuit of Fig. 46-1, a binary 1 is equivalent to +5 V, and a binary 0 is equivalent to ground potential. Studying the truth table of Fig. 46-1 reveals that for a two-input OR gate, the circuit is off only when points *A* and *B* are both at ground potential. Another way of explaining the results of the truth table is to say that the OR gate is a mostly on device. In the case of the truth table of Fig. 46-1, the two-input OR gate is on 75 percent of the time. The OR gate has a symbol, and it is shown in Fig. 46-2.

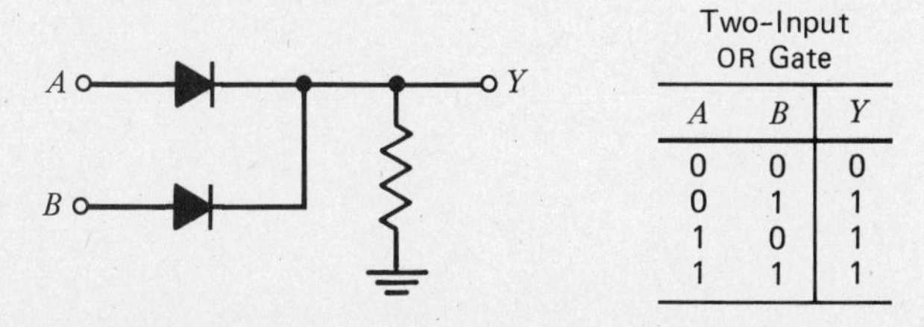

Two-Input OR Gate

A	B	Y
0	0	0
0	1	1
1	0	1
1	1	1

Fig. 46-1. Two-input OR gate and truth table.

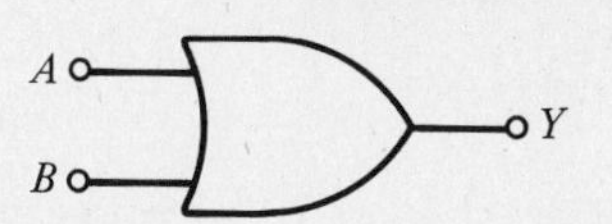

Fig. 46-2. OR gate symbol.

The AND gate has two or more input signals but creates only one output signal. All inputs must be high to get a high output signal. AND gates can also be of either discrete or IC design. Figure 46-3 shows a discrete AND gate and its truth table.

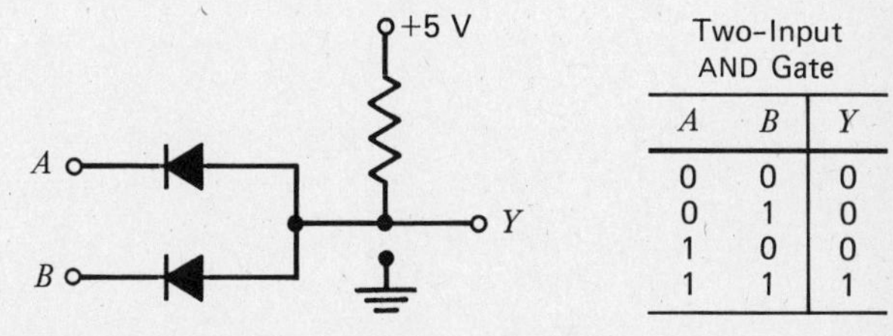

Two-Input AND Gate

A	B	Y
0	0	0
0	1	0
1	0	0
1	1	1

Fig. 46-3. Two-input AND gate and truth table. Note that 0 = low = 0 V and 1 = high = 5 V.

The AND gate circuit is on only when points *A* and *B* are both at a +5 V level, or at a binary 1. Another way of describing the results found in the truth table is to say that the AND gate is a mostly off device. In the case of the truth table of Fig. 46-3, the two-input AND gate is off 75 percent of the time. The AND gate has a symbol, and it is shown in Fig. 46-4.

EQUIPMENT

DC power supply, 0–10 V
VTVM
Protoboard or springboard
Test leads

COMPONENTS

(3) 1N4004 silicon diodes
(1) 1-kΩ 0.5-W resistor

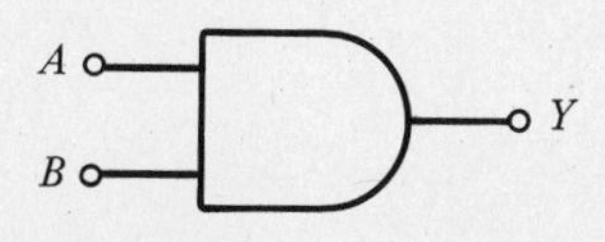

Fig. 46-4. AND gate symbol.

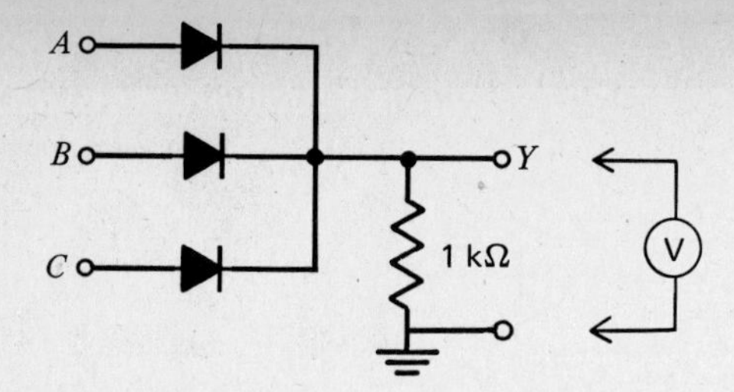

Fig. 46-5. Three-input OR gate.

PROCEDURE

1. Connect the circuit in Fig. 46-5. Apply the correct polarity of voltage to the circuit.
2. Connect points *A, B,* and *C* to ground. Measure and record in Table 46-1 the voltage level measured from point *Y* to ground.

Note: Remember that a 1 = +5 V and a 0 = ground potential.

3. Complete Table 46-1 for the remaining combinations of *A, B,* and *C.*

4. Connect the circuit of Fig. 46-6. Apply the correct polarity of voltage to the circuit.
5. Connect points *A, B,* and *C* to ground. Measure and record in Table 46-2 the voltage level measured from point *Y* to ground.
6. Complete Table 46-2 for the remaining combinations of *A, B,* and *C.*

Note: Remember that a 1 = +5 V and a 0 = ground potential.

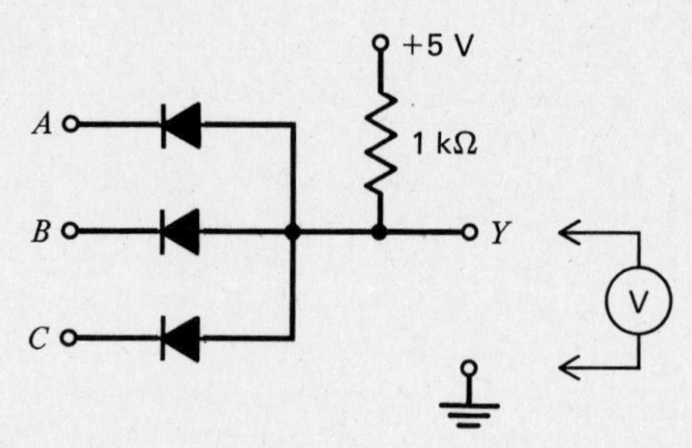

Fig. 46-6. Three-input AND gate.

NAME

DATE

RESULTS FOR EXPERIMENT 46

QUESTIONS

1. In the above circuits, what voltage level would a binary 1 represent? A binary 0? Are the answers the same for both circuits of Figs. 46-5 and 46-6?

2. What is an OR gate? What percentage is the three-input circuit on?

3. What is an AND gate? What percentage is the three-input circuit off?

4. What function do the diodes perform in the OR and AND gates?

5. What are the fundamental differences between the two circuits shown in Figs. 46-5 and 46-6? Do the differences significantly affect overall outcomes? Explain.

REPORT

Write a complete report. Discuss the results. Discuss the three most significant aspects of the experiment and write a conclusion.

TABLE 46–1

A	*B*	*C*	*Y*
0	0	0	________
0	0	1	________
0	1	0	________
0	1	1	________
1	0	0	________
1	0	1	________
1	1	0	________
1	1	1	________

TABLE 46–2

A	*B*	*C*	*Y*
0	0	0	________
0	0	1	________
0	1	0	________
0	1	1	________
1	0	0	________
1	0	1	________
1	1	0	________
1	1	1	________

EXPERIMENT 47

OTHER LOGIC CIRCUITS

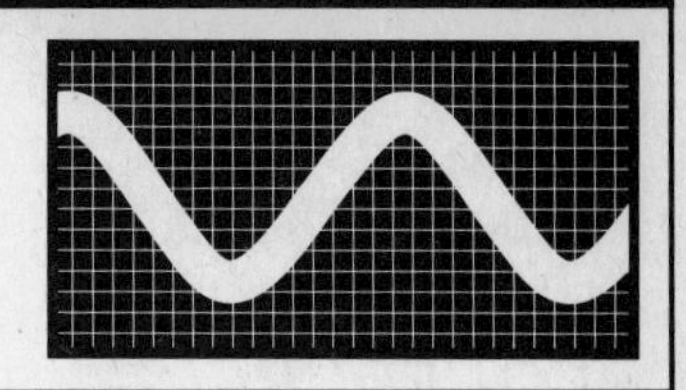

OBJECTIVES

At the completion of this experiment, you will be able to:

- Build a diode NOR gate.
- Build a diode NAND gate.
- Construct a truth table for gate circuits.

SUGGESTED READING

Chapter 31, *Basic Electronics,* B. Grob, Seventh Edition

INTRODUCTION

Two fundamental circuits are the NOR and NAND gates. The NOR gate has two or more input signals but creates only one output signal. If any input signal is high (binary 1), the output signal is low. NOR gates can be discrete, that is, made from several components such as diodes and resistors. NOR gates can also come packaged as integrated circuits (ICs). The NOR gate circuit constructed in this experiment is discrete. Figure 47-1 shows a discrete NOR gate and its truth table.

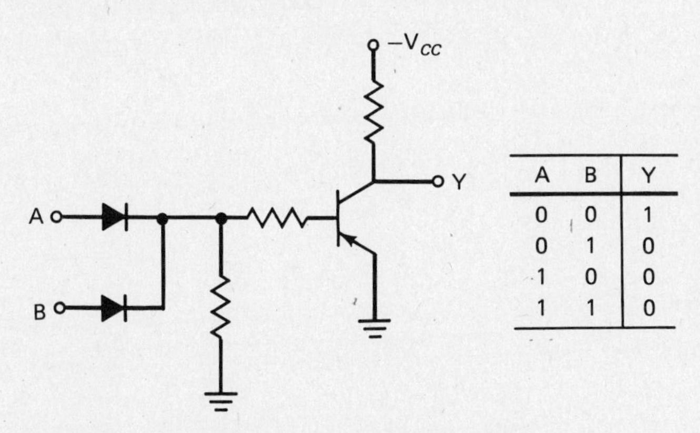

A	B	Y
0	0	1
0	1	0
1	0	0
1	1	0

Fig. 47-1. Two input NOR gate and truth table.

Truth tables are data tables that quickly show the overall circuit action of the gate under investigation. For the circuit of Fig. 47-1, a binary 1 is equivalent to +5 V, and a binary 0 is equivalent to ground potential. Studying the truth table of Fig. 47-1 reveals that for a two-input NOR gate, the circuit is on only when points A and B are both at ground potential. Another way of explaining the results of the truth table is to say that the NOR gate is a mostly off device. In the case of the truth table of Fig. 47-1, the two-input NOR gate is off 75 percent of the time. The NOR gate has a schematic symbol, and it is shown in Fig. 47-2.

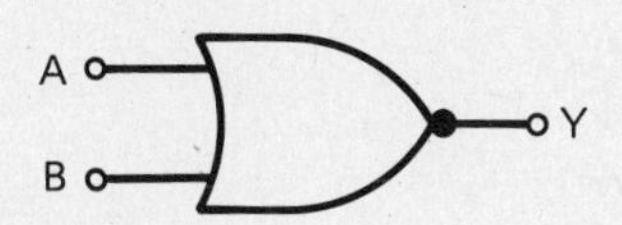

Fig. 47-2. OR gate symbol.

The NAND gate has two or more input signals but creates only one output signal. NAND gates can also be either of discrete or IC design. Figure 47-3 shows a discrete NAND gate and its truth table. The NAND gate circuit is off only when points A and B are both at a +5-V level, or at a binary 1. Another way of describing the results found in the truth table is to say that the NAND gate is a MOSTLY on device. In the truth table of Fig. 47-3, the two-input NAND gate is on 75 percent of the time. The NAND gate has a schematic symbol, and it is shown in Fig. 47-4.

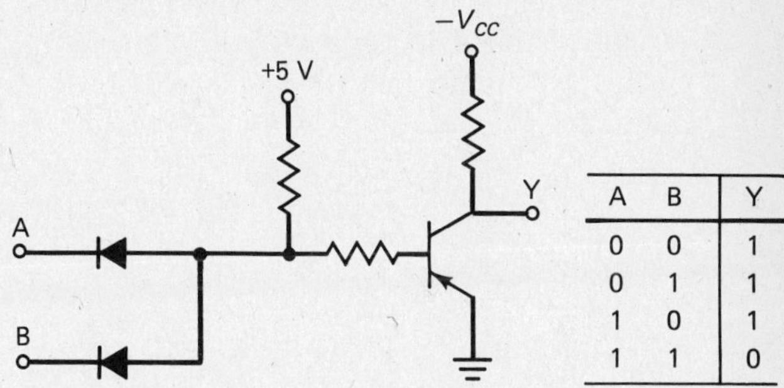

A	B	Y
0	0	1
0	1	1
1	0	1
1	1	0

Fig. 47-3. Two input and gate and truth table.

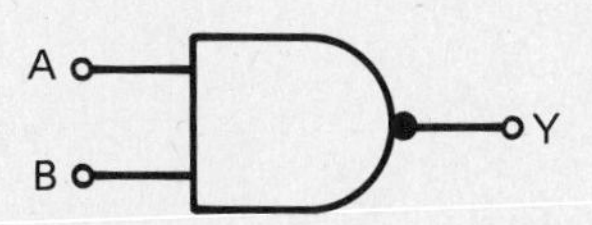

Fig. 47-4. NAND gate symbol.

EQUIPMENT

Power supply 0–10 V dc
VTVM
Protoboard/springboard
Test leads

COMPONENTS

(3) 1N4004 silicon diodes
(1) 2N3904 transistor
(2) 1-kΩ $\frac{1}{2}$-W resistors
(1) 10-kΩ $\frac{1}{2}$-W resistor

PROCEDURE

1. Connect the circuit in Fig. 47-5. Apply the correct polarity of voltage to the circuit.

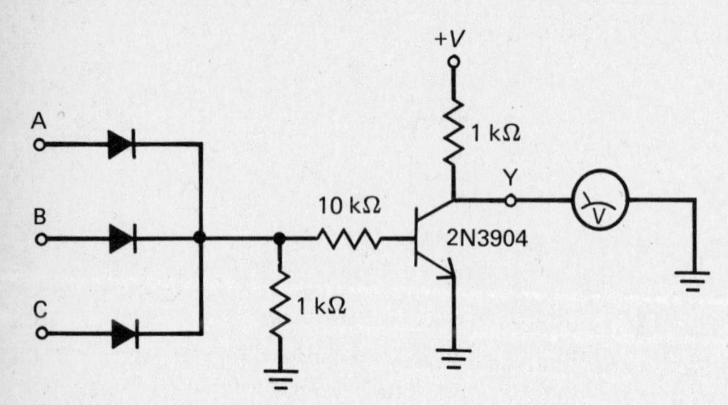

Fig. 47-5. Three input NOR gate.

2. Connect points A, B, and C to ground. Measure and record in Table 47-1 the voltage level measured from point *Y* to ground. Remember that 1 = +5 V and 0 = ground potential.

3. Complete Table 47-1 for the remaining combinations of A, B, and C.

4. Connect the circuit of Fig. 47-6. Apply the correct polarity of voltage to the circuit.

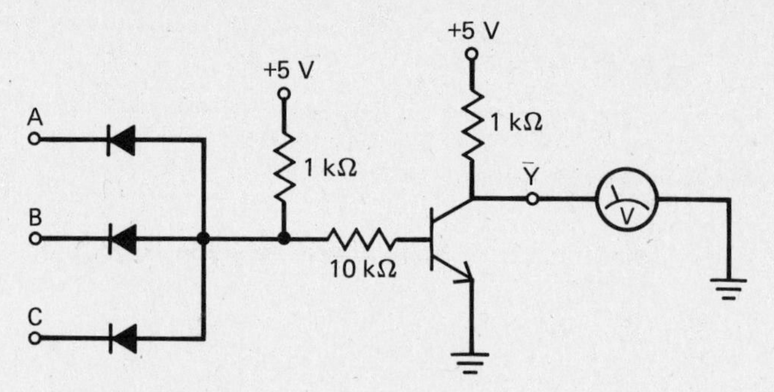

Fig. 47-6. Three input NAND gate.

5. Connect points A, B, and C to ground. Measure and record in Table 47-2 the voltage level measured from point *Y* to ground. Again remember that 1 = +5 V and 0 = ground potential.

6. Complete Table 47-2 for the remaining combinations of A, B, and C.

NAME DATE

RESULTS FOR EXPERIMENT 47

QUESTIONS

1. In the above circuits what voltage level would a binary 1 represent? Binary 0? Are the answers the same for both the circuits of Figs. 47-5 and 47-6?

2. What is a NOR gate? What percentage of the time is the three-input circuit off?

3. What is a NAND gate? What percentage of the time is the three-input circuit on?

4. What function do the diodes perform in the NOR and NAND gates?

5. What are the fundamental differences between the two circuits shown in Figs. 47-3 and 47-4? Do the differences significantly affect overall outcomes? Explain.

REPORT

Write a complete report. Discuss the measured and calculated results. Discuss the three most significant aspects of the experiment and write a conclusion.

TABLE 47–1

A	B	C	*Y*
0	0	0	______
0	0	1	______
0	1	0	______
0	1	1	______
1	0	0	______
1	0	1	______
1	1	0	______
1	1	1	______

TABLE 47–2

A	B	C	*Y*
0	0	0	______
0	0	1	______
0	1	0	______
0	1	1	______
1	0	0	______
1	0	1	______
1	1	0	______
1	1	1	______

EXPERIMENT 48
MULTIVIBRATOR CIRCUITS

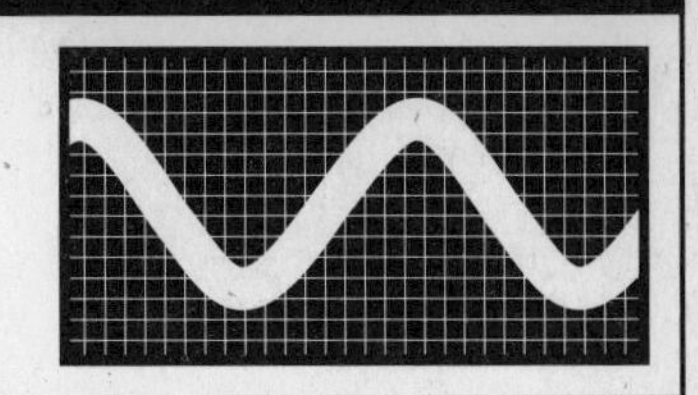

OBJECTIVES

At the completion of this experiment, you will be able to:

- Construct a basic multivibrator circuit.
- Test for proper operation in an astable multivibrator circuit.

SUGGESTED READING

Chapter 29, *Basic Electronics,* B. Grob, Seventh Edition

INTRODUCTION

Multivibrators are also known as *latches, registers,* and *flip-flops.*

A *flip-flop* is a circuit that receives an input signal that is either 5 V or at ground potential. The circuit then produces and remembers a predictable output signal which is also either 5 V or at ground potential. The circuit can maintain its memory even after the input signal has been removed from the device.

The simplest astable multivibrator is formed by connecting two inverting gates back to back through two capacitors, as shown in Fig. 48-1. This connection produces a circuit which switches from low to high repeatedly. The time between switching, or the frequency of operation, is determined by the value of the capacitors.

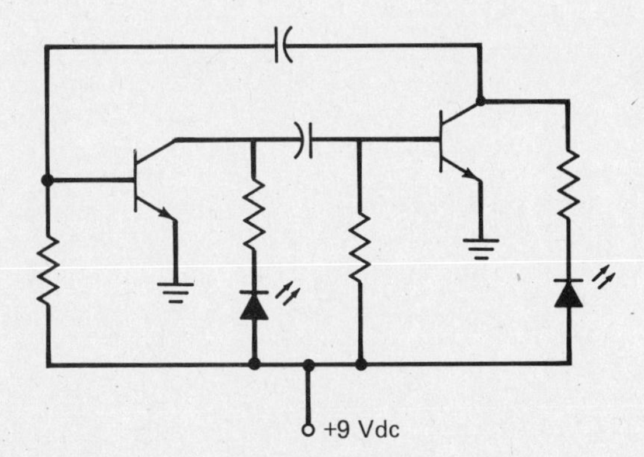

Fig. 48-1. Basic astable-multivibrator circuit.

EQUIPMENT

DC power supply, 0–10 V
VTVM
Protoboards
Test leads

COMPONENTS

(2) 2N3904
(2) LEDs
(2) 220-Ω $\frac{1}{2}$-W resistors
(2) 82-kΩ $\frac{1}{2}$-W resistors
(2) 25-μF 10-V dc electrolytic capacitors
(2) 47-μF 10-V dc electrolytic capacitors

PROCEDURE

1. Construct the astable multivibrator circuit of Fig. 48-2 using the two 25-μF 10-V dc electrolytic capacitors.

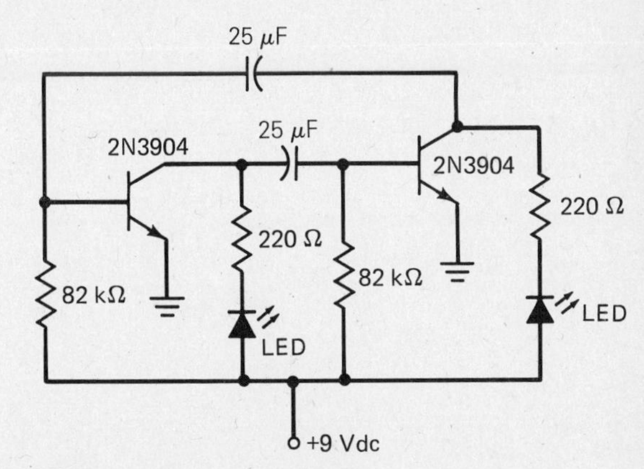

Fig. 48-2. Astable multivibrator.

2. Turn on the power supply. Determine the frequency of the circuit.

3. Turn off the power supply.

4. Change the value of the capacitors to 47 μF. Turn on the power supply. Determine the output frequency and note the difference in circuit action. Describe the circuit action of this step in the "Results" section of this experiment. How do the results of steps 2 and 4 differ?

NAME DATE

RESULTS FOR EXPERIMENT 48

QUESTIONS

1. What components determine the frequency of operation?

2. What does a multivibrator do?

REPORT

Write a complete report. Discuss the measured and calculated results. Discuss the three most significant aspects of the experiment, and write a conclusion.

EXPERIMENT 49

OPERATIONAL AMPLIFIERS

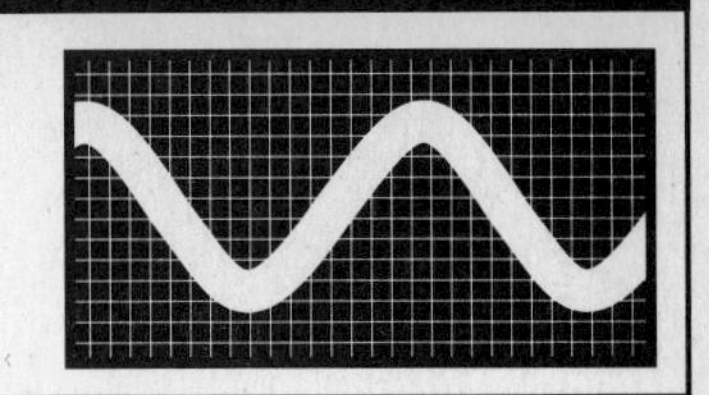

OBJECTIVES

At the completion of this experiment, you will be able to:

- Demonstrate the operation of an inverting amplifier.
- Demonstrate the operation of a noninverting amplifier.
- Demonstrate the operation of a buffer amplifier.

SUGGESTED READING

Chapter 32, *Basic Electronics*, B. Grob, Seventh Edition

INTRODUCTION

The first effort of this experiment will be directed toward the definition of terms used in conjunction with the operational amplifier. The operational amplifier, sometimes shortened to op amp, is a highly sophisticated linear integrated circuit (IC) direct-current amplifier, demonstrating high gain, high input impedance, and low output impedance. Originally, the term referred to high-gain, high-performance vacuum tube direct-current amplifiers that were designed to perform mathematical operations with predetermined voltage levels. Operational amplifiers were the basic building blocks of analog computers because they can perform the mathematical operations of amplification, addition, subtraction, integration, and differentiation. The operational amplifier used today can still be used to perform these mathematical operations; however, more useful circuit designs have been created. In combination with nonlinear elements such as diodes, they may be used as limiters, level detectors, and nonlinear function generators. By designing operational amplifier circuits which include other active components such as transistors, it is even possible to multiply and divide analog voltages by taking the logarithms and antilogarithms of input voltages.

The modern-day device tends to operate at lower voltages and does not have any of the common problems associated with vacuum tubes. Today's operational amplifier is in an IC format and still resembles the high-gain, direct-current amplifier which uses external feedback for controlled responses. When working with operational amplifiers, the user will find that they lend themselves easily and adapt well to a variety of industrial applications. They can be designed to function as filters, oscillators, pulse modulators, peak detectors, signal-function generators, small-signal rectifiers, instrumentation amplifiers, and a seemingly endless variety of specialized circuit applications. It has been determined that the operational amplifier is the most commonly used IC found in industry today. Figure 49-1 shows an IC comparison chart for commonly used linear devices.

When the schematic diagram of an op amp IC is displayed, it will take the form of a simple triangle. Some European and Japanese manufacturers will use a modified triangle schematic symbol which looks very similar to the symbol used for a digital gate circuit. We will not be using the modified symbol for this text; instead, we will be using a simple triangle symbol, and it is used in all schematics presented for your use in this experiment. The symbol that will be used is represented in Fig. 49-2.

To correctly address input signal information to the operational amplifier, two input terminals are made available. They are traditionally drawn on the left-hand side of the schematic diagram, as represented in Fig. 49-3. The input terminals are connected internally to a differential amplifier located inside the IC casing. These terminals are referred to as the *inverting* and *noninverting* inputs and carry the symbols of − and +, respectively. In addition to the input terminals, the device must make use of an output terminal. The operational amplifier will use only one terminal for this function, and it is drawn at the apex of the triangle, located on the right-hand side of the schematic diagram. Again direct your attention toward Fig. 49-3 and notice the placement of the output terminal.

Under normal operating conditions, if an ac signal is applied to the inverting terminal (−) with reference to ground, a 180° inversion, out-of-phase, signal would be seen at the output terminal. This inversion may be difficult to see with the use of a conventional single-trace scope, but when a dual-trace oscilloscope is used to compare the input versus the output signal, the 180° out-of-phase signal is easy to see. An ac signal applied to the noninverting terminal (+) with reference to ground would cause no inversion or phase shift in the signal being measured at the output terminal. A comparison of ac input signals of sine, triangle, and square waves applied to the inverting and noninverting terminals of a typical operational amplifier is shown in Fig. 49-4.

Device	Operating Temp. Range °C Min	Max	A_{VOL} Min K	R_i Min Ω	P_D mW	$\lvert I_{io}\rvert$ Max nA	I_b Max nA	CMV_i Min V	†Typ. Slew Rate SR V/μSec	V_o Min V	$\lvert V_{io}\rvert$ Max mV	Offset Adjust	Internal Compen-sation	Output Protec-tion	Input Protec-tion	JEDEC Package Type
709A	−55	+125	25	350 K	108	50	200	±8	0.3	±12	1	no	no	no	no	TO-91, 99, 116
709B	−55	+125	25	150 K	165	200	500	±8	0.3	±12	5	no	no	no	no	TO-91, 99, 116
709C	0	+70	15	50 K	200	500	1500	±8	0.3	±12	10	no	no	no	no	TO-91, 99, 116
739C (Dual)	0	+70	6.5	37 K	420	1000	2000	±10	1.0	+12, −14	6	no	no	yes	yes	TO-116
741B	−55	+125	50	300 K	85	200	500	±12	0.5	±12	5	yes	yes	yes	yes	TO-91, 99, 116
741C	0	+70	20	150 K	85	200	500	±12	0.5	±12	6	yes	yes	yes	yes	TO-91, 99, 116
747B (Dual)	−55	+125	50	300 K	85	200	500	±12	0.5	±12	5	yes	yes	yes	yes	TO-101, 116
747C (Dual)	0	+70	20	150 K	85	200	500	±12	0.5	±12	6	yes	yes	yes	yes	TO-101, 116
748B	−55	+125	50	300 K	85	200	500	±12	0.5	±12	5	yes	no	yes	yes	TO-99
748C	0	+70	20	150 K	85	200	500	±12	0.5	±12	6	yes	no	yes	yes	TO-99
749B (Dual)	−55	+125	25	100 K	220	400	750	±11	1.5	+12, −14.5	3	no	no	yes	yes	TO-116
749C (Dual)	0	+70	15	70 K	330	500	1000	±11	1.5	+12, −14.5	6	no	no	yes	yes	TO-116
800B	−55	+125	10	250 K	180	100	1000	±4	—	±6	50	no	no	no	no	TO-101
800D	−55	+125	10	100 K	180	200	2000	±4	—	±12	—	no	no	no	no	TO-101
801B	−55	+125	10	250 K	180	100	1000	±4	—	±12	50	no	no	no	no	TO-100
801D	−55	+125	10	100 K	180	200	2000	±4	—	±12	—	no	no	no	no	TO-100
805B	−55	+125	30	500 K	225	50	500	±8	2.5	±12	5	no	no	no	yes	TO-91, 99
805C	0	+100	10	100	225	100	1000	±8	2.5	±12	10	no	no	no	yes	TO-91, 99
806B	−55	+125	30	500	225	50	500	±8	2.5	±9	5	no	no	no	yes	TO-91, 99
806C	0	+100	10	100	225	100	1000	±8	2.5	±9	10	no	no	no	yes	TO-91, 99
807B	−55	+125	30	500	225	50	500	±8	2.5	±12	2.5	no	no	no	yes	TO-91, 99
808A	−55	+125	25	1 M	225	15	50	±8	2.5	±12	5	no	no	no	yes	TO-91, 99
808B	−55	+125	25	1 M	225	30	50	±8	2.5	±12	10	no	no	no	yes	TO-91, 99
809B	−55	+125	10	100	150	100	500	±10	—	±10	10	no	no	yes	yes	TO-99, 116
809C	0	+100	10	50	150	350	1000	±10	—	±10	10	no	no	yes	yes	TO-99, 116
810B (Dual)	−55	+125	10	100	150	100	500	±10	—	±10	10	no	no	yes	yes	TO-116
810C (Dual)	0	+100	10	50	150	350	1000	±10	—	±10	10	no	no	yes	yes	TO-116
715B	−55	+125°C	15	IM (typ)	210	250	750	±15	18	±10	±5	yes	yes	yes	yes	TO-100
715C	0	+70°C	10	IM (typ)	300	1500	250	±15	18	±10	±7.5	yes	no	yes	yes	TO-100
846B	−55	+125	100	25 M	90	5	30	±12.5	2.0	±12	±3	yes	no	yes	yes	TO-99
846C	0	+70	50	15 M	75	15	50	±12	2.0	±12	±5	yes	no	yes	yes	TO-99
LM101A	−55	+125	50	1.5 M	120	10	25	±12	0.5	±12	±2	yes	no	yes	yes	TO-99
LM101B	−55	+125	50	300 K	120	200	500	±12	0.5	±12	±5	yes	no	yes	yes	TO-99
LM201A	−25	+85	50	1.5 M	120	10	75	±12	0.5	±12	±2	yes	no	yes	yes	TO-99
LM201C	−25	+85	20	300 K	90	200	500	±12	0.5	±12	±7.5	yes	no	yes	yes	TO-99
LM301A	0	+70	25	500 K	120	50	250	±12	0.5	±12	±7.5	yes	no	no	yes	TO-99
LM307D	0	+70	25	500 K	120	50	250	±12	0.5	±12	±7.5	yes	no	no	yes	TO-99
811B	−55	+125	10	100	150	100	500	±10	—	±10	10	no	no	yes	yes	TO-99, 116
811C	0	+100	10	50	—	350	1000	±10	—	±10	10	no	no	yes	yes	TO-99, 116
813C	0	+70	6	—	120	2000	5000	±5	—	—	4	no	no	yes	yes	TO-99, 116
819B	−55	+125	5	50 K	25	100	500	±4	—	±4	10	no	no	yes	yes	TO-99
841B	−55	+125	50	300	85	200	500	±12	0.5	±12	5	no	no	yes	yes	TO-99
841C	0	+100	20	150 K	85	200	500	±12	0.5	±12	6	no	no	yes	yes	TO-99
844B	−55	+125	100	25 M	75	5	30	±12.5	2.0	±12	±3	yes	yes	yes	yes	TO-99
844C	0	+70	50	15 M	90	15	50	±12	2.0	±12	±5	yes	yes	yes	yes	TO-99
LM107B	−55	+125	50	1.5 M	120	10	75	±12	0.5	±12	±2	yes	yes	yes	yes	TO-99
LM207C	−25	+85	50	1.5 M	120	10	75	±12	0.5	±12	±2	yes	yes	yes	yes	TO-99
MC1437C (Dual)	0	+70	15	50 K	200	500	1500	±8	0.3	±12	10	no	no	no	no	TO-116
MC1439C	0	+70	15	100	200	100	1000	±11	4.2	±10	7.5	no	no	yes	yes	TO-99, 116
MC1458C (Dual)	0	+70	20	150 K	85	200	500	±12	0.5	±12	6	no	yes	yes	yes	TO-99, 116
MC1537B (Dual)	−55	+125	25	150 K	165	200	500	±8	0.3	±12	5	no	no	no	no	TO-91, 99, 116
MC1539B	−55	+125	50	150	150	60	500	±11	4.2	±10	3	no	no	yes	yes	TO-99, 116
MC1558B (Dual)	−55	+125	50	300 K	85	200	500	±12	0.5	±12	5	yes	yes	yes	yes	TO-101, 116

†Unity Gain

Fig. 49-1. IC comparison chart for linear devices.

Of course, the operational amplifier will need to have supply voltages and other components added in order to operate efficiently.

Circuit Configurations

Circuits using operational amplifiers commonly display properties radically different from those of the individual devices themselves. For example, you will find that the circuit's closed-loop gain, A_{CL}, is only a fraction of the device's internal open-loop gain, A_{OL}. In addition, the circuit input impedance is often much different from the operational amplifier's internal input impedance (although for some circuits you will find it to be of the same magnitude). Output impedance of the operational amplifier circuit is usually less than that of the op amp device, but the bandwidth is usually greater. It is therefore advisable to make the first analysis of this circuit by assuming an ideal op amp situation and then modifying the analysis for the imperfections existing in the real world.

An ideal operational amplifier would display the following five characteristics:

Fig. 49-2. Schematic symbol for an op amp.

1. Infinite open-loop gain, A_{OL} = infinite
2. Infinite input impedance, Z_{in} = infinite
3. Zero output impedance, $Z_{out} = 0$
4. Zero offset voltage, $V_{OS} = 0$
5. Zero bias current, $I_b = 0$

Although these approximations are by no means conclusive, they are the values which influence most other characteristics. The approximations will simplify the analysis of operational circuitry. For example, the assumption of an infinite input impedance allows us to ignore the loss of any signal current into the amplifier's input terminal. The lack of bias current enables us to neglect the effect of this variable. These assumptions of the ideal amplifier can now be applied to the following circuits.

The Noninverting Amplifier

The noninverting operational amplifier circuit will usually be drawn as the schematic configuration shown in Fig. 49-5. Notice that both schematics are basically the same and are just redrawn for easier interpretation.

The voltage appearing at the inverting input terminal is at the same potential as V_{in}. Therefore, the voltage across R_1 will be the same as the potential of V_{in}, or

$$V_{R_1} = V_{in}$$

Knowing the above relationship, we can determine that the amount of current flowing through R_1 will be equal to the current flowing through R_F, the feedback resistor; it would then follow that

$$I_{R_1} = \frac{V_{in}}{R_1}$$

Voltage across R_F would be determined by the current flowing through itself, and this can be related back to the input voltage:

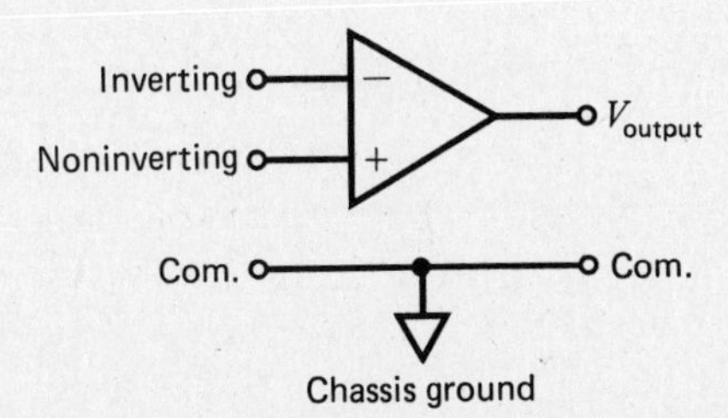

Fig. 49-3. The inverting and noninverting terminals for an op amp.

$$V_{R_F} = (I)(R_F) = \frac{(V_{in})(R_F)}{R_I} = V_{in}\frac{R_F}{R_I}$$

Since the left end of resistor R_F is at the input potential, the voltage across R_F is

$$V_{R_F} = V_{in}\frac{R_F}{R_1}$$

This signal appears to be noninverting at the left end of the resistor, and the output voltage V_{out} can be calculated as

$$V_{out} = V_{in} + V_S\frac{R_F}{R_I} = V_{in}\left(1 + \frac{R_F}{R_I}\right)$$

We now have two dependable methods for determining the voltage gain of a noninverting operational amplifier.

The first method makes use of the formula which states that the voltage gain of an amplifier is equal to the ratio of the output voltage V_{out} to the input voltage V_{in}, or

$$A_V = \frac{V_{out}}{V_{in}}$$

Combining this equation with one previous one, we can now realize that

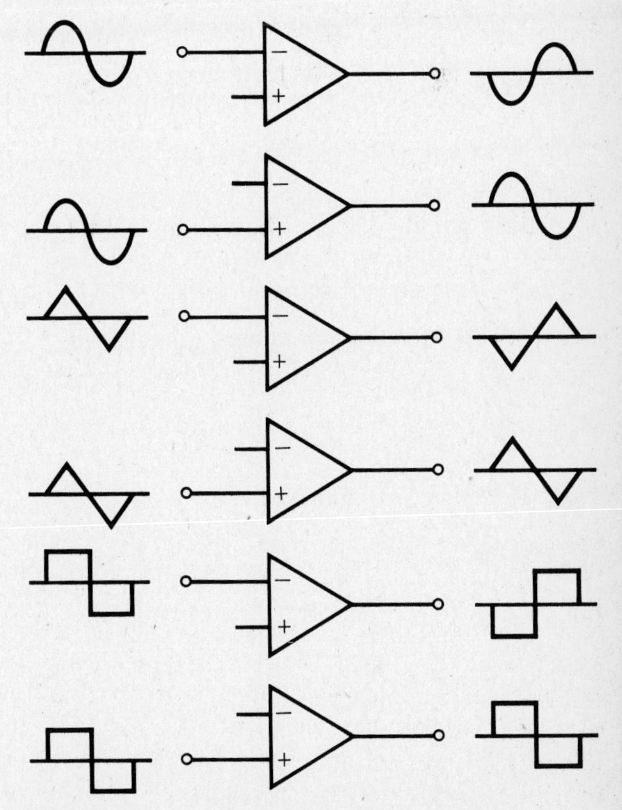

Fig. 49-4. Op amp input-output signal comparison.

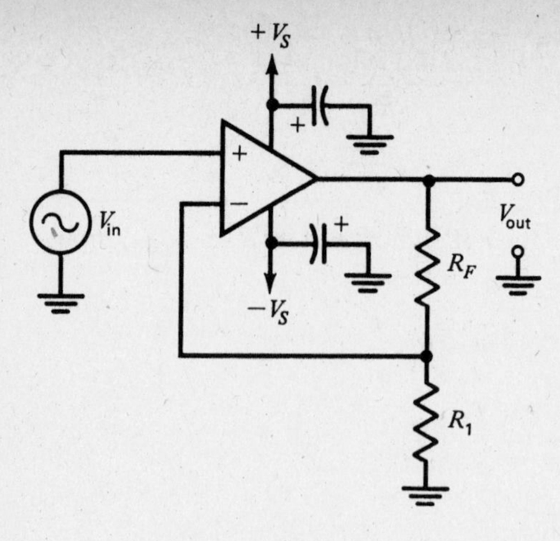

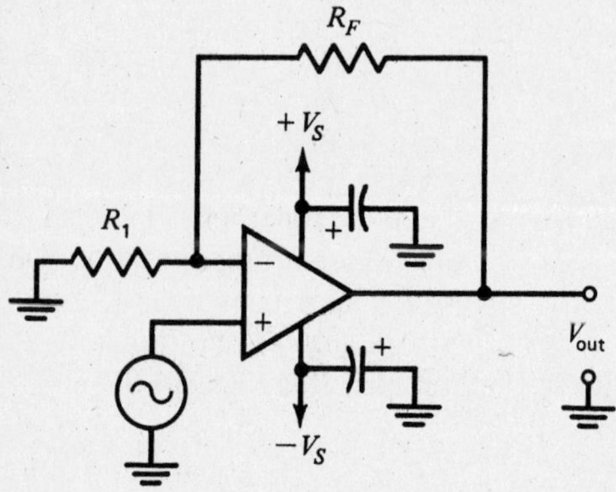

Fig. 49-5. Op amp noninverting schematic diagrams.

$$A_V = 1 + \frac{R_F}{R_1}$$

This demonstrates a unique aspect of the noninverting operational amplifier, which is that the voltage gain of this particular amplifier will always be greater than 1, or greater than that of a unity-gain amplifier. Therefore,

$$A_V > 1 \quad \text{(for a noninverting amplifier)}$$

Sample Problem 1. In a noninverting amplifier, where $R_1 = 10\ \text{k}\Omega$ and $R_F = 50\ \text{k}\Omega$, what will the output signal level be with an input signal (V_{in}) of 0.5 V? Draw the schematic diagram and show all work.

Given: A noninverting operational amplifier of

$$R_F = 50\ \text{k}\Omega$$
$$R_1 = 10\ \text{k}\Omega$$
$$V_{in} = 0.5\ \text{V} = 500\ \text{mV}$$

Find: Schematic diagram and V_{out}.

Solution: See Fig. 49-6.

$$A_V = 1 + \frac{R_F}{R_1}$$
$$= 1 + \frac{50\ \text{k}\Omega}{10\ \text{k}\Omega}$$
$$= 6.00$$
$$V_{out} = (V_{in})(A_V)$$
$$= (0.5\ \text{V})(6.00)$$
$$= 3.00\ \text{V}$$

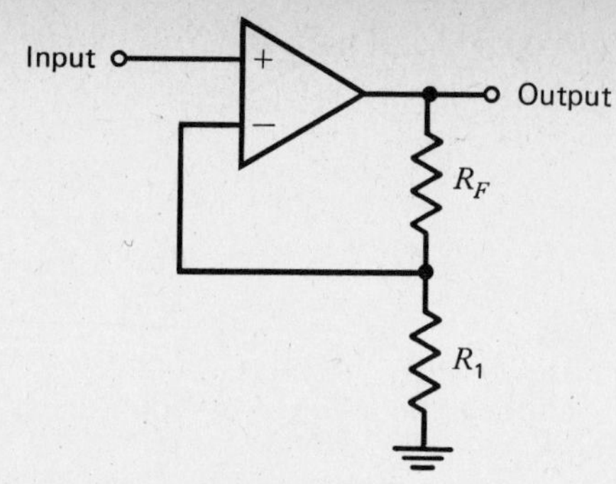

Fig. 49-6. Noninverting op amp.

The Inverting Amplifier

As previously stated, we can calculate the voltage gain of any amplifier by noting the relationship existing between the input and output signal magnitudes. The following equation is valid for any amplifier:

$$A_V = \frac{V_{out}}{V_{in}}$$

Operational amplifiers usually use feedback in the design circuitry in order for them to perform linearly. Negative feedback enables the designer to easily select and control the voltage gain. The inverting operational amplifier schematic diagram is shown in Fig. 49-7. Note in Fig. 49-7 that the power supply connections, V+ and V−, are implied but not shown.

Voltage at the inverting input terminal is at the same potential as the noninverting input terminal, or zero. This point is also said to be at a virtual ground reference level. The current through R_{in} will be equal in magnitude to the power supply voltage level divided by the resistance of R_{in}. This relationship is represented by the formula

$$I_{R_{in}} = \frac{V_{in}}{R_{in}}$$

The current of $I_{R_{in}}$ also flows through R_F since the current will not enter the operational amplifier, therefore the voltage potential across R_F is related to the current through R_F. This is also related to the parameters revolving around the input resistor R_{in}. Therefore,

$$V_{R_F} = I_{R_{in}} \times R_F$$
$$= \frac{V_{in}}{R_{in}} R_F$$

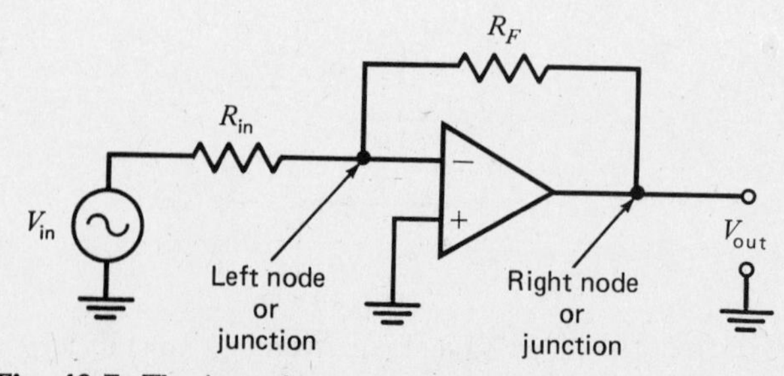

Fig. 49-7. The inverting op amp circuit.

The left side, or node, of R_F is at a zero reference point, and the right node of R_F must be equal to the output voltage V_{out}. But in reality the voltage across R_F is related to

$$V_{in}\frac{R_F}{R_{in}} = V_{out}$$

with the right node negative with respect to the input voltage V_{in}, represented by the formula

$$\frac{V_{out}}{V_{in}} = -\frac{R_F}{R_{in} = A_{V_{CL}}}$$

where

$$Z_{in} = R_{in}$$

Through rearranging, we have

$$V_{out} = -\frac{R_F}{R_{in}} V_{in}$$

and solving for the voltage gain,

$$A_V = \frac{V_{out}}{V_{in}} = \frac{R_F}{R_{in}}$$

In summary, it can now be realized that the voltage gain can be equal to, less than, or greater than 1:

$$A_V = 1$$
$$A_V < 1$$
$$A_V > 1$$

Typically, it will be found that R_{in} is at least 1 kΩ and that for the operational amplifiers the device's input impedance is equal to the input resistance of R_{in}. Therefore,

$$Z_{in} = R_{in}$$

Sample Problem 2. Notice in the circuit of Fig. 49-8 that a 1.00-mV peak ac signal is applied to the circuit and that an oscilloscope is connected to monitor the output signal. What would the expected magnitude and phase of the output signal be?

Given:

$$R_F = 100\ \text{k}\Omega$$
$$R_{in} = 1\ \text{k}\Omega$$
$$V_{in} = 1.00\ \text{mV}$$

Find: Phase of V_{out}.

Solution:

$$A_V = -\frac{R_F}{R_{in}} = -\frac{100\ \text{k}\Omega}{1\ \text{k}\Omega} = -100$$

$$V_{out} = (A_V)(V_{in}) = (-100)(1.00\ \text{mV}) = -0.1\ \text{V}$$

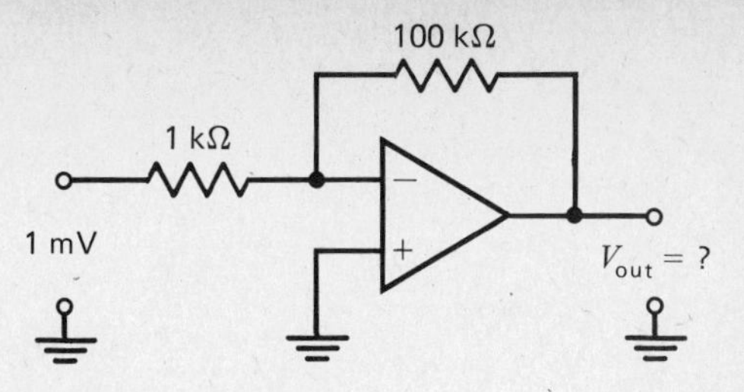

Fig. 49-8. Inverting op amp.

The output signal displays a phase shift of 180° out of phase from its input voltage V_{in} reference potential, hence the name *inverting amplifier*.

Sample Problem 3. In Sample Problem 2, what loading effect does this circuit present to the ac signal generator feeding it?

Given:

$$R_F = 100\ \text{k}\Omega$$
$$R_{in} = 1\ \text{k}\Omega$$
$$V_{in} = 1.00\ \text{mV peak}$$
$$A_V = -100$$
$$V_{out} = -0.1\ \text{V}$$

Find: Z_{in}.

Solution:

$$Z_{in} = R_{in} = 1\ \text{k}\Omega$$

The Buffer Amplifier

Buffering is the process by which the output load does not affect the operating conditions of the previous stage. The circuit employed to provide buffering is known as a buffer stage of a voltage follower (output voltage follows the input) stage. Operational amplifiers are commonly used for this function when a certain degree of isolation is needed between audio amplifier stages. The buffer amplifier will always yield a voltage gain close to 1. This type of op amp circuit results in a very high input impedance Z_{in} (typically over 100 kΩ) and a low output impedance Z_{out} (usually less than 100 Ω). The primary function of the operational amplifier in these types of circuits is to isolate the input source from the output load. For the buffer amplifier, the ability to amplify is not as important a factor as its ability to match a resistive source to a load. Figures 49-9 and 49-10 show a typical buffer circuit.

EQUIPMENT

Audio-frequency signal generator
Oscilloscope
DC power supply ± 15 V
Springboard or protoboard
Test leads

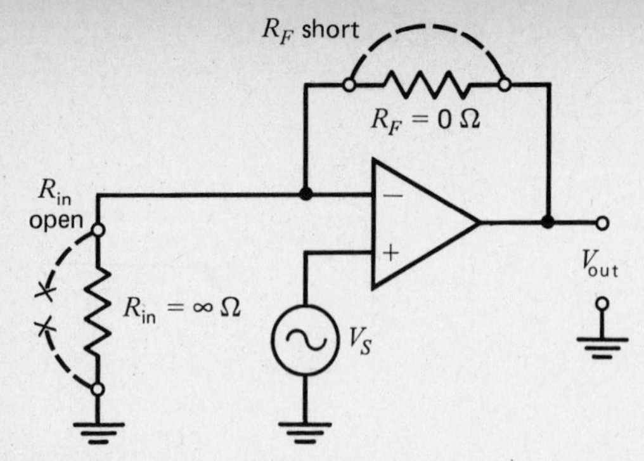

Fig. 49-9. The noninverting amplifier used as a voltage follower (buffer) circuit.

COMPONENTS

(1) 741 operational amplifier

Resistors (all 0.5 W):

(2) 10 kΩ	(1) 47 kΩ
(1) 27 kΩ	(1) 82 kΩ
(1) 39 kΩ	

PROCEDURE

1. Connect the circuit shown in Fig. 49-11. Apply power and observe and record the input and output traces on the oscilloscope.
2. Adjust the amplitude of the signal generator to $1.00\ V_{p\text{-}p}$ at 1.00 kHz. Notice the difference between the input and output wave shapes of the amplifier.
3. Verify that the gain of the amplifier is equal to 1.
4. Connect the circuit shown in Fig. 49-12.
5. Apply power and adjust the amplitude of the signal generator to $1.00\ V_{p\text{-}p}$ at 1.00 kHz. Compare the input to the output signal. Record the input and output traces on graticule paper.
6. Determine the voltage gain of the amplifier such that

$$A_V = 1 + \frac{R_F}{R_1}$$

Record this information.

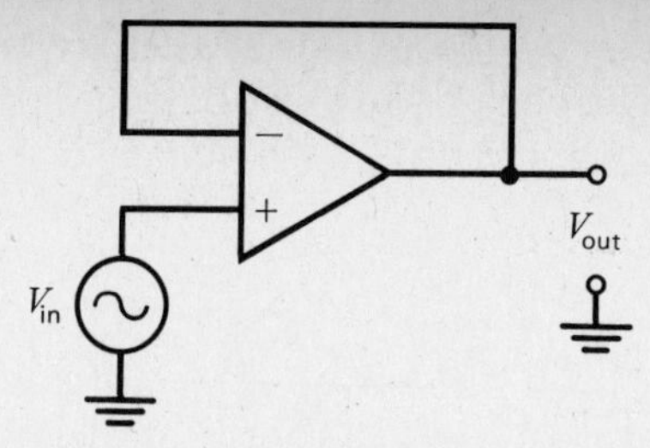

Fig. 49-10. Voltage follower stage with the elimination of $R_F + R_{in}$.

7. Maintain the input amplitude from the signal generator at $1.00\ V_{p\text{-}p}$ and change the value of R_1 to 27 kΩ, 39 kΩ, 47 kΩ, and 82 kΩ. Determine the results of the output signal level and the voltage gain for each change of R_F in Fig. 49-12 and record the results in Table 49-1.
8. Connect the circuit shown in Fig. 49-13. Apply power to the circuit and adjust the amplitude of the signal generator to $1.00\ V_{p\text{-}p}$ at a frequency of 1.00 kHz.
9. Compare the input to output signal of the amplifier. Record the input and output traces upon graticule paper.
10. Determine the voltage gain of the amplifier.
11. Repeat step 7 (for Fig. 49-13), and record this information in Table 49-2.

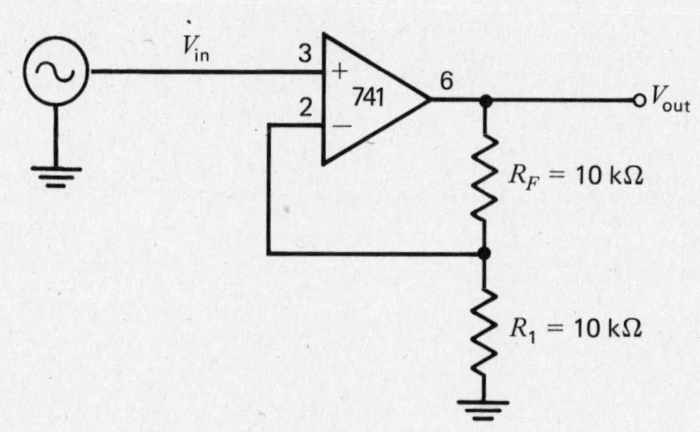

Fig. 49-12. Noninverting amplifier.

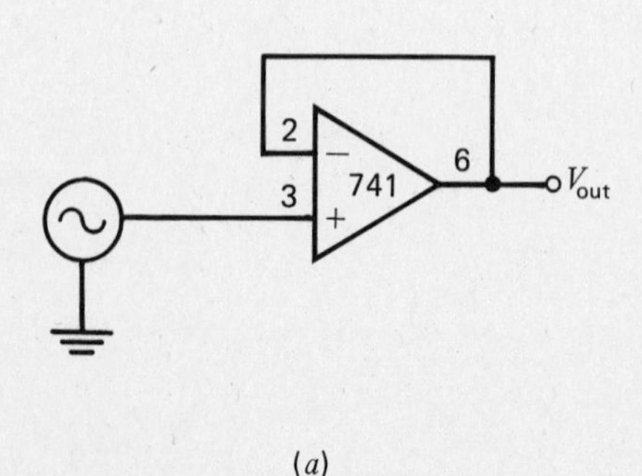

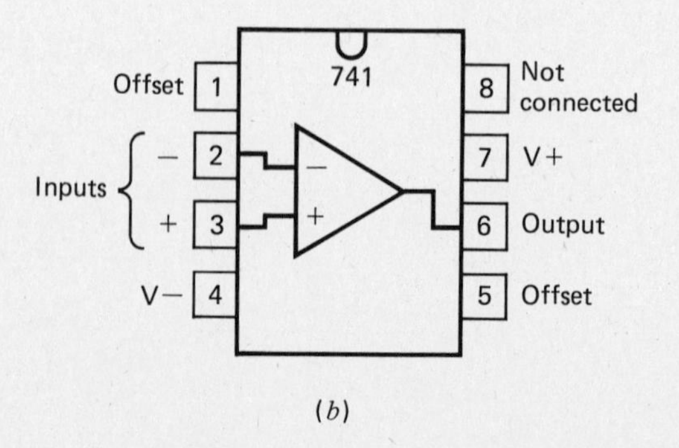

Fig. 49-11. (*a*) Voltage follower and (*b*) pinouts for a 741C.

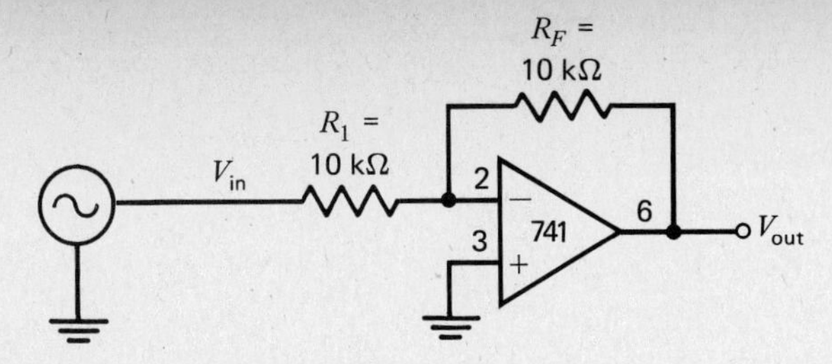

Fig. 49-13. Inverting amplifier.

NAME DATE

RESULTS FOR EXPERIMENT 49

QUESTIONS

1. For a voltage follower circuit, what is the difference between the input and output signals for phase and gain?

2. For the noninverting amplifier, how does the measured voltage gain compare with the formula for determining gain? What is the percentage of error between these two values?

3. For the inverting amplifier, do the experimental results agree with the gain equation?

4. What would the purpose be for a buffer amplifier?

5. What voltage gain can be expected from a noninverting amplifier?

REPORT

Write a complete report. Discuss the results. Discuss the three most significant aspects of the experiment and write a conclusion.

TABLE 49–1. $V_{in} = 1\ V_{p\text{-}p}$

R_1, kΩ	V_{out}	A_V
27	______	______
39	______	______
47	______	______
82	______	______

TABLE 49–2. $V_{in} = 1\ V_{p\text{-}p}$

R_1, kΩ	V_{out}	A_V
27		
39		
47		
82		

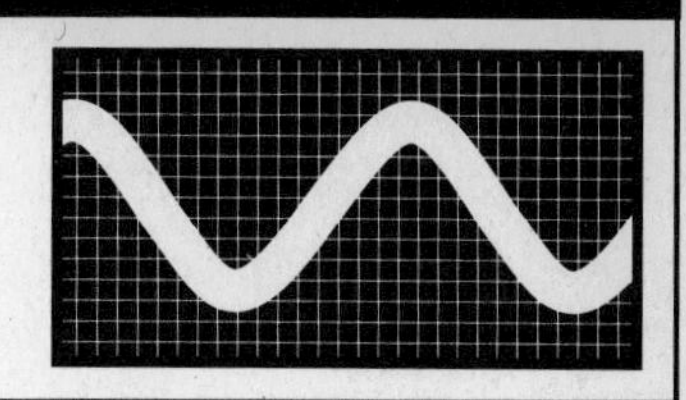

EXPERIMENT 50

INTEGRATED LOGIC CIRCUITS

OBJECTIVES

At the completion of this experiment, you will be able to:

- Verify the proper operation of an integrated circuit (IC) OR gate.
- Verify the proper operation of an IC AND gate.

SUGGESTED READING

Chapter 32, *Basic Electronics,* B. Grob, Seventh Edition

INTRODUCTION

The computers of today do not process numbers in base 10 (that is, 0, 1, 2, 3, . . . , 9). Computers instead use binary logic, base 2 (0 and 1), to perform their functions. Two fundamental circuits are the IC OR gate and the IC AND gate.

Integrated circuits contain a complete circuit, consisting of active and passive components connected in a unique circuit configuration contained in a space no larger than a transistor. The active components are transistors and diodes; the passive components are the resistors and capacitors. The components of an IC are not discrete parts wired together to form a complete circuit. Instead the IC is a complete circuit formed by a photographic process, on a small slab of silicon.

The purpose of this experiment is to introduce integrated logic circuits. The IC OR gate has two or more input signals but creates only one output signal. If any input signal is high, the output signal is high. The OR gate circuit constructed in this experiment has an IC orientation. Figure 50-1 shows an IC OR gate and its truth table.

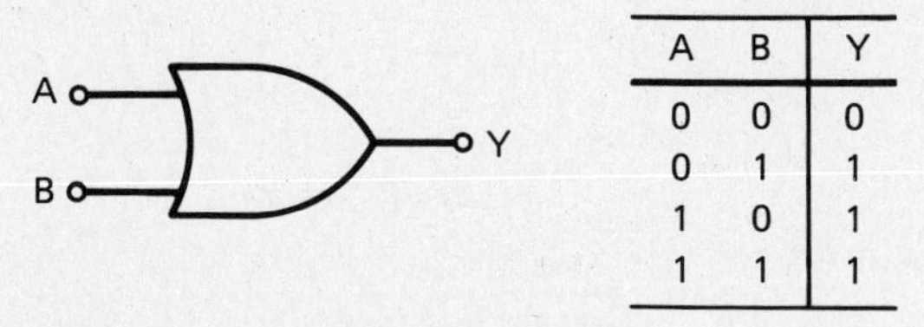

A	B	Y
0	0	0
0	1	1
1	0	1
1	1	1

Fig. 50-1. IC OR gate and truth table.

The IC AND gate has two or more input signals but creates only one output signal. All inputs must be high to get a high output signal. Figure 50-2 shows an IC AND gate and its truth table.

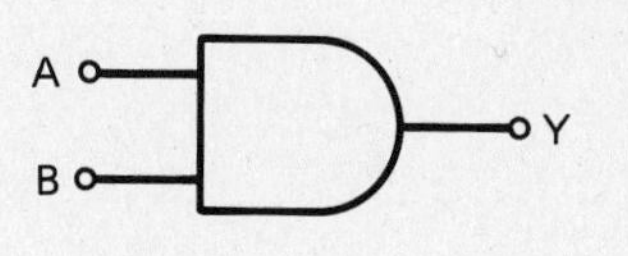

A	B	Y
0	0	0
0	1	0
1	0	0
1	1	1

Fig. 50-2. IC AND gate and truth table.

It is important to check the manufacturer's specification sheets for proper pin identification before use. The 7400 series ICs are available in the 14- or 16-pin format. The 7400 family of logic ICs uses pins 7 and 8 for the ground connection and pins 14 and 16 for +5 V.

EQUIPMENT

DC power supply, 0–10 V
VTVM
Protoboard
Test leads

COMPONENTS

(1) 7432 IC quad OR gate
(1) 7408 IC quad AND gate

PROCEDURE

1. Construct the circuit of Fig. 50-3. Be sure to connect the ground and +5 V to the proper pins (see your instructor).

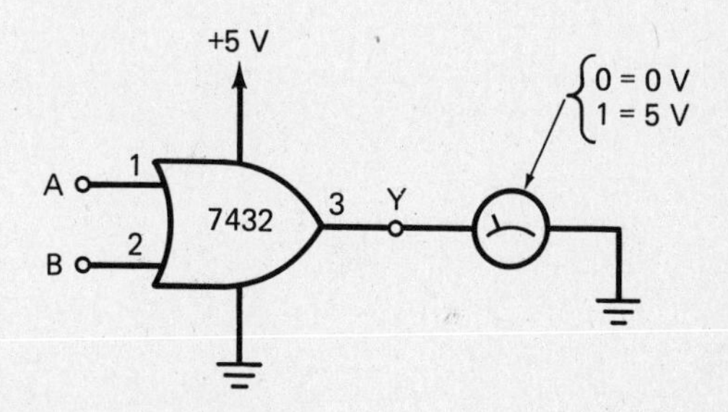

Fig. 50-3. OR gate circuit.

2. Connect inputs A and B to 0 (ground level), and measure the output level with the VTVM. Record these measurements in Table 50-1.

3. Connect inputs A and B to the states indicated in Table 50-1, and record the measurements of the VTVM in Table 50-1.

4. Construct the circuit of Fig. 50-4. Be sure to connect the ground and +5 V to the proper pins (see your instructor).

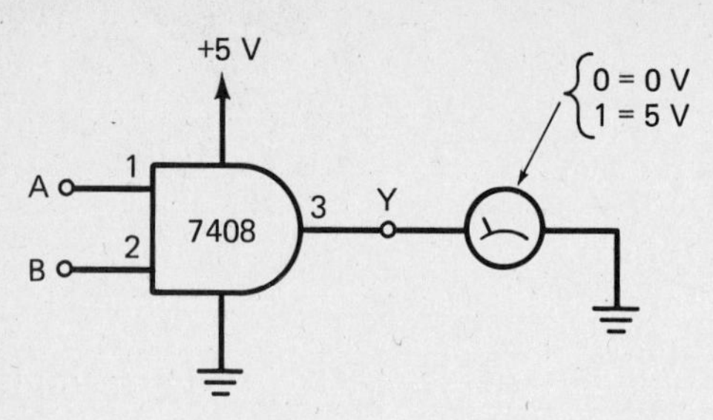

Fig. 50-4. AND gate circuit.

5. Connect inputs A and B to 0 (ground level), and measure the output level with a VTVM. Record these measurements in Table 50-2.

6. Connect inputs A and B to the states indicated in Table 50-2, and record the measurements of the VTVM in Table 50-2.

NAME

DATE

RESULTS FOR EXPERIMENT 50

QUESTIONS

1. In the above circuits, what voltage level would binary 1 represent? Binary 0? Are the answers the same for both circuits?

2. What is an OR gate?

3. In a three-input OR gate circuit, what percentage of the time is the circuit on?

4. What is an AND gate?

5. In a three-input AND gate circuit, what percentage of the time is the circuit off?

6. What are the fundamental differences between the two circuits constructed in this experiment?

REPORT

Write a complete report. Discuss the three most significant aspects of the experiment, and write a conclusion.

TABLE 50–1

A	B	Y
0	0	
0	1	
1	0	
1	1	

TABLE 50–2

A	B	Y
0	0	
0	1	
1	0	
1	1	

EXPERIMENT 51

TROUBLESHOOTING POWER SUPPLIES

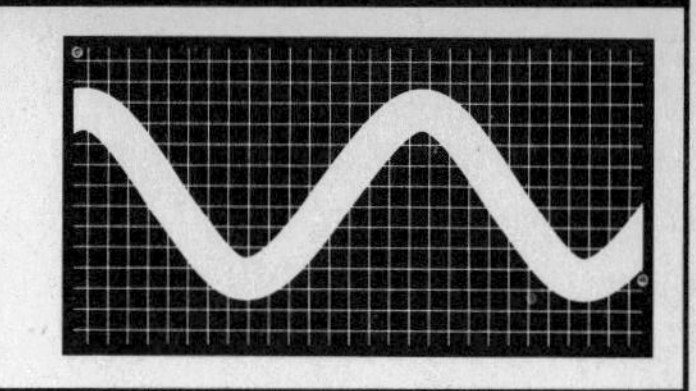

OBJECTIVES

At the completion of this experiment, you will be able to:

- Review the design of a basic power supply.
- Evaluate power supply design.
- Evaluate other power supply designs.

SUGGESTED READING

Chapters 11, 27, 28, and 29, *Basic Electronics,* B. Grob, Seventh Edition

INTRODUCTION

Most modern electronic devices require a dc voltage source or dc power supply, like a battery, to operate correctly. The power supply modifies the existing power in some way. There are ac power supplies as well as dc supplies.

Almost all power supplies that today's electronic technicians troubleshoot and repair have some form of regulation. This means that their outputs are designed to provide a constant level of voltage, regardless of how much current is needed to operate the equipment. Remember that connecting a load of any kind to a power supply is like connecting a resistor in series with a battery. Therefore, current will flow, and a voltage drop will be present across the load (resistor) and the battery (its internal resistance). A regulated power supply is designed to eliminate the effects of the supply's loading (voltage drop), so that the load will always have a constant fixed voltage across it, no matter how much current flows in the circuit. Although this explanation is extremely simple, it provides a foundation to get started.

In summary, regulation in a power supply means the ability to maintain the output voltage level under various load conditions. However, before the supply voltage can be regulated, an *unregulated* voltage must be provided. To do this, the dc power supply will have to use a rectifier and filter system which helps convert the alternating current, like the 110-V 60-Hz line voltage, to a relatively smooth direct current, as a battery provides, for the product (load) to operate.

There are many different types of power supply circuits. However, most supplies have some basic components that perform fundamental tasks required of all power supplies. Therefore, let's review these tasks and the components that perform them.

Transformers

Most power supplies use a transformer to transform or change the line voltage from one level to another. Usually, this is a reduction in voltage from the large line voltage (110 V) to a smaller voltage used for the power supply. This is usually referred to as a *step-down transformer* because it steps down or decreases the line voltage. Also remember that a transformer consists of two separate coils or inductors (a primary and a secondary) that are actually lengths of wire wrapped around a core. Besides reducing the voltage, the transformer isolates the secondary voltage from the line voltage.

In troubleshooting, it is common practice to check the continuity of the transformer's windings with an ohmmeter, to be sure that the wires have not shorted or opened. Always do this with the power off, and be sure to unplug the equipment.

Fuses

Most power supplies have line fuses connected in series with the transformer. Typically, fuses protect the product's circuitry from any great changes in ac line voltage or in the amount of current drawn by the load. Although it may seem obvious to check a fuse whenever a product fails to operate, several common problems are often overlooked by many technicians. For example, fuse ratings must be checked and adhered to for proper operation. If a fuse is supposed to be 1 A and a 0.5-A fuse is used, the product may work for a while and suddenly appear to break. Also if a slow-blow fuse is specified but not used, the product may continually blow the fuse, no matter what rating is used. *Slow-blow fuses cannot or should not be replaced with regular fuses, and regular fuses should not be replaced with slow-blow fuses.* This simple rule cannot be overemphasized. Always check the fuse when you are troubleshooting.

Rectification

The term *rectification* refers to the concept of changing ac voltage to dc voltage. Therefore, it is common to have a rectifier, typically made of diodes that conduct in only one direction, at the output of the secondary. A wide variety of rectifiers are used in dc power supplies. The type of rectification system employed depends on the design and power demands of the circuit used.

Ripple

Remember that the purpose of any filtering system is to remove the ac component of the rectified ac line voltage. The ideal dc output would be like a battery: pure direct current with no trace of alternating current. However, because no filter can remove 100 percent of the ac component, there is always some amount of what is called *ripple* on the direct current. This ripple on the dc level is usually represented by a percentage:

$$\text{Ripple factor, \%} = \frac{V \text{ ripple}}{V \text{ dc, out}} \times 100$$

For troubleshooting purposes, the ripple factor of any filtered output should never exceed 1 percent. In fact, 0.2 percent or less is a good value.

Safety

There is nothing more important for a technician than safety. Regardless of how much knowledge or experience a person has, one small mistake or accident can lead to injury or even death from electric shock.

Electricians are trained to measure high voltages with one hand. As in the example above, you are putting your body (especially your heart and lungs) in the middle—between your two hands—when this occurs. So be careful when you are dealing with line voltages. And never stay near water when high voltages are present.

If someone receives electric shock, always remove the person from the source as quickly as possible. You can do this by shutting off the product's line switch or by simply pulling the plug. Then proper medical attention should be administered as soon as possible. In the case of heart failure or severe shock, CPR or artificial respiration should be administered immediately.

However, the best way to avoid injury or worse is to follow these simple rules:

1. Never work near water.
2. Use a separate isolation transformer with a 1:1 voltage ratio. Plug the transformer into the ac line, then plug the product and/or equipment into the transformer. Even a fused power distribution bar serves as protection.
3. Never work on equipment while you have metal jewelry on your fingers, wrists, neck, or arms.
4. Never walk away from equipment or products that are exposed. Always turn off the power, and put some cover over the opening to prevent accidental discharge of high-voltage capacitors.
5. Never use equipment that is underrated for the work you are doing. This is especially true for high-voltage probes used to test CRT anodes.
6. Always refer to the service manual or call the manufacturer for advice if no manual is available.
7. Always replace components with recommended parts. Do not use components that are out of tolerance or underrated. And always discharge capacitors with an insulated-handle screwdriver before you try to remove them.
8. Never work on equipment that you are not qualified to troubleshoot or repair. Even if you do not injure yourself, you could injure or ruin the equipment.

Basic Power Supply Troubleshooting

Although experience and factory-type training courses are the best way to learn how to test and repair equipment, two troubleshooting summaries are presented below. One is a general set of principles, many of which are reinforced in the experiments in this manual.

Power supplies are often the cause of most consumer product failures because their circuits use high currents and large voltages and dissipate a lot of heat, which can cause component damage.

1. Check Fuses. Obviously, this is the easiest repair you can make. But don't be too quick to judge a bad fuse. Remember that slow-blow fuses cannot be replaced by regular quick-blow types. Always check the amperage rating.
2. Check the Transformer Windings. This is not often a problem with newer equipment. However, older products typically have had more time running under higher temperatures. Therefore, a greater chance exists that the coating around windings has melted or worn away, causing a short, or there is a chance that heat has caused an open. Disconnect the primary and any secondaries from the circuit, and use an ohmmeter to check for continuity. Don't forget that you should always check the input and output voltages to verify their levels with the schematic.
3. Check the Rectifier. A defective diode can cause lowered voltage or no voltage. Use the voltmeter to check ac input and dc output voltages.
4. Check Filter Capacitors. Excessive audible noise, called *hum,* and low output voltage can be caused by a bad filter capacitor. Check the capacitors out of the circuit for open resistance. And check their input and output voltages with an oscilloscope, to verify their filtering (smoothing) ability.
5. Check the Regulator Circuitry. This is often the most difficult part of troubleshooting a regulated power supply. If the rectifier and filter elements are good, then the regulator portion is suspected. You will have to check all components, one at a time, to

verify their nominal values or their operation. There is no standard way to check this circuitry except to use a variable high-wattage load resistance and to simulate operation with the real load disconnected. Voltages around the transistor and any other components can then be measured. Use an oscilloscope or voltmeter and any schematic provided. Remember that the theory you just read in the previous section provides a basis for analyzing the measurement results. For example, if you decrease the load and draw more current, you should see some kind of change in the bias around the pass regulator. If there is no change, you can be sure that the regulator circuit is faulty. At that point, you may have to replace components, one at a time, if you cannot determine the faulty one. Many experienced technicians use this method.

6. Check the Load. Connect sections of the load circuit, one at a time, if possible. Remember that a short in the load can cause the power supply to fail. Also if the load is faulty, the power supply operation may appear to be faulty but is not.

Typical Power Supply Problem

No Voltage: This problem could be caused by a blown fuse or a failure in the filter circuitry.

Solutions: Measure the primary voltage V_{in}. Visually inspect the supply for heat damage—look for darkened components. Replace the fuses. Use the ohmmeter to check for continuity or opens where applicable. If the supply has a total resistance that is low, it is suggested that the load be removed. If the total resistance is high, the problem is then due to the load. After this, reconnect the supply and power on again. If the fuse blows, the supply may have a short. The procedure discussed how to check all the discrete components for typical resistance and voltage readings. It then discussed how to connect each part of the circuit at a time to determine where the short was. Finally, the procedure discussed more possibilities, including the use of a variac (vary input voltage), before coming to a conclusion.

The circuit you will develop is a low-voltage dc power supply. As Fig. 51-1 shows, it is a full-wave bridge rectifier with a pi filter.

The output of this power supply is approximately 12 V. The load is indicated as purely resistive and is designated as R_1. Resistor R_1 will draw a maximum current of 150 mA; R_b is a bleeder resistor and is always connected.

Resistor R_1 is designed to draw 150 mA at 12 V. From Ohm's law R_1 is calculated at 80 Ω. Given the value of R_1, V_{out}, and the load current, the power delivered to the load P_1 can be calculated at 1.8 W. Note that P_1 is not equivalent to peak power P_{peak}. And P_{peak} can be determined if V_{peak} and I_{peak} are known. So V_{peak} equals $V_{rms}/0.707 = 17.8$ V. And I_{peak} equals V_{peak}/R_1, which is 0.222 A. Therefore P_{peak} is equivalent to the product of V_{peak} and I_{peak} and is 3.95 W. Normally a power supply is derated in terms of power by a factor of 2, so an 8-W power rating is advised.

In determining a value of R_b, I_b must be calculated. Bleeder currents are normally one-tenth (0.10) of the load current. In this case I_b is 0.015 A. Now R_b can be determined from Ohm's law as 12 V/0.015 A, or 800 Ω. Power dissipated by R_b can be calculated again as V_{peak}/R_b and is 0.0178 A. Also P_{peak} is determined as the product of V_{peak} and I_{peak} and is 0.316 W.

In the schematic, the filter capacitors have been specified at 50 μF. Their dc working voltage rating must be at least 12 V. A standard value of 15 WV dc makes a good choice.

The next component to be selected is the filter inductor. The inductance value has been specified as 1 H. The current rating must be at least 165 mA and should be derated to at least 300 mA. Note that using an inductor with a higher than required current rating reduced dc losses due to the larger size of the wire used in the windings.

The diodes used in the full-wave rectifier have two main ratings: maximum current and peak inverse voltage. The normal average current for this circuit approaches 200 mA. However, when it is first energized, a large surge exists because the uncharged state of the capacitor provides little opposition to current. Allowing for a margin of safety, diodes with a 1-A rating should be sufficient.

The last component to be selected is the transformer, and it has a specified secondary voltage of 12.6 V. We have already estimated an average power supply current of approximately 200 mA, and most transformers with a specified voltage far exceed that capacity. Therefore, a 12.6-V transformer rated at 0.5, 1.0, or 1.5 A will be satisfactory.

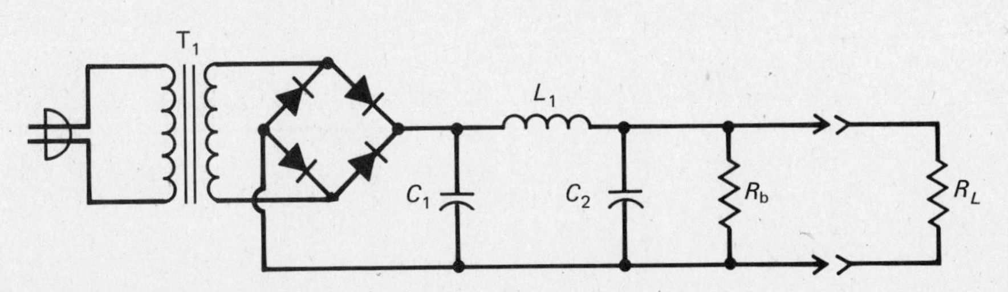

Fig. 51-1. Full-wave bridge rectifier circuit.

EQUIPMENT

Oscilloscope
VTVM
Protoboard
Test leads
Filament transformer (120 V:12.6 V)

COMPONENTS

(4) 1N4004 diodes or equivalent
(2) 47-μF capacitors
(1) 1-H inductor (choke) or largest available
(4) 330-Ω 1-W resistors connected in parallel
(1) 1000-Ω 1-W resistor
(1) 100 kΩ 1-W resistor
(1) SPST switch

PROCEDURE

1. Build the circuit shown in Fig. 51-2. Do not energize. Doublecheck all component connections.
2. When you are satisfied that the circuit is connected correctly, energize it.
3. Measure the unloaded dc and ac peak-to-peak ripple voltages. Record these measurements in Table 51-1 in the "Results" section of this experiment.
4. Disconnect the power to the supply.
5. Connect the load resistor to the output terminals of the circuit, as shown in Fig. 51-3.
6. Measure and record the fully loaded dc and ac peak-to-peak ripple voltages in Table 51-1.
7. Disconnect the power to the supply.
8. From the information measured, calculate the percentage of regulation by the formula

$$\% \text{ regulation} = \frac{V_{\text{no load}} - V_{\text{full load}}}{V_{\text{full load}}} \times 100$$

Record the percentage of regulation in Table 51-1. However, note that a low percentage value would indicate an output voltage that does not vary much from full-load to no-load condition. A low percentage is preferable and indicates a more constant output voltage.
9. With the power disconnected to the supply, remove R_L from the circuit. Connect a voltmeter across R_b. Energize and deenergize the supply circuit. Record how long it takes for C_2 to discharge.
The RC time constant for the R_b and C_2 components should be totally discharged in 5 time constants. The length of time required can be calculated by

$$\begin{aligned} \text{Time} &= 5T \qquad \text{where} \qquad T = RC \\ &= 5R_bC_2 \\ &= 0.25 \text{ s} \end{aligned}$$

10. With the circuit deenergized, change R_b to 100 kΩ. Energize and deenergize the circuit, and note the discharge time. Calculate the discharge time for this circuit.

Note: If there were no bleeder resistor, how long would circuit discharge take? In some circuits, a discharge time of days is possible. For this reason, never operate a power supply without a bleeder resistor.

11. When the capacitor has fully discharged (power off), replace the 100-kΩ resistor with the original 1-kΩ resistor.
12. With the circuit deenergized and R_1 disconnected, connect an oscilloscope across the power supply output terminals. Energize the circuit. With the oscilloscope correctly adjusted, measure the peak-to-peak ripple voltage across R_b. Record this value and the calculated value of V_{rms} in Table 51-1.
The ripple voltage and ripple percentage provide an indication of the quality or smoothness of a dc

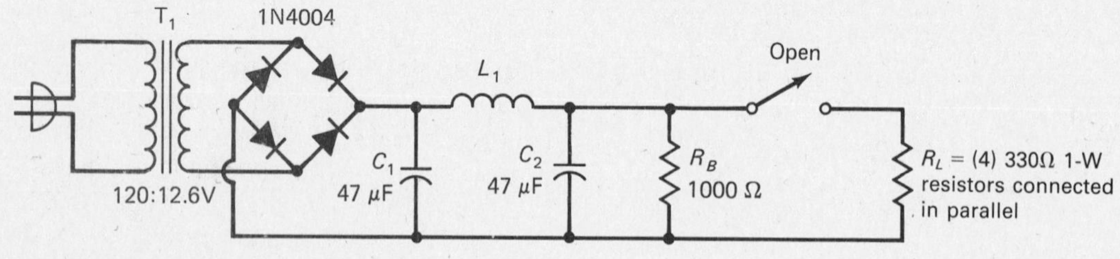

Fig. 51-2. Full-wave test circuit.

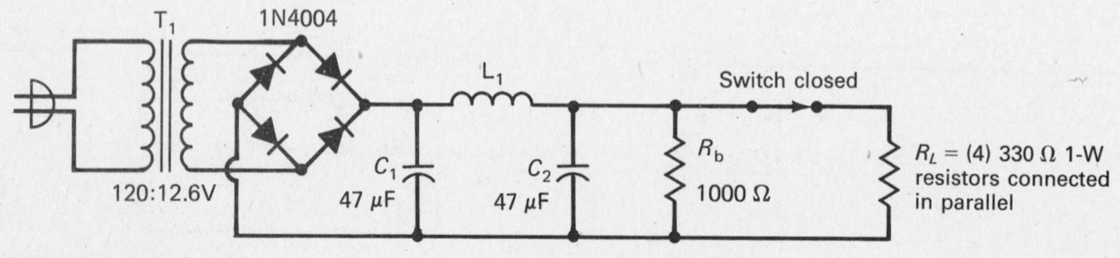

Fig. 51-3. Full-wave circuit loaded.

voltage. Ripple must be measured under both load and no-load conditions. Ripple is measured with an oscilloscope. The results will be in peak-to-peak values; after the measurement is taken, convert to rms values.

13. Deenergize the circuit.

14. With the circuit deenergized, connect R_1 across R_b. Energize and record the peak-to-peak ripple voltage in Table 51-1. Calculate the V_{rms} equivalent and record it in Table 51-1. Deenergize the circuit.

The full-load ripple should be larger and have a different waveshape than no-load ripple. The ripple waveshape should change from a sawtooth to a sine wave. Under no load, the capacitor remains charged longer, and the inductor provides little effect. When the load is increased the capacitor discharges faster and the inductor begins filtering more.

15. Determine and record in Table 51-1 the ripple percentage from

$$\text{Ripple, \%} = \frac{V_{rms}}{V_{dc}} \times 100$$

A low ripple percentage indicates a high-quality filter. When the load is increased, the ripple voltage increases, and the dc voltage level decreases.

Power Supply Defect

The most common defect in a power supply is an open fuse or circuit breaker. The next most probable problem is an open diode.

Shorted, open, and leaking capacitors are common. The symptoms produced by these defects are usually recognizable to a skilled technician.

16. With the power supply deenergized, remove one diode, as shown in Fig. 51-4. Making sure that the circuit is under full load, energize the circuit. Measure and record the dc and ripple levels; note the ripple waveshape.

17. Compare the output characteristics of the open-diode circuit in step 16 with the original circuit characteristics.

A power supply overload or a shorted filter capacitor will result in excessive diode current. This excessive diode current can cause a diode to open. If this happens in a full-wave bridge-type circuit, the circuit will operate as a half-wave circuit. This defect will result in obvious symptoms. The output voltage will drop, the ripple voltage will change, and the ripple frequency should decrease from 120 to 60 Hz.

18. With the circuit energized, measure the dc voltage across the open diode at points *A* to *B*. As soon as you are finished examining the circuit and taking measurements, deenergize and replace the diode.

19. With the circuit deenergized and the diode replaced, remove capacitor C_1, as shown in Fig. 51-5. This simulates an open-capacitor condition. The capacitor is supposed to charge to the peak value of the input signal. Therefore an open capacitor should cause an output voltage drop. Removal of the capacitor will also cause the ripple to increase.

20. With the load R_1 connected, energize; then measure, and record the dc output voltage, ripple voltage, and ripple frequency in Table 51-1.

21. Deenergize the circuit and reconnect capacitor C_1.

22. With the circuit deenergized, remove C_2, as shown in Fig. 51-6.

23. Energize the circuit and take measurements of V_{dc} and V_{ripple}. Capacitor C_2 is important in removing ripple. Its effect on the dc voltage level may not be as noticeable compared to the effects of the input capacitor C_1. Both capacitors have an effect on the ripple and dc average; but when compared, the input capacitor C_1 has the greatest effect overall.

24. Deenergize the circuit and disconnect. The effect on short circuits was not discussed in this experiment. It was not presented because electric shorts

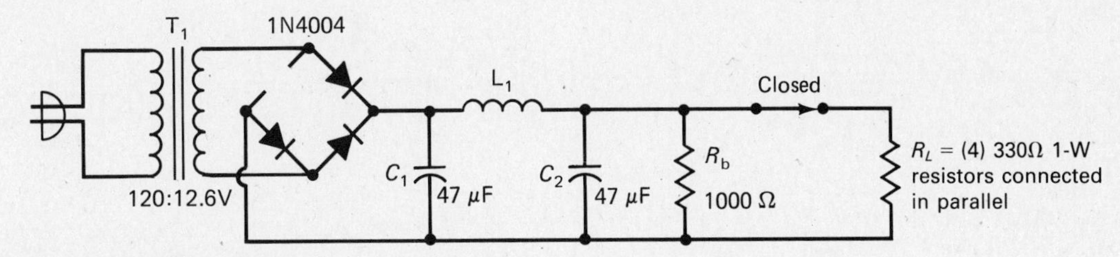

Fig. 51-4. Full-wave open diode test circuit.

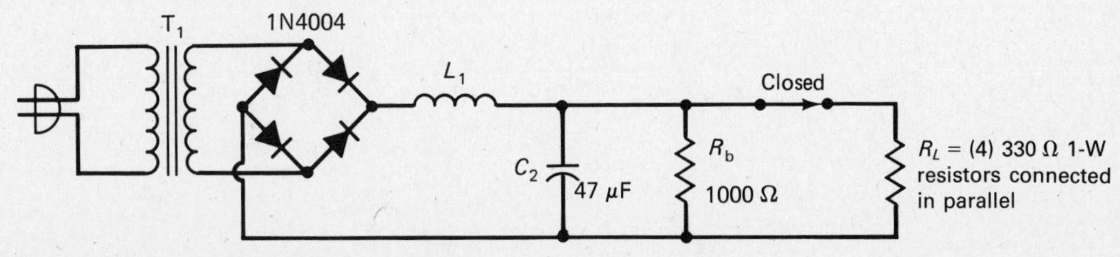

Fig. 51-5. Full-wave open input capacitor test circuit.

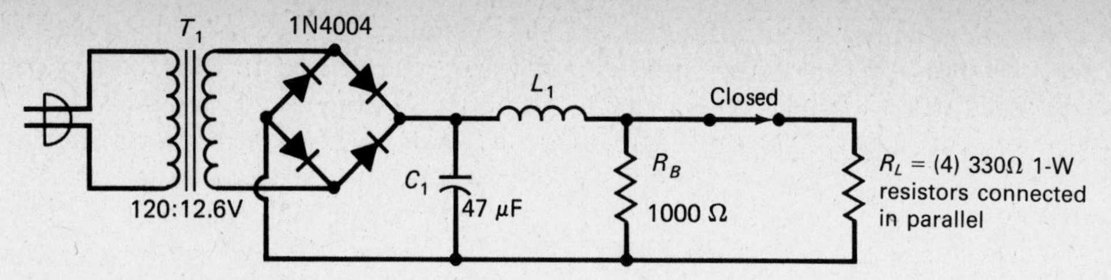

Fig. 51-6. Full-wave open output capacitor test circuit.

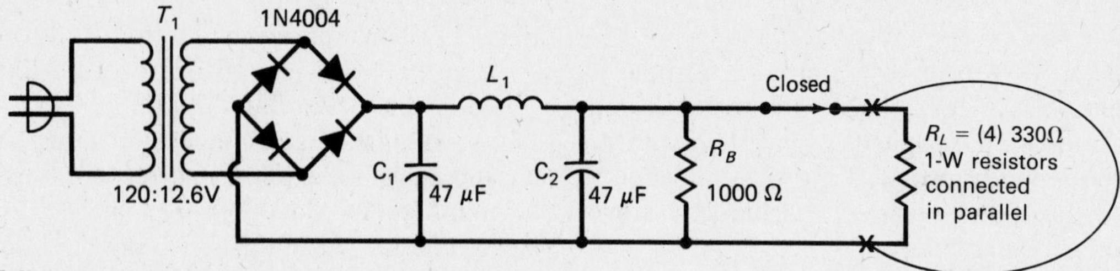

Fig. 51-7. Full-wave shorted output circuit (do not construct!).

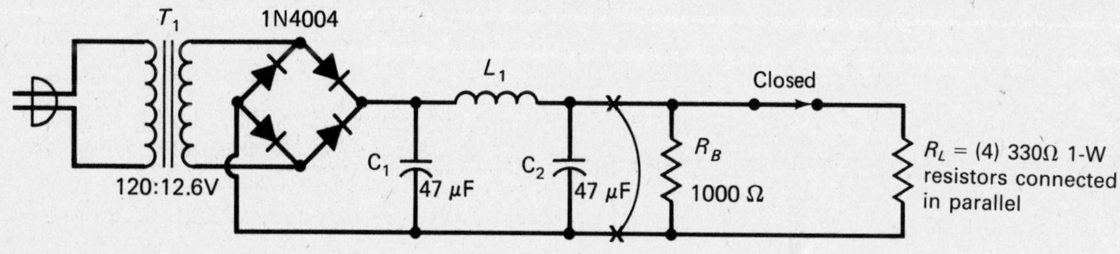

Fig. 51-8. Full-wave shorted capacitor circuit (do not construct!).

can cause damage. Consider the shorted circuit in Fig. 51-7. The overall effect would be an increase in output current.

The sudden increase in current would continue until the maximum ratings of the components were reached. Circuit opens would occur, most likely a diode.

Another circuit condition (see Fig. 51-8) is a shorted or leaky capacitor. A leaky capacitor occurs when the dielectric strength deteriorates and allows more current to flow through the capacitor. Again increased current flow manifests itself in opened components such as a diode.

NAME DATE

RESULTS FOR EXPERIMENT 51

QUESTIONS

1. What are the circuit characteristics of a power supply with an open output capacitor?

2. What are the circuit characteristics of a power supply with an open input capacitor?

3. What are the circuit characteristics of a power supply with an open rectifier diode?

4. What does the percentage of full-load ripple indicate?

5. Why is a bleeder resistor important in a power supply?

6. What safety factors must be observed when you are working with a power supply?

REPORT

Write a complete report. Discuss the measured and calculated results. Discuss the three most significant aspects of the experiment, and write a conclusion.

TABLE 51–1

Procedure Step	Value	Measured	Calculated
3	V_{dc}	______	
3	V_{ripple}	______	
6	V_{dc}	______	
6	V_{ripple}	______	
8	% Regulation		______
9	T for C_2	______	______
10	T for C_2	______	______
12	V_{ripple}	______	
12	V_{rms}		______
14	V_{ripple}	______	
14	V_{rms}		______
15	% Ripple		______
16	V_{dc}	______	
	V_{ripple}	______	
18	V_{dc}	______	
20	V_{dc}	______	
	V_{ripple}	______	
	f_{ripple}	______	
23	V_{dc}	______	
	V_{ripple}	______	

OPTIONAL EXPERIMENT 52

VACUUM TUBE AMPLIFIER

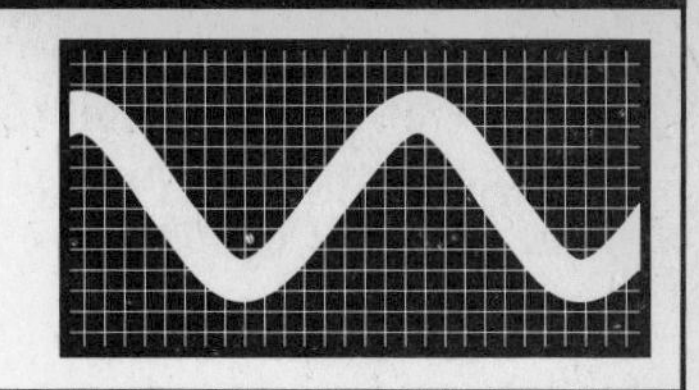

OBJECTIVES

At the completion of this experiment, you will be able to:

- Study the operation of a tube as an amplifier.
- Measure the voltage gain of a vacuum tube amplifier.
- Study a family of characteristic curves for a vacuum tube.

SUGGESTED READING

Chapter 28, *Basic Electronics,* B. Grob, Seventh Edition

INTRODUCTION

Vacuum tubes are able to operate efficiently over a wide range of frequencies. Their operation is based upon their ability to control, almost instantaneously, the flow of millions of electrons. The main difference between vacuum tubes and transistors is that vacuum tubes are controlled by voltage, whereas transistors, typically, are controlled by current.

Vacuum tubes can be used as (1) voltage amplifiers, (2) power amplifiers, (3) detectors, (4) oscillators, (5) frequency detectors, (6) regulators, (7) modulators, and (8) rectifiers. As a rectifier, a vacuum tube, like a silicon diode, will allow current to flow in only one direction. The vacuum tube operates as follows:

Within the evacuated (void of air) tube, there are electric elements called a *cathode* (an electron emitter, given the sign −) and a *plate* (the electron acceptor, given the sign +). If a positive voltage is applied to the plate, electrons emitted from the cathode are attracted to the plate. In this case, the cathode must be heated in order to release the electrons. One method of heating the cathode so that it will release electrons is through the use of a filament supply, typically 6.3 V_{rms}. With the cathode heated, and a positive voltage applied to the plate (positive electrode), electrons will flow from the cathode (−) to the plate (+), creating the same results as if a PN junction diode was forward-biased.

When a third electrode, called a *control grid,* is added inside a vacuum tube, the tube is then called a *triode.* It is located between the plate and the cathode, as shown in Fig. 52-1. The control grid actually controls the number of electrons allowed to flow from the cathode to the plate. Again, voltage is the controlling factor. If the control grid is made more negative than the cathode, some of the electrons going toward the plate will be repelled by the grid, thus reducing the plate current. If the plate is made as positive as possible and the maximum number of electrons given off by the cathode is attracted to the plate, the triode is said to be *saturated.* That is, plate current has reached a saturation point, and any increase in the plate voltage will not increase plate current.

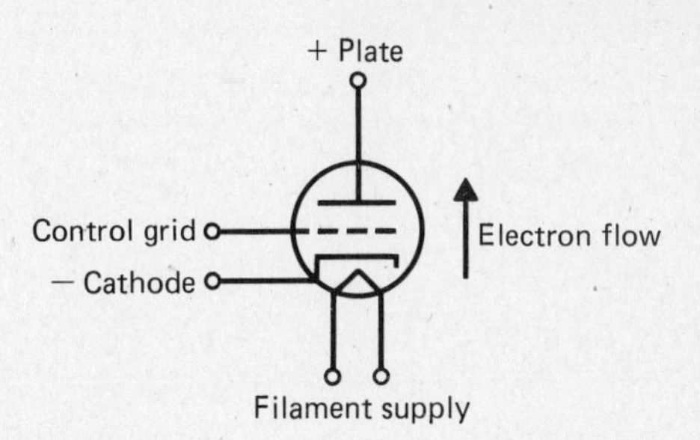

Fig. 52-1. Triode.

Conversely, if the control grid is biased so negative as to prohibit the flow of electrons from cathode to plate, the triode is said to be a *cut off* triode. In this case, the flow of current is stopped, or cut off.

The operation of a triode tube can be plotted on a load line similar to that of a transistor, with the plate voltage on the x axis, plate current on the y axis, and differing values of grid voltage as a family of curves.

Vacuum tubes have characteristics that are denoted by particular standard symbols. The important characteristics and their symbols are listed below. Because a vacuum tube has nonlinear characteristics (that, is, characteristics not bounded by constant limits), the characteristics are referred to as *parameters.* If the parameter is denoted by an uppercase symbol, it refers to a dc parameter based on a fixed value of current and voltage. When a lowercase symbol is used, it is referred to as a dynamic or an ac parameter.

Plate Resistance

The plate resistance is the measure of opposition to the flow of current through the tube. The formula is basically Ohm's law. For dc parameters,

$$R_p = \frac{E_p}{I_p}$$

where E_p = dc plate voltage
I_p = dc plate current

For ac parameters,

$$r_p = \frac{\Delta e_p}{\Delta i_p} \qquad \text{(with } e_g \text{ constant)}$$

where r_p = ac plate resistance
Δe_p = small change in plate voltage
Δi_p = small change in plate current

Amplification Factor

The amplification factor is a measure of how much more control the grid voltage exerts over plate current compared to the plate voltage. The symbol used to denote amplification is the Greek letter mu (μ). The formula for this calculation is, for dc parameters,

$$\mu = \frac{E_p}{E_g}$$

and for ac parameters,

$$\mu = \frac{\Delta e_p}{\Delta e_g} \qquad \text{(with } i_p \text{ constant)}$$

where μ = amplification factor (a ratio)
Δe_p = small change in plate voltage
Δe_g = small change in grid voltage

Transconductance

The transconductance is a factor which relates the amount of plate current change which is caused by a grid voltage change. The formula is basically one of conductance, where $G = I/E$. For dc parameters,

$$G_m = \frac{I_p}{E_g}$$

For ac parameters,

$$g_m = \frac{\Delta i_p}{\Delta e_g} \qquad \text{(with plate voltage held constant)}$$

where g_m = transconductance, in siemens
Δi_g = small change in plate current
Δe_g = small change in grid voltage

EQUIPMENT

High-voltage dc power supply (0–300 V)
Low-voltage dc power supply (0–10 V)
Filament power supply
Audio signal generator
Test leads
Tube socket for the 6SN7 or 6J5
Test leads

COMPONENTS

(1) Tube 6J5 or 6SN7
(2) 0.01-μF capacitor (disc)
(1) 22-kΩ 2-W resistor
(1) 100-kΩ 0.5-W resistor

PROCEDURE

1. Connect the circuit shown in Fig. 52-2. Be careful with the high-voltage power supply. Do not keep on the circuit while the plate voltage is supplied.

Note: B+ is the standard symbol for the plate supply voltage. Here, B+ = 300 V dc. The 22-kΩ resistor is referred to as R_p, plate resistor. The 100-kΩ resistor is referred to as R_g, grid resistor. Directly below the grid resistor is the variable power supply, V_{gg}. This is a separate power source that controls the grid voltage, which in turn controls the flow of electrons from the heated cathode (−) to the plate, also referred to as the *anode* (+). Also, note that the filament is not drawn on the schematic of Fig. 52-2 (it is assumed to be connected).

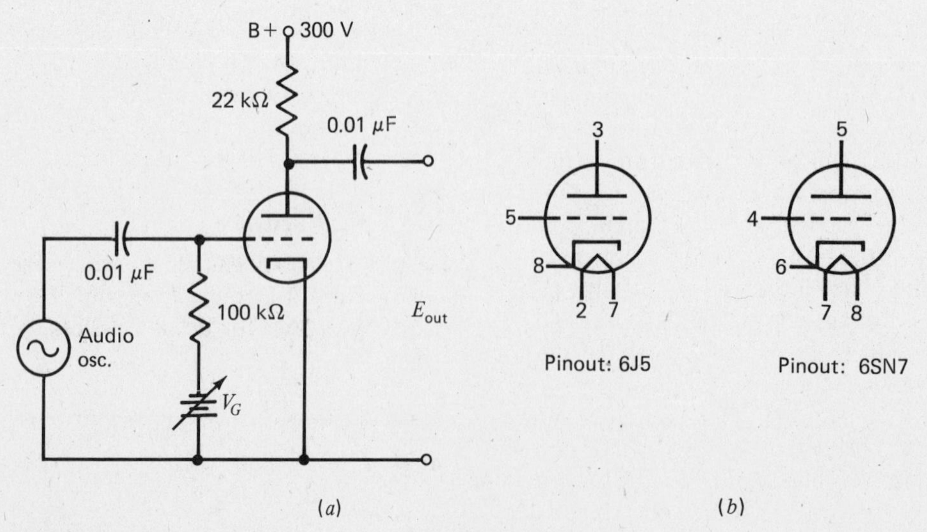

Fig. 52-2. (*a*) Triode amplifier. (*b*) Pinouts for a 6J5 and a 6SN7 tube.

2. With the circuit of Fig. 52-2 connected, adjust the grid voltage to −2 V. Also, adjust the audio signal generator to 1 kHz at 2 $V_{p\text{-}p}$ output.
3. Measure and record in Table 52-1 the signal voltage from grid to cathode.
4. Measure and record in Table 52-1 the input and output signals for grid voltages of −2 V, −4 V, and −6 V. Calculate and record the voltage gain (A_V) for each one.
5. With the grid voltage set at −4 V, decrease the frequency of oscillation to 20 Hz and record the peak-to-peak output voltage in Table 52-1. Also record the peak-to-peak output voltages with the frequency at 15 kHz and 50 kHz.
6. Calculate the voltage gain (A_V) of the amplifier by using the formula below, where $\mu = 20$ for the 6J5 or 6SN7 vacuum tube. Also, $r_p = 7.0\ \text{k}\Omega$ and is the ac plate resistance. Record this in Table 52-1.

$$A_V = \frac{\mu \times R_p}{R_p + r_p}$$

7. Compare the calculated voltage gain with the measured voltage gain. Calculate and record in Table 52-1 the percentage of error between the measured and calculated voltage gains.
8. Replace the 22-kΩ plate resistor with an 8.2-kΩ resistor. For the corresponding grid voltages of −2 V, −4 V, −6 V, and −8 V, measure and record in Table 52-1 the corresponding plate voltages.
9. With the 8.2-kΩ resistor as R_p, repeat steps 2 to 8. Record results in a separate data table of your own, similar to Table 52-1.

NAME DATE

RESULTS FOR OPTIONAL EXPERIMENT 52

QUESTIONS

1. Define the purpose and function of each of the elements found in a triode.

2. In this experiment, what step had the largest effect on gain?

3. For this tube experiment, define the condition causing saturation.

4. For this tube experiment, define the condition causing cutoff.

5. How do the calculated values of voltage gain compare to the actual measured values? How are differences accounted for?

REPORT

Write a complete report. Discuss the results. Discuss the three most significant aspects of the experiment and write a conclusion.

TABLE 52–1

Procedure Step	Measurement	Value V	Calculated A_V	Measured A_V
3	V_{GK}	________		
4	V_{in} at $V_G = -2$ V	________	________	________
	V_{out} at $V_G = -2$ V	________		
	V_{in} at $V_G = -4$ V	________	________	________
	V_{out} at $V_G = -4$ V	________		
	V_{in} at $V_G = -6$ V	________	________	________
	V_{out} at $V_G = -6$ V	________		
5	V_{out} at 20 Hz	________		
	V_{out} at 15 kHz	________		
	V_{out} at 50 kHz	________		
6			________	
7	Percent of error = ________%			
8	V_P with $V_G = -2$ V	________		
	V_P with $V_G = -4$ V	________		
	V_P with $V_G = -6$ V	________		
	V_P with $V_G = -8$ V	________		

1N4001 thru 1N4007

Designers▲Data Sheet

"SURMETIC"▲ RECTIFIERS

. . . subminiature size, axial lead mounted rectifiers for general-purpose low-power applications.

Designers Data for "Worst Case" Conditions

The Designers▲ Data Sheets permit the design of most circuits entirely from the information presented. Limit curves — representing boundaries on device characteristics — are given to facilitate "worst case" design.

LEAD MOUNTED SILICON RECTIFIERS

50-1000 VOLTS DIFFUSED JUNCTION

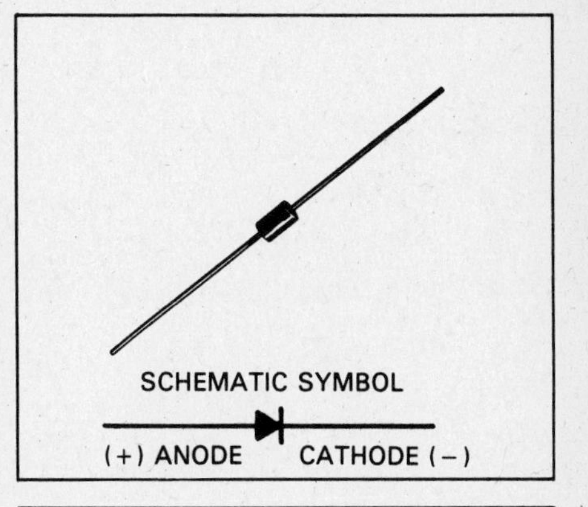

*MAXIMUM RATINGS

Rating	Symbol	1N4001	1N4002	1N4003	1N4004	1N4005	1N4006	1N4007	Unit
Peak Repetitive Reverse Voltage Working Peak Reverse Voltage DC Blocking Voltage	V_{RRM} V_{RWM} V_R	50	100	200	400	600	800	1000	Volts
Non-Repetitive Peak Reverse Voltage (halfwave, single phase, 60 Hz)	V_{RSM}	60	120	240	480	720	1000	1200	Volts
RMS Reverse Voltage	$V_{R(RMS)}$	35	70	140	280	420	560	700	Volts
Average Rectified Forward Current (single phase, resistive load, 60 Hz, see Figure 8, $T_A = 75^oC$)	I_O	1.0							Amp
Non-Repetitive Peak Surge Current (surge applied at rated load conditions, see Figure 2)	I_{FSM}	30 (for 1 cycle)							Amp
Operating and Storage Junction Temperature Range	T_J, T_{stg}	-65 to +175							^oC

*ELECTRICAL CHARACTERISTICS

Characteristic and Conditions	Symbol	Typ	Max	Unit
Maximum Instantaneous Forward Voltage Drop ($i_F = 1.0$ Amp, $T_J = 25^oC$) Figure 1	v_F	0.93	1.1	Volts
Maximum Full-Cycle Average Forward Voltage Drop ($I_O = 1.0$ Amp, $T_L = 75^oC$, 1 inch leads)	$V_{F(AV)}$	–	0.8	Volts
Maximum Reverse Current (rated dc voltage)	I_R			μA
$T_J = 25^oC$		0.05	10	
$T_J = 100^oC$		1.0	50	
Maximum Full-Cycle Average Reverse Current ($I_O = 1.0$ Amp, $T_L = 75^oC$, 1 inch leads	$I_{R(AV)}$	–	30	μA

*Indicates JEDEC Registered Data.

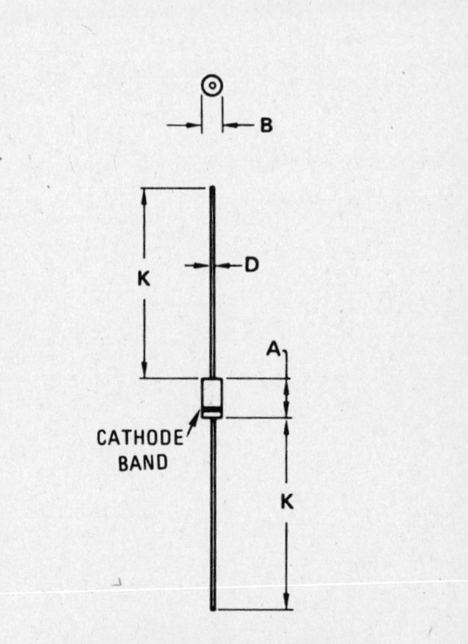

DIM	MILLIMETERS MIN	MILLIMETERS MAX	INCHES MIN	INCHES MAX
A	5.97	6.60	0.235	0.260
B	2.79	3.05	0.110	0.120
D	0.76	0.86	0.030	0.034
K	27.94	–	1.100	–

CASE 59-04

Does Not Conform to DO-41 Outline.

MECHANICAL CHARACTERISTICS

CASE: Void free, Transfer Molded
MAXIMUM LEAD TEMPERATURE FOR SOLDERING PURPOSES: 350^oC, 3/8" from case for 10 seconds at 5 lbs. tension
FINISH: All external surfaces are corrosion-resistant, leads are readily solderable
POLARITY: Cathode indicated by color band
WEIGHT: 0.40 Grams (approximately)

▲Trademark of Motorola Inc.

DS 6015 R3

2N3903 (SILICON)
2N3904

NPN SILICON ANNULAR TRANSISTORS

. . . designed for general purpose switching and amplifier applications and for complementary circuitry with types 2N3905 and 2N3906.

- Collector-Emitter Breakdown Voltage – BV_{CEO} = 40 Vdc (Min)
- Current Gain Specified from 100 μA to 100 mA
- Complete Switching and Amplifier Specifications
- Low Capacitance – C_{ob} = 4.0 pF (Max)

NPN SILICON SWITCHING & AMPLIFIER TRANSISTORS

*MAXIMUM RATINGS

Rating	Symbol	Value	Unit
Collector-Base Voltage	V_{CB}	60	Vdc
Collector-Emitter Voltage	V_{CEO}	40	Vdc
Emitter-Base Voltage	V_{EB}	6.0	Vdc
Collector Current – Continuous	I_C	200	mAdc
Total Power Dissipation @ T_A = 25°C Derate above 25°C	P_D	350 2.8	mW mW/°C
Total Power Dissipation @ T_C = 25°C Derate above 25°C	P_D	1.0 8.0	Watts mW/°C
Junction Operating Temperature	T_J	150	°C
Storage Temperature Range	T_{stg}	-55 to +150	°C

THERMAL CHARACTERISTICS

Characteristic	Symbol	Max	Unit
Thermal Resistance, Junction to Ambient	$R_{\theta JA}$	357	°C/W
Thermal Resistance, Junction to Case	$R_{\theta JC}$	125.	°C/W

*Indicates JEDEC Registered Data

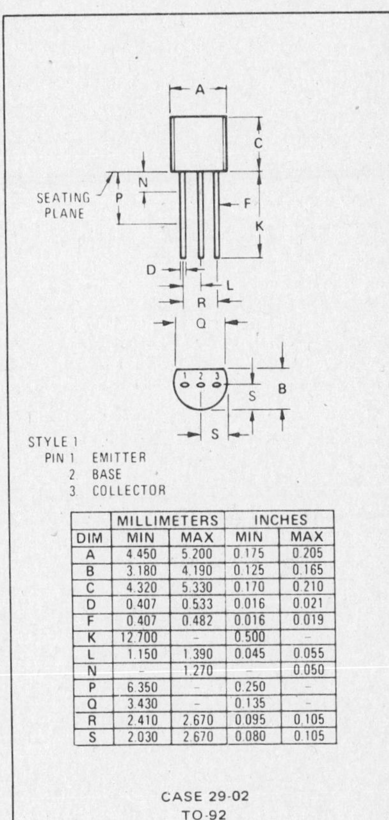

	MILLIMETERS		INCHES	
DIM	MIN	MAX	MIN	MAX
A	4.450	5.200	0.175	0.205
B	3.180	4.190	0.125	0.165
C	4.320	5.330	0.170	0.210
D	0.407	0.533	0.016	0.021
F	0.407	0.482	0.016	0.019
K	12.700	–	0.500	–
L	1.150	1.390	0.045	0.055
N	–	1.270	–	0.050
P	6.350	–	0.250	–
Q	3.430	–	0.135	–
R	2.410	2.670	0.095	0.105
S	2.030	2.670	0.080	0.105

CASE 29-02
TO-92

Appendix C Applicable Color Codes

TABLE C-1 Color Code for Carbon Composition Resistors

COLOR	DIGIT 1st	DIGIT 2nd	MULTIPLIER	TOLERANCE
Black	0	0	1	—
Brown	1	1	10	—
Red	2	2	100	—
Orange	3	3	1,000	—
Yellow	4	4	10,000	—
Green	5	5	100,000	—
Blue	6	6	1,000,000	—
Violet	7	7	10,000,000	—
Gray	8	8	100,000,000	—
White	9	9	1,000,000,000	—
Gold	—	—	0.1	± 5%
Silver	—	—	0.01	±10%
No band	—	—	—	±20%

TABLE C-2 Color Code for Carbon Film Resistors

COLOR	DIGITS 1st	DIGITS 2nd	DIGITS 3rd	MULTIPLIER	TOLERANCE
Black	0	0	0	1	
Brown	1	1	1	10	±1%
Red	2	2	2	100	±2%
Orange	3	3	3	1,000	
Yellow	4	4	4	10,000	
Green	5	5	5	100,000	±5%
Blue	6	6	6	1,000,000	±0.25%
Violet	7	7	7	10,000,000	±0.10%
Gray	8	8	8		±0.05%
White	9	9	9		±5%
Gold	—	—	—	0.1	±10%
Silver	—	—	—	0.01	

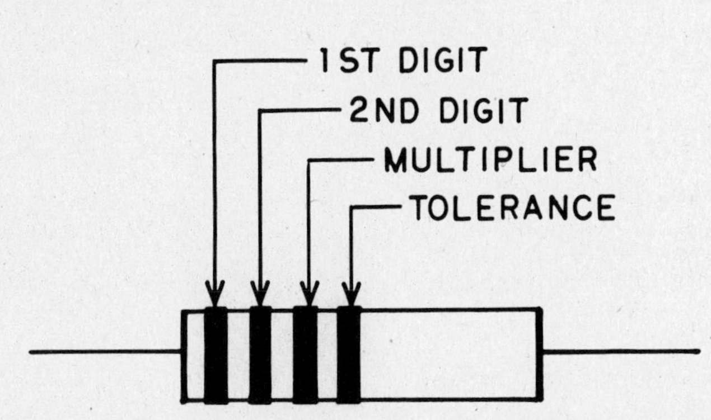

FIG. C-1 Color coding on a carbon composition resistor.

1ST DIGIT
2ND DIGIT
3RD DIGIT
MULTIPLIER
TOLERANCE

FIG. C-2 Color coding on a carbon film resistor.

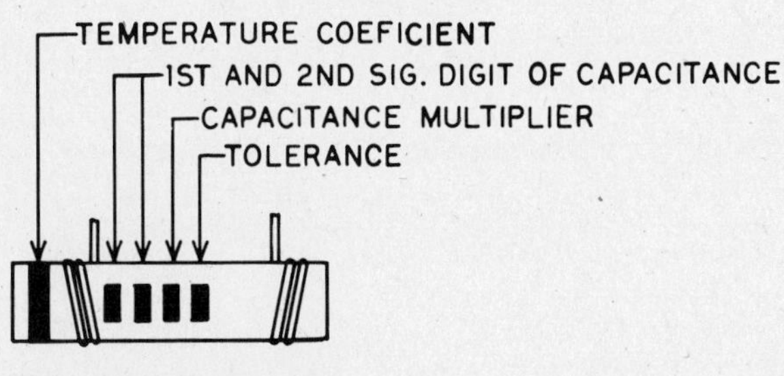

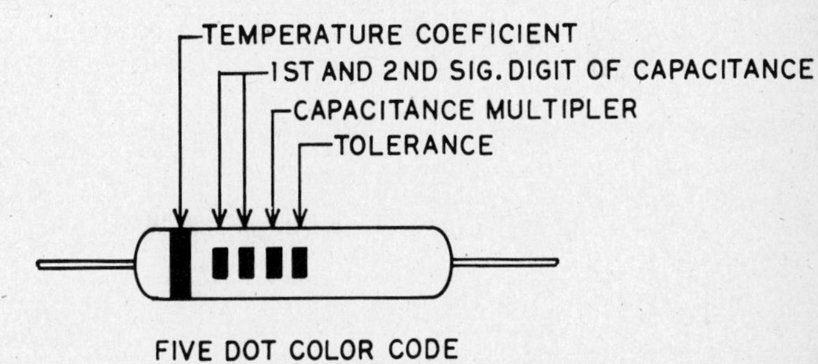

TEMPERATURE COEFICIENT
TEMPERATURE COEFICIENT MULTIPLIER
1ST AND 2ND SIG. DIGIT OF CAPACITANCE
CAPACITANCE MULTIPLIER
TOLERANCE

SIX DOT COLOR CODE
RADIAL LEAD

FIG. C-3 Color coding on tubular ceramic capacitors.

Reproduced from Baer/Ottaway, *Electrical and Electronic Drawing*, Fifth Edition, McGraw-Hill Book Company, 1986.

TABLE C-3 Color Codes for Tubular and Disc Ceramic Capacitors

BAND OR DOT COLOR	CAPACITANCE IN PICOFARADS: SIGNIFICANT DIGITS 1st	SIGNIFICANT DIGITS 2nd	CAPACITANCE MULTIPLIER	TOLERANCE ≤10 pF	TOLERANCE >10 pF	TEMPERATURE COEFFICIENT pp °C (5-DOT SYSTEM)	TEMPERATURE COEFFICIENT 6-DOT SYSTEM SIG. FIG.	TEMPERATURE COEFFICIENT MULTIPLIER
Black	0	0	1	±2.0 pF	±20%	0	0.0	−1
Brown	1	1	10	±0.1 pF	±1%	−33		−10
Red	2	2	100		±2%	−75	1.0	−100
Orange	3	3	1000		±3%	−150	1.5	−1000
Yellow	4	4				−230	2.0	−10000
Green	5	5		±0.5 pF	±5%	−330	3.3	+1
Blue	6	6				−470	4.7	+10
Violet	7	7				−750	7.5	+100
Gray	8	8	0.01	±0.25 pF		+150 to −1500		+1000
White	9	9	0.1	±1.0 pF	±10%	+100 to −75		+10000
Silver	—	—						
Gold	—	—						

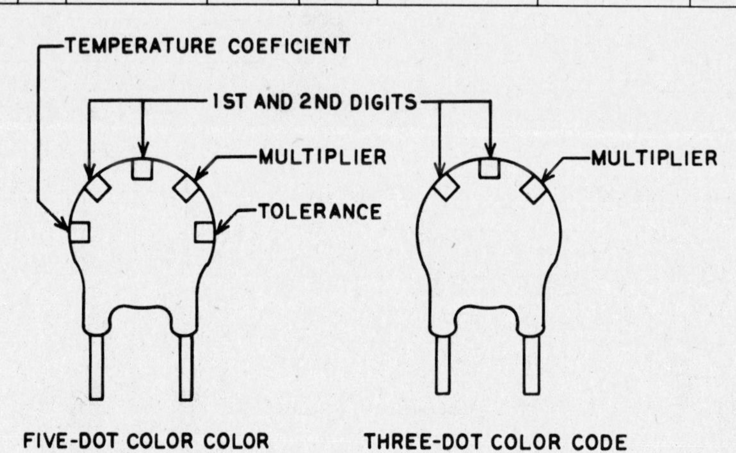

FIG. C-4 Color coding on a ceramic disk capacitor.

TABLE C-4 Color Code for Molded Paper Tubular Capacitors

COLOR	(CAPACITANCE IN PICOFARADS) SIGNIFICANT DIGITS 1st	SIGNIFICANT DIGITS 2nd	MULTIPLIER	TOLERANCE	VOLTAGE
Black	0	0	1	±20%	—
Brown	1	1	10		100
Red	2	2	100		200
Orange	3	3	1000	±30%	300
Yellow	4	4	10000		400
Green	5	5			500
Blue	6	6			600
Violet	7	7			700
Gray	8	8			800
White	9	9			900
Gold	—	—			1000
Silver	—	—		±10%	—

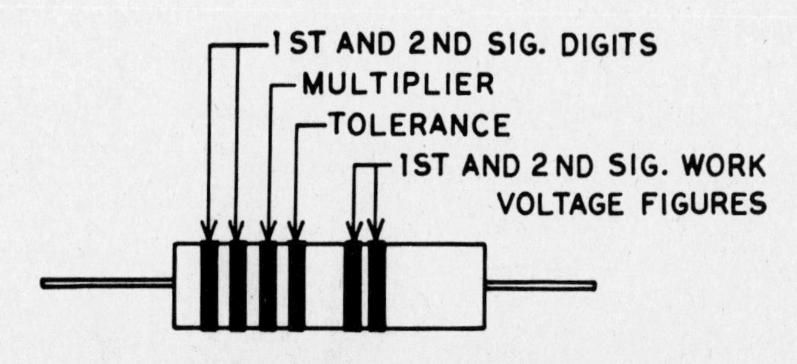

FIG. C-5 Color coding on a paper tubular capacitor.

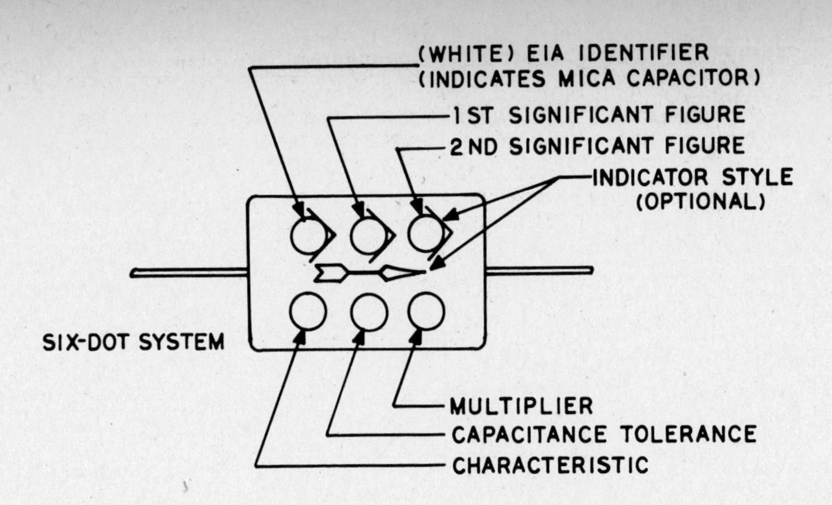

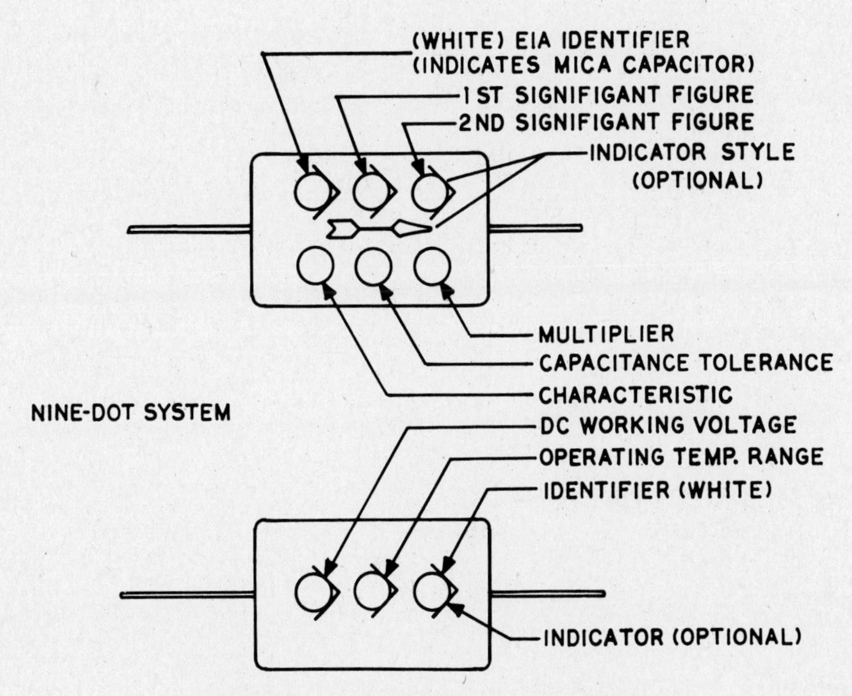

FIG. C-6 Color coding on mica capacitors.

TABLE C-5 Color Code for Mica Capacitors

COLOR	CHARAC-TERISTIC	DIGITS 1st	DIGITS 2nd	MULTI-PLIER	TOLER-ANCE	DC WORKING VOLTAGE	OPERATING TEMPERATURE RANGE
Black		0	0	1	±20%	100	
Brown	B	1	1	10	±1%		−55°C to +85°C
Red	C	2	2	100	±2%	300	
Orange	D	3	3	1,000			−55°C to +125°C
Yellow	E	4	4	10,000		500	
Green	F	5	5		±5%		
Blue		6	6				
Violet		7	7				
Gray		8	8				
White		9	9				
Gold		—	—	0.1	±½%	1000	
Silver		—	—	0.01	±10%		

TABLE C-6 Mica Capacitor Characteristics

CHARAC-TERISTIC	TEMPERATURE COEFFICIENT OF CAPACITANCE (ppm/°C)	MAXIMUM CAPACITANCE DRIFT
B	Not specified	Not specified
C	±200	±(0.5% + 0.5 pF)
D	±100	±(0.3% + 0.1 pF)
E	−20 to +100	±(0.1% + 0.1 pF)
F	0 to +70	±(0.05% + 0.1 pF)

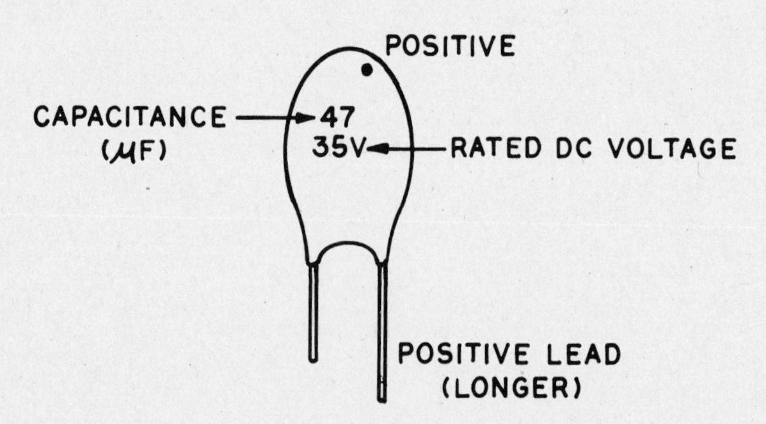

FIG. C-7 Marking of a tantalum capacitor.

Appendix D Lab Report Preparation

Appendix D-1: How to Write Lab Reports

1. Write short sentences whenever possible.
2. Use a dictionary and spell all words correctly.
3. Write neatly in blue or black ink.
4. Avoid personal pronouns: I, you, we, etc.
5. Never discuss results unless they are part of your data.
6. Do not copy or paraphrase textbook theory.
7. Label all data tables and graphs with titles, numbers, proper units, and column headings.
8. Never write in the margins of your paper.
9. Leave enough space between sentences for the instructor to make corrections.
10. Technical accuracy and completeness are the most important part of a lab report. Unless the report is well organized and easy to read, it is of little value.

Every instructor will have different standards and different ideas about report writing. However, most lab reports reflect the scientific method as follows:

- A hypothesis is formulated. This is like a statement of purpose.
- Data is collected and analyzed. This is like the procedure and results.
- The hypothesis is proven or disproven based upon the results. This is like the discussion and conclusion.

Refer to the two sample reports that follow in Appendixes D-3 and D-4. One is a poor report that does not follow these suggestions. The other is a good report that does use the suggestions.

Appendix D-2: Blank Style Sheet

Experiment No.

Name: ____________________

Date: ____________________

Title:

Class: ____________________

Instr: ____________________

Purpose:

Procedure:

Results:

Discussion of Results:

1.

2.

3.

Notes:

Conclusion:

Appendix D-3: Sample Poor Report

Experiment No. 3

Name: J. Doe

Date:

Title: Ohm's law

Class: Elec.

Instr: Jones

Purpose: To do Ohm's law experiment.

Procedure: 1-16 in manual.

Results:

Discussion of Results:

1. The more voltage I had the more current I had when I measured.

2. When we changed the resistors I saw more current but it was hard to read the meter because the needle was bent a little.

3. $I = \frac{V}{R}$ is ohm's law.

Notes: This experiment is good to learn about ohm's law but I had trouble hooking up the circuit.

Conclusion: Ohm's law works good.

Appendix D-4: Sample Good Report

Experiment No. 3

Name: John Doe

Date:

Title: Ohm's Law

Class: Elec. 1-A

Instr: Mr. R. Jones

Purpose: To validate Ohm's Law: $I = \frac{V}{R}$

Procedure: Experiments in Basic Electronics, pages 16-20, steps 1-16.

Results: Data tables 3-1 and 3-2, attached.

Discussion of Results:

1. With resistance held constant, the current varied in direct proportion to any changes in applied voltage.

2. With voltage held constant, the current was inversely proportional to any changes in circuit resistance.

3. Current can be held constant as long as voltage and resistance are kept in proportion. This is consistent with the formula $I = \frac{V}{R}$.

Notes: The pointer of meter number 16 was bent.

Conclusion: Ohm's Law is valid based upon the results of this experiment. Current is directly proportional to voltage and inversely proportional to resistance.

Appendix E Blank Graph Paper

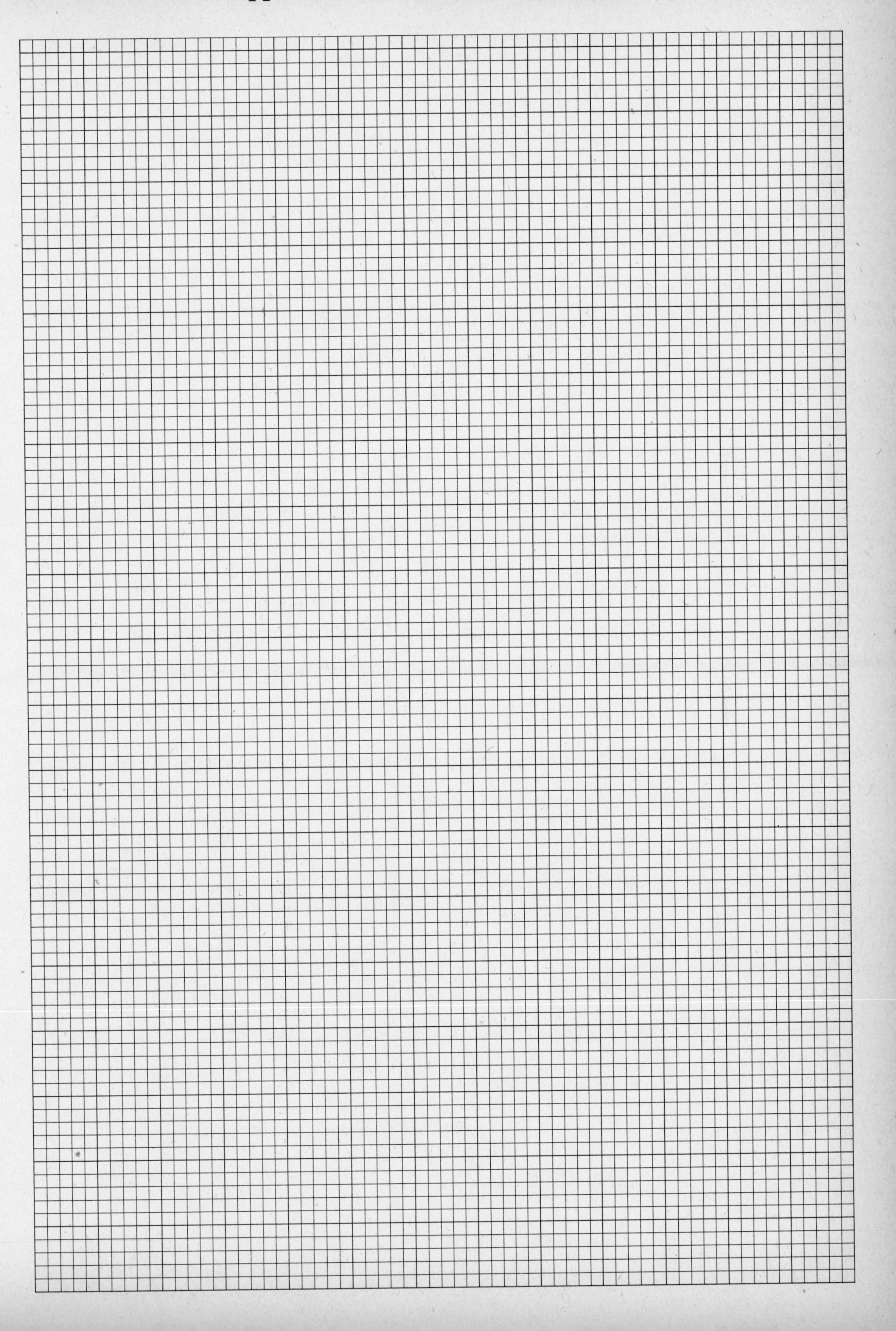

10
9
8
7
6
5
4
3
2
EUGENE DIETZGEN CO.
1
9
8
7
6
5
4
3
2
1

Appendix F Oscilloscope Graticules

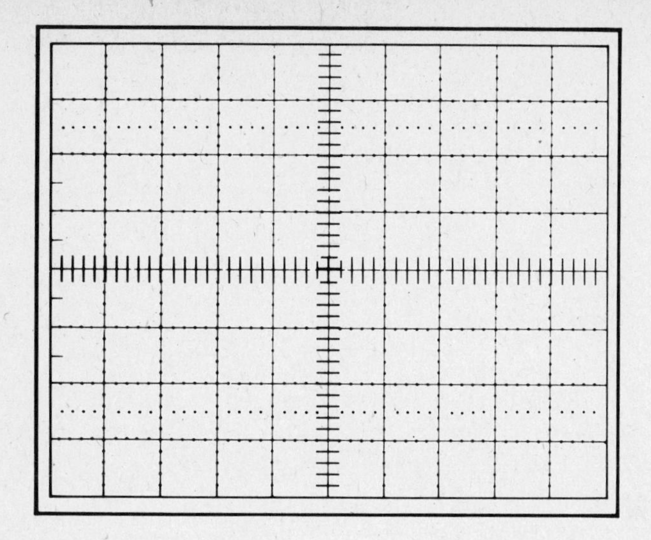

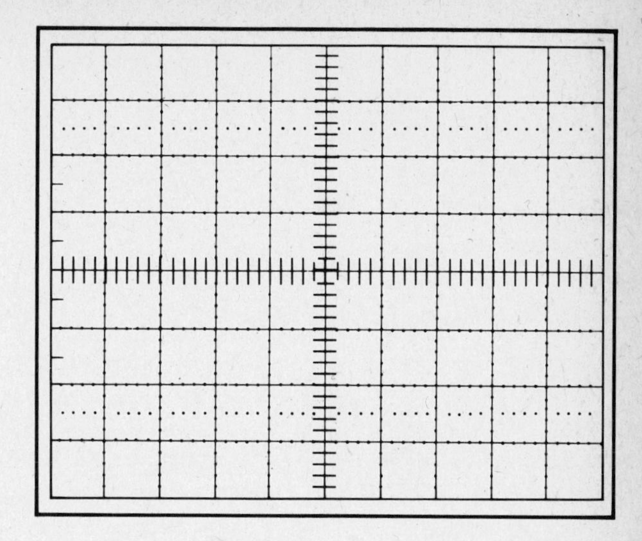

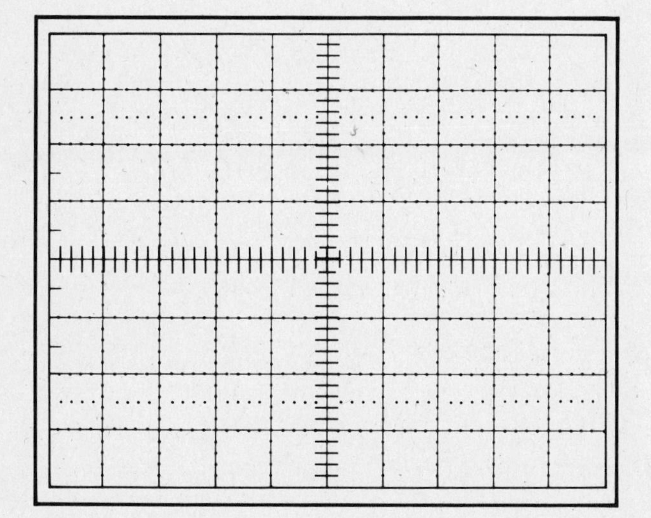

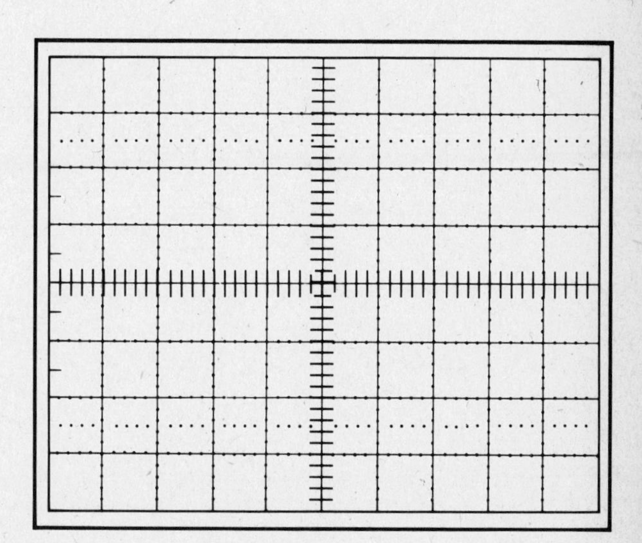

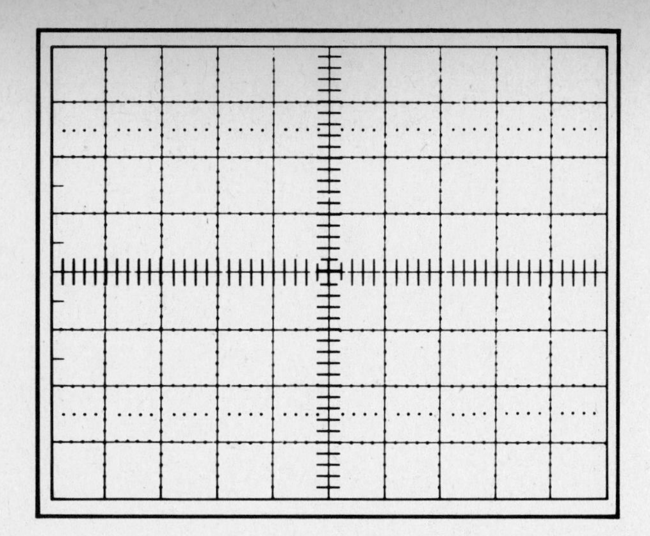
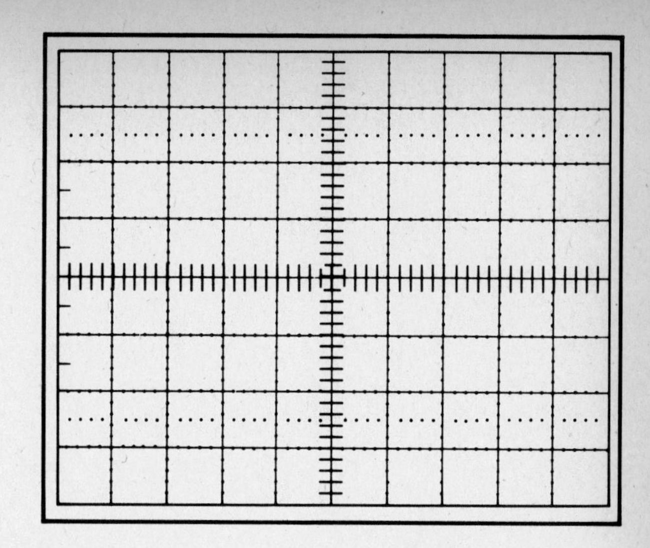
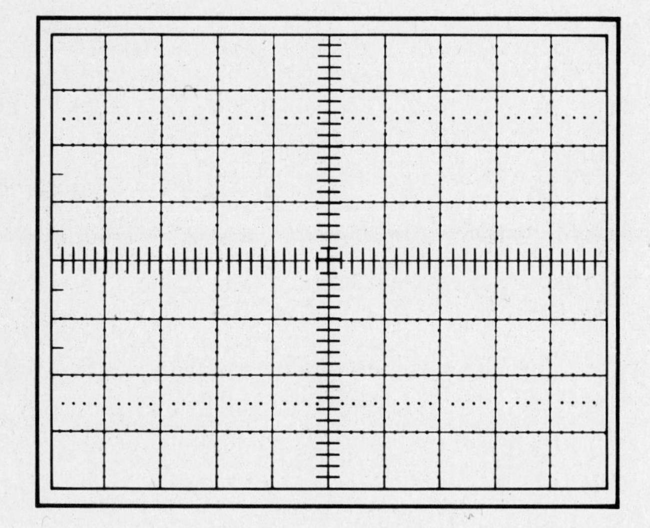
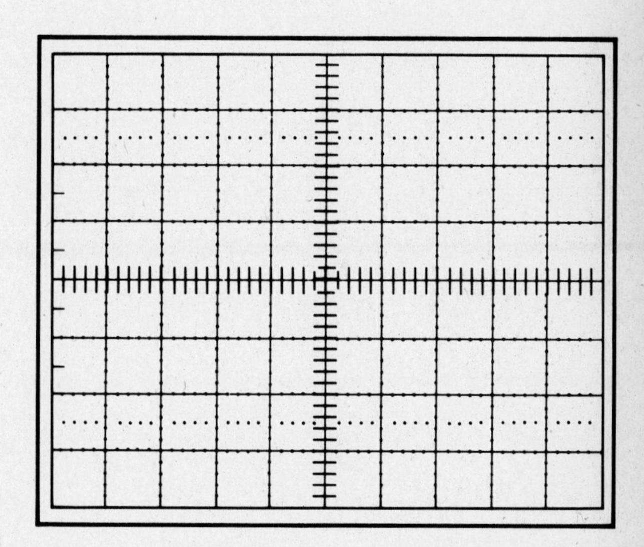

Appendix G How To Make Graphs

Graphs usually show the relationship of two or more variables. The relationship is actually the curve or line that results. Here are some things to remember when making graphs.

1. Be neat and complete.
2. Never connect points. Always show the characteristic of the curve.
3. There should be room in the margins to title the graph.
4. Use the fullest scales possible. Do not confine a graph to one corner of the paper.
5. Use semilog graph paper for exponential quantities. Label the X axis as 1×10^2 for 100s, 1×10^3 for 1000s etc.

Examples

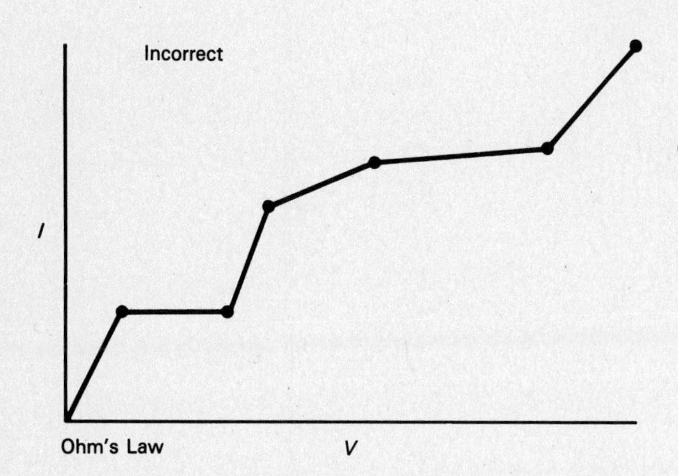

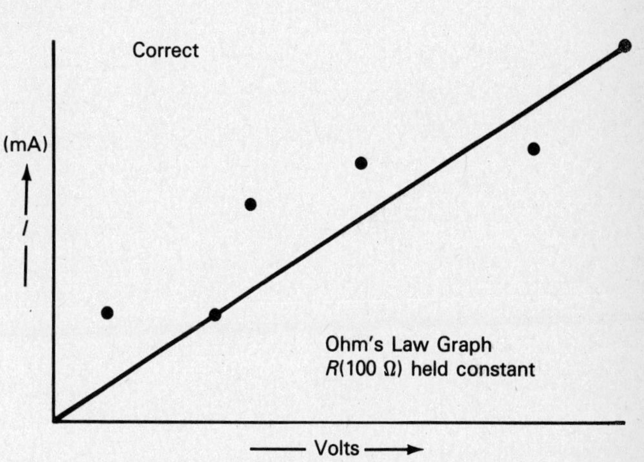

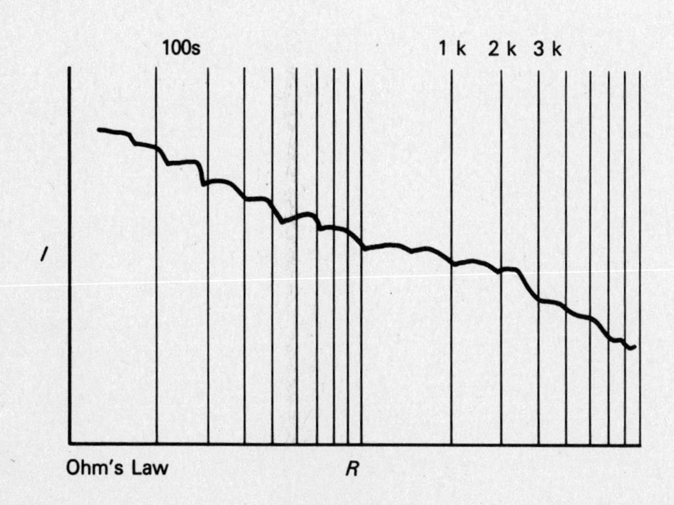

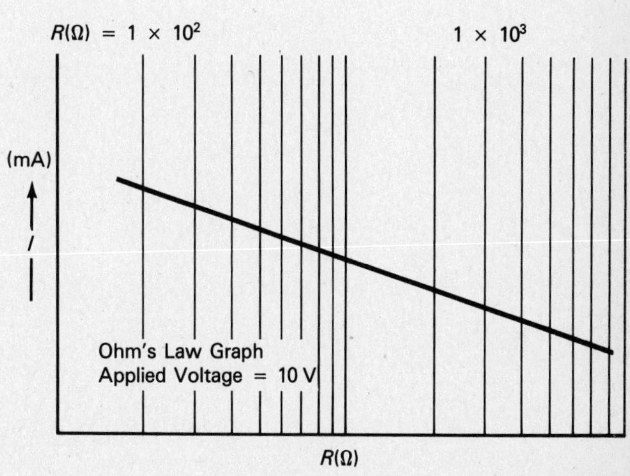

Appendix H The Oscilloscope

The purpose of this appendix on oscilloscopes is to help you learn to use this measurement tool accurately and efficiently. This introduction is divided into two sections: the first section describes the functional parts of the basic oscilloscope, and the second section describes probes.

The oscilloscope is the most important tool to an experienced electronics technician (see Fig. H-1). While working through this appendix, it is best to have your laboratory oscilloscope (or the one you will be using in your studies of electronics) in front of you so that you can learn, practice, and apply your newly acquired knowledge. This appendix will discuss fundamentals which apply to any oscilloscope; your instructor or laboratory aid will help you with the fine details and special provisions that apply to your particular oscilloscope.

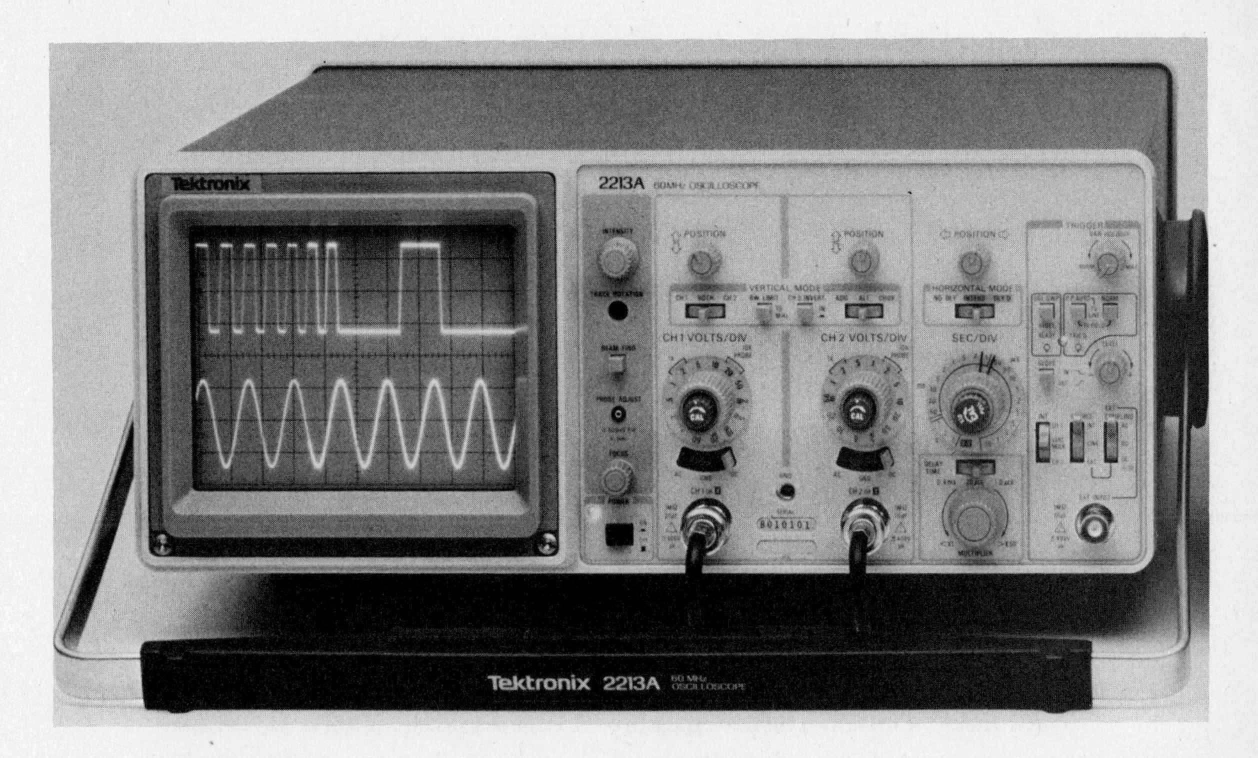

FIG. H-1 The oscilloscope.

Part 1: Functional Oscilloscope Sections

There are four functional parts to the basic oscilloscope: the display system, the vertical system, the horizontal system, and the trigger system.

Display System. The display system is a coordinated system of controls. It includes a cathode-ray tube (CRT), intensity control, and focus control. The CRT has a phosphor coating inside it. As an electron beam is moved across the phosphor coating, a glow is created which follows the beam and persists for a short time. A grid, also known as a graticule, is etched or painted on the screen of the CRT. This grid serves as a reference for taking measurements. Figure H-2 shows a typical oscilloscope graticule. Note the major and minor divisions.

The intensity control adjusts the amount of glow emitted by the electron beam and phosphor coating. The beam-trace should be adjusted so that it is easy to see and produces no halo. The focus control adjusts the beam for an optimum trace.

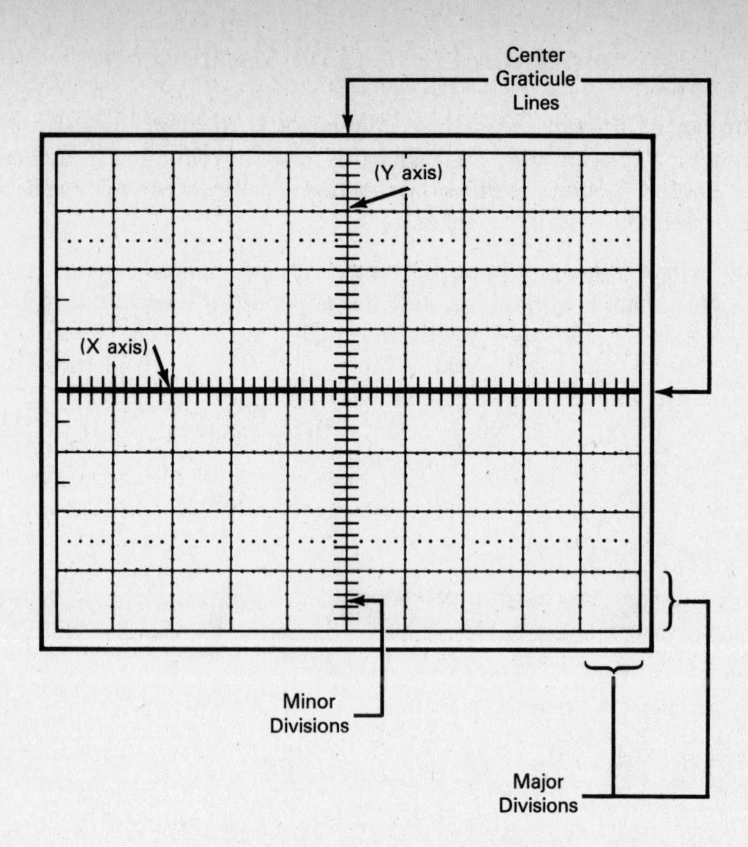

FIG. H-2 Oscilloscope CRT.

Vertical System. The vertical system controls and develops the deflection voltages which are displayed on the CRT. In this system, you typically find controls for vertical position, vertical sensitivity, and input coupling. The vertical position controls the placement of the trace. Adjusting this control will move the entire trace either up or down along the Y axis of the CRT. This control is adjusted as needed to help the operator make accurate measurements. The vertical-sensitivity control, also known as the volts-per-division switch, controls the Y axis sensitivity. The range is usually controlled from 1 mV to 50 V per division. For example, if you were to observe a trace occupying 4 divisions on the CRT, and the volts-per-division switch was turned to 1 mV/div., then the measured voltage would equal 1 mV × 4 divisions, or 4 mV. If the switch were turned to the 50 V/div. position and a 4 division trace was observed, then the oscilloscope would be measuring 200 V. The input-coupling switch lets the operator determine how the circuit under examination is connected to the oscilloscope. The three positions of this switch are ground, dc coupling, and ac coupling. When the switch is in the ground position, the operator can adjust the position of the trace with no input signal applied to the oscilloscope. This function is used primarily to align the oscilloscope to a reference point prior to taking a measurement. When in the dc-coupling position, the oscilloscope allows the operator to see the entire signal. However, when the input-coupling switch is in the ac-coupling position, only the ac signal components are displayed on the CRT. All dc components are blocked when the switch is in the ac position.

Horizontal System. As the oscilloscope trace is moved across the CRT (from left to right), it moves at a rate of speed which is related to frequency. The horizontal system is dominated by two main controls: the horizontal-position control and the time-base control. The horizontal-position control performs the same task as the vertical-position, but utilizes the X axis. The time-base, or seconds-per-division control, is used to select the appropriate sweep necessary to see the input signal. Ranges typically found on the time-base control extend from 0.1 μs to 0.5 μs per division on the CRT.

Trigger System. The trigger system allows the operator to select a part of the input signal and synchronize it with the trace displayed on the CRT. Normally, a trigger-level control is available. The position of the trigger-level control determines where on the selected trace the oscilloscope triggering will occur.

Each oscilloscope has different features. Your instructor is the best source for varying operational procedures.

Part 2: Oscilloscope Probes

Probes should accurately reproduce the signal for your oscilloscope. Probes can be divided by function into two main areas: current sensing and voltage sensing. Voltage-sensing probes can be further divided into passive and active types. For most applications, the probes that were supplied with your oscilloscope are the ones you should use. An operator picks the type of probe based on the voltage intended to be measured. For example, if you are measuring a 50 V signal, and the largest vertical sensitivity available is 5 V, then that particular signal will occupy 10 divisions on the CRT. This is a situation where attenuation is needed, and a $\times 10$ probe would reduce the amplitude of your signal to a reasonable proportion. The best way to ensure that your oscilloscope and probe measurement system have the least effect on the accuracy of your measurements is to use the probe recommended for your oscilloscope.

Appendix I Component List

Resistors

All resistors 0.5 W, 5% unless indicated otherwise.

(1) 10 Ω
(1) 15 Ω
(1) 56 Ω
(1) 68 Ω
(3) 100 Ω
(1) 100 Ω, 1 W
(1) 120 Ω
(3) 150 Ω
(1) 150 Ω, 1 W
(2) 220 Ω
(1) 270 Ω
(1) 330 Ω
(1) 330 Ω, 1 W
(1) 390 Ω
(3) 470 Ω
(2) 560 Ω
(1) 680 Ω
(1) 820 Ω
(3) 1 kΩ
(1) 1 kΩ, 1 W
(1) 1.2 kΩ
(2) 1.5 kΩ
(1) 2.2 kΩ
(1) 2.7 kΩ
(1) 3.3 kΩ
(2) 4.7 kΩ
(1) 5.6 kΩ
(1) 8.2 kΩ
(3) 10 kΩ
(1) 22 kΩ
(1) 22 kΩ, 2 W
(1) 27 kΩ
(1) 33 kΩ
(1) 39 kΩ
(2) 47 kΩ
(1) 68 kΩ
(2) 82 kΩ
(1) 86 kΩ
(2) 100 kΩ
(1) 100 kΩ, 1 W
(1) 150 kΩ
(1) 220 kΩ
(1) 470 kΩ
(1) 1 MΩ
(1) 1.2 MΩ
(1) 3 MΩ
(1) 3.3 MΩ

Capacitors

All capacitors 25 V or greater.

(1) 0.0068 μF
(2) 0.01 μF
(1) 0.1 μF
(3) 10 μF
(2) 25 μF
(2) 47 μF electrolytic
(2) 47 μF
(2) 4 μF or 1 μF

Inductors

(2) 33 mH
(1) 100 mH
(2) 1 H (or optional value)

Potentiometers

(1) 1 kΩ
(1) 5 kΩ
(1) 100 kΩ
(1) 1 MΩ

Batteries

(4) D cells
(4) D-cell holders

Diodes

(4) 1N4004 or equivalent
(2) LEDs
(1) Zener, 5 V (1 W)
(2) IR#S1M solar cells or equivalent photo diode

Transistors

(1) 2N3638
(2) 2N3904

FET

(1) 2N3823 or equivalent

Op Amps and ICs

(1) 741
(1) 7408
(1) 7432

Vacuum Tube (Optional)

6J5 or equivalent

Bench Equipment

Ammeter with 30-mA capacity
Voltmeter: DVM, VTVM, or VOM
DC power supply, 0–30 V (± 15 V for op amps experiment)
High-voltage power supply with 120:12.6 V center-tap @ 6.3 V filament transformer
Galvanometer or microammeter movement
Signal generator (sine wave, to 1 MHz preferred)
Oscilloscope (solid-state, auto-trigger, dual-trace, with operator's manual preferred)
Frequency counter

Miscellaneous Parts

(1) Circuit board; proto springboard or breadboard
(2) SPST switch
(1) SPDT switch
(1) 0–1 mA meter movement
(1) 50 μA meter movement
6 Test leads
(1) Decade box
(1) Magnetic compass
(1) Heavy-duty horseshoe magnet, 20 lb plus pull
(1) Sheet-metal shield, 6 × 6 in
(1) Grease pencil
2–3 ft, thin insulated wire
Iron filings
No. 18 steel nail
(4) light bulbs: 25, 60, 100, 150 W